DESIGN OF STEEL STRUCTURES

DESIGN OF STEEL STRUCTURES

Elias G. Abu-Saba

Associate Professor of Architectural Engineering
North Carolina Agricultural and
Technical State University

CHAPMAN & HALL
I(T)P An International Thomson Publishing Company

New York • Albany • Bonn • Boston • Cincinnati
• Detroit • London • Madrid • Melbourne • Mexico City
• Pacific Grove • Paris • San Francisco • Singapore
• Tokyo • Toronto • Washington

This edition published by
Chapman & Hall
A division of International Thomson Publishing Inc.
I(T)P The ITP logo is a trademark under license

Printed in the United States of America

For more information, contact:

Chapman & Hall
One Penn Plaza
New York, NY 10119

International Thomson Publishing
Berkshire House 168-173
High Holborn
London WC1V 7AA
England

Thomas Nelson Australia
102 Dodds Street
South Merlbourne, 3205
Victoria, Australia

Nelson Canada
1120 Birchmount Road
Scarborough, Ontario
Canada, M1K 5G4

Chapman & Hall
2-6 Boundary Row
London SE1 8HN

International Thomson Editores
Campos Eliseos 385, Piso 7
Col. Polanco
11560 Mexico D.F. Mexico

International Thomson Publishing Gmbh
Königwinterer Strasse 418
53228 Bonn
Germany

International Thomson Publishing Asia
221 Henderson Road
#05-10 Henderson Building
Singapore 0315

International Thomson Publishing-Japan
Hirakawacho-cho Kyowa Building, 3F
1-2-1 Hirakawacho-cho
Chiyoda-ku, 102 Tokyo
Japan

1 2 3 4 5 6 7 8 9 10 XXX 01 00 99 97 96 95

Library of Congress Cataloging-in-Publication Data

Abu-Saba, Elias G.,
Design of steel structures / by Elias G. Abu-Saba.
p. cm.
Includes bibliographical references and index.
ISBN 0-412-98491-1
1. Building, Iron and steel. 2. Structural design. 3. Load factor design. I. Title.
TA684.A144 1995
693'.71—dc20 94-48546
CIP

Please send your order for this or any Chapman & Hall book to **Chapman & Hall, 29 West 35th Street, New York, NY 10001, Attn: Customer Service Department.** You may also call our Order Department at 1-212-244-3336 or fax your purchase order to 1-800-248-4724.

For a complete listing of Chapman & Hall's titles, send your requests to **Chapman & Hall, Dept. BC, One Penn Plaza, New York, NY 10119.**

I dedicate this book to the memory of my parents, Jurjis and Sabat Abu-Saba, who did not have the privilege to go to school. Yet they believed in the power of knowledge and provided us, their children, with the opportunity to learn and grow.

The infinite spans the human mind.
The spirit spins free of space and time.
The joy and sadness of life are a wink
In the eternal flow.
The stream cascades and meanders
To merge and be lost in the greater sea.

Elias G. Abu-Saba
November 6, 1994

Contents

Preface

This book is intended for classroom teaching in architectural and civil engineering at the graduate and undergraduate levels. Although it has been developed from lecture notes given in structural steel design, it can be useful to practicing engineers. Many of the examples presented in this book are drawn from the field of design of structures.

Design of Steel Structures can be used for one or two semesters of three hours each on the undergraduate level. For a two-semester curriculum, Chapters 1 through 8 can be used during the first semester. Heavy emphasis should be placed on Chapters 1 through 5, giving the student a brief exposure to the consideration of wind and earthquakes in the design of buildings. With the new federal requirements vis a vis wind and earthquake hazards, it is beneficial to the student to have some understanding of the underlying concepts in this field. In addition to the class lectures, the instructor should require the student to submit a term project that includes the complete structural design of a multi-story building using standard design procedures as specified by AISC Specifications. Thus, the use of the *AISC Steel Construction Manual* is a must in teaching this course. In the second semester, Chapters 9 through 13 should be covered. At the undergraduate level, Chapters 11 through 13 should be used on a limited basis, leaving the student more time to concentrate on composite construction and built-up girders.

Chapters 6, 11, 12, 13, and 14 can be used for a graduate course. As a prerequisite for the graduate course, the student must have a minimum of three credit hours in the design of steel structures. The instructor will go into more depth in the presentation of Chapters 6, 11, 12, and 13. The student should be required to submit a term project using rigid frames in multi story buildings. Chapter 14 provides a simple method which can easily be computerized, allowing the student the facility of designing medium- to high-rise buildings using steel frames. U. S. customary units are used throughout, although some examples are presented with S. I. units. To help students convert from U. S. units to S. I. units, tables of conversion are provided in Chapter 1.

The allowable stress design method (ASD) is used predominantly in this book. To help the student appreciate the load and resistance factored design method (LRFD), the text includes discussions and examples utilizing this method.

Fundamental principles of steel design are presented in logical order to encourage the student to tackle the problem of design with a more general perspective. Tables, graphs and other design aids are introduced to help facilitate the design process. The selection of sections in the design of composite construction is made easier by a short-cut approach, thus eliminating the tediousness of the trial-and-error method.

The author gratefully acknowledges Robert M/ Hackett of the University of Mississippi, Joseph P. Colaco of CBM Engineers, Inc., Ramesh Malla of the University of Conneticut, and Mahesh Pendey of the University of Waterloo, Ontario for their comments during the development of the book.

The author expresses his appreciation to the hundreds of students who have attended his classes at North Carolina Agricultural and Technical State University. Without them this book could not have been materialized.

Finally, the author wishes to thank his wife, Dr. Mary Bentley Abu-Saba, for her continued support and encouragement.

DESIGN OF STEEL STRUCTURES

1

Introduction

1.1 INTRODUCTION

Conversion Factors

The building industry in the United States is gradually adopting the new metric-based system referred to as SI units (for System International). For many years, the Congress of the United States has tried to legislate the use of the metric system without success. At present, scientific and technical periodicals and journals in this country are requiring the use of both systems in their publications. In this book, illustrations and example problems use both the SI and U.S. systems. Table 1.1 lists the standard units for both systems.

To help the reader make the transition from one system to another, conversion factors are provided in Table 1.2.

Design Philosophy

The design of structures is a creative process. At the same time, the structural designer must have a basic understanding of the concepts of solid mechanics and be able to work harmoniously with the architect who is in charge of the project, the contractor who will perform the construction, and the owner who will bear the cost of the project and use it. The principal goals of the structural designer are to provide a safe and reliable structure that will serve the function for which it was intended, an

TABLE 1.1 Units of Measurement

Name of Unit	Abbreviation	Use
	U.S. System	
Length		
Foot	ft	Large dimensions, building plans, beam spans
Inch	in.	Small dimensions, size of member cross sections
Area		
Square feet	ft^2	Large areas
Square inches	$in.^2$	Small areas, properties of cross sections
Volume		
Cubic feet	ft^3	Large volumes, quantities of materials
Cubic inches	$in.^3$	Small volumes
Force, Mass		
Pound	lb	Specific weight, force, load
Kip	k	1000 lb
Pounds per foot	lb/ft	Linear load as on a beam
Kips per foot	k/ft	Linear load as on a beam
Pounds per square foot	lb/ft^2 or psf	Distributed load on a surface
Kips per square foot	k/ft^2 or ksf	Distributed load on a surface
Pounds per cubic foot	lb/ft^3 or pcf	Relative density, weight
Stress		
Pounds per square inch	$lb/in.^2$ or psi	Stress in structures
Kips per square inch	$k/in.^2$ or ksi	Stress in structures
	SI System	
Length		
Meter	m	Large dimensions, building plans, beam spans
Millimeter	mm	Small dimensions, size of member cross sections
Area		
Square meter	m^2	Large areas
Square millimeter	mm^2	Small areas, properties of cross sections

Name of Unit	Abbreviation	Use
Volume		
Cubic meter	m^3	Large volumes, quantities of materials
Cubic millimeter	mm^3	Small volumes
Mass		
Kilogram	kg	Mass of materials
Kilograms per cubic meter	kg/m^3	Density
Force (Load on Structures)		
Newton	N	Force or load
Kilonewton	kN	1000 N
Stress		
Pascal	Pa	Stress or pressure ($1 \text{ Pa} = 1 \text{ N/m}^2$)
Kilopascal	kPa	1000 Pa
Megapascal	MPa	1,000,000 pascal
Gigapascal	GPa	1,000,000,000 pascal

economical structure that can be built and maintained within the specified budget, and a structure that is aesthetically acceptable.

Design Loads

Design loads for buildings include dead and live loads. Dead loads consist of the weight of all permanent constructions, including fixed equipment that is placed on the structure. In bridges, it includes the weight of decks, sidewalks, railings, utility posts and cables, and the bridge frame. Live loads are dynamic and vary in time. They include vehicles, snow, personnel, movable machinery, equipment, furniture, merchandise, wind and earthquake forces, and the like. Once a building frame has been selected on the basis of dead and live loads, a check must be made using a combination of these loads. In some regions, a check must include the seismic forces. Member sizes may have to be modified from the initial selection to meet the wind and seismic forces.

Live load tables are provided by almost all building codes. In unusual cases, the design load intensity is established to the satisfaction of a building official. However, the actual distribution of the live load for maximum effect is the responsibility of the design engineer. Table 1.3 lists live load intensities for various occupancies.

TABLE 1.2 Conversion Factors for Measurement Units

To Convert from U.S. Units to SI Units, Multiply by	U.S. Units	SI Units	To Convert from SI Units to U.S. Units, Multiply by
25.4	in.	mm	0.03937
0.3048	ft	m	3.281
645.2	$in.^2$	mm^2	1.550×10^{-3}
16.39×103	$in.^3$	mm^3	61.02×10^{-6}
416.2×103	$in.^4$	mm^4	2.403×10^{-6}
0.09290	ft^2	m^2	10.76
0.02832	ft^3	m^3	35.31
0.4536	lb (mass)	kg	2.205
4.448	lb (force)	N	0.2248
4.448	k (force)	kN	0.2248
1.356	ft-lb (moment)	N-m	0.7376
1.356	k-ft (moment)	kN-m	0.7376
1.488	lb/ft (mass)	kg/m	0.6720
14.59	lb/ft (force)	N/m	0.06853
14.59	k/ft (force)	kN/m	0.06853
6.895	$lb/in.^2$ (force/unit area)	kPa	0.1450
6.895	$k/in.^2$ (force/unit area)	MPa	0.1450
0.04788	lb/ft^2 (force/unit area)	kPa	20.93
47.88	k/ft^2 (force/unit area)	MPa	0.02093
16.02	lb/ft^3 (density)	kg/m^3	0.6242

1.2 STRUCTURAL STEEL AND ITS PROPERTIES IN CONSTRUCTION

Historically, the use of steel in construction for commercial buildings has been widely adopted in the United States. The availability of steel makes it much easier to use. In addition, steel has many characteristics that make it more advantageous than concrete. Structural steel takes less time to erect. The combination of high strength, light weight, ease of fabrication and erection, and many other favorable properties makes steel the material of choice for construction in this country. These properties will be discussed briefly in the following paragraphs.

High Strength

The strength of a construction material is defined by the ratio of the weight it carries to its own weight. When compared with other building

TABLE 1.3 Minimum Uniformly Distributed Live Loads

Occupancy or Use	Live Load (lb/ft^2)
Apartments (see residential)	
Armories and drill rooms	150
Assembly halls and other places of assembly	
Fixed seats	60
Movable seats	100
Balcony (exterior)	100
Bowling alleys, poolrooms, and similar recreational areas	75
Corridors	
First floor	100
Other floors, same as occupancy served except as indicated	
Dance halls	100
Diningrooms and restaurants	100
Dwellings (see residential)	
Garages (passenger cars)[a]	100
Grandstands (see reviewing stands)	
Gymnasiums, main floors and balconies	100
Hospitals	
Operating rooms	60
Private rooms	40
Wards	40
Hotels (see residential)	
Libraries	
Reading rooms	60
Stackrooms	150
Manufacturing	125
Marquees	75
Office buildings	
Lobbies	100
Offices	80
Penal institutions	
Cell blocks	40
Corridors	100
Residential	
Multifamily Houses	
Corridors	60
Private apartments	40
Public rooms	100
Dwellings	
First floor	40
Second floor and habitable attics	30
Unhabitable attics	20

[a] Floors shall be designed to carry 150% of the maximum wheel load anywhere on the floor.

(*cont'd.*)

TABLE 1.3 (*cont'd.*)

Occupancy or Use	Live Load (lb/ft^2)
Hotels	
Corridors serving public rooms	100
Guest rooms	40
Private corridors	40
Public rooms	100
Public corridors	60
Reviewing stands and bleachers[b]	
Schools	
Classrooms	40
Corridors	100
Sidewalks, vehicular driveways, and yards subject to trucking	250
Skating rinks	10
Stairs' fire escapes and exitways	100
Storage warehouse	
Heavy	250
Light	125
Stores	
Retail	
First floor, rooms	100
Upper floors	75
Wholesale	125
Theaters	
Aisles, corridors, and lobbies	100
Balconies	60
Orchestra floors	60
Stage floors	150
Yards and terraces, pedestrians	100

[b] For detailed recommendations, see American Standard Places of Outdoor Assembly, Grandstands and Tents, Z20.3 1950, or the latest revision thereof approved by the American Standard Association, Inc., National Fire Protection Association, Boston, MA.

Source: Minimum Uniformly Distributed Live Loads ASCE 7-88 (Formerly ANSI A58.1), Table 2, p. 4.

materials, steel is considered to have a high strength ratio. This is important in the construction of long-span bridges, tall buildings, and buildings that are erected on relatively poor soil.

Strength and ductility are the two properties that make steel suitable for building structures that otherwise could not have been possible. The strength of steel provides buildings with a minimum number of columns and relatively small members. Its ductility relieves overstressing in certain

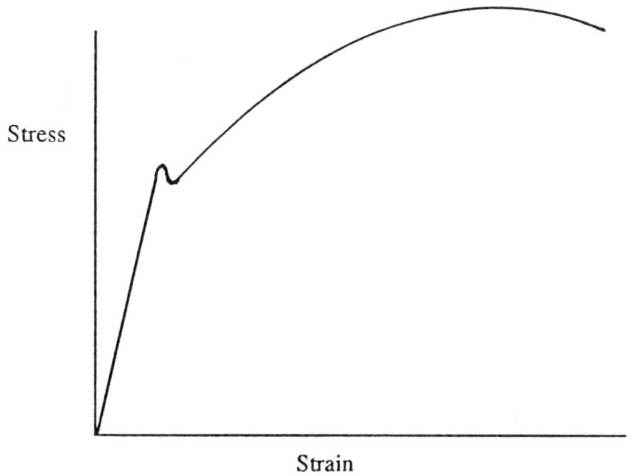

Figure 1.1 Typical stress–strain curve.

members in a given structure by allowing redistribution of stresses due to yielding.

A typical stress–strain diagram for structural steel is shown in Figure 1.1. If we look at this diagram, it can be observed that steel obeys Hooke's law up to the first yield. In this region, stress is directly proportional to strain. Beyond this point, steel experiences a plastic condition momentarily and then enters into the strain-hardening state. At failure, the strain ranges from 150 to 200 times the elastic strain. During strain hardening, the stress continues to increase to a maximum and then drops slightly before failure. At the end of the strain-hardening state, the cross section of the tension specimen is reduced. This characteristic is referred to as *necking*.

1.3 APPLICATIONS

The diagram shown in Figure 1.2 represents a typical interior panel of a library stack room's framing system. It has a reinforced concrete floor slab $4\frac{1}{2}$ in. thick. Tiling weighs 1 lb/ft^2, and ceiling loads are equivalent to 10 lb/ft^2.

Example 1.1

(a) Determine the uniform load on a typical beam and express it in pounds per linear foot.

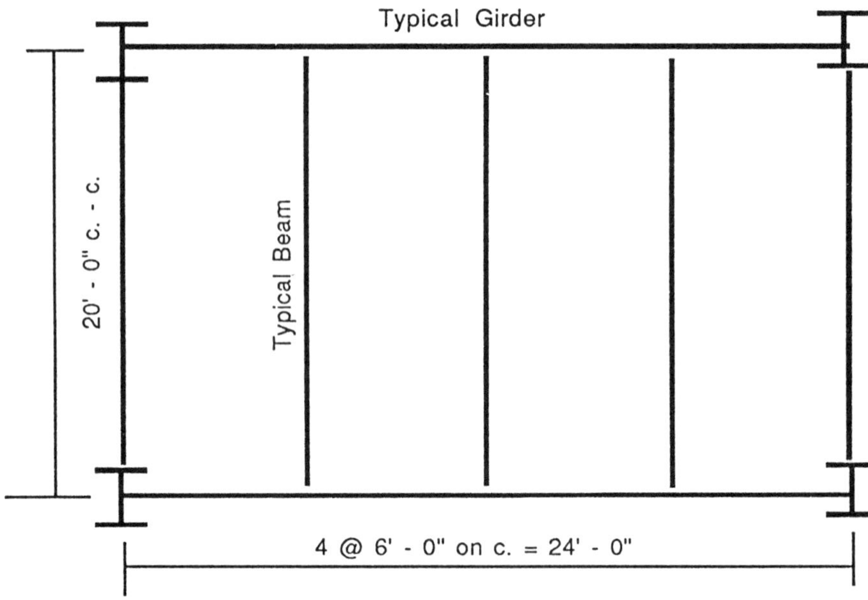

Figure 1.2 Floor plan: typical interior bay.

(b) Assuming that the typical beam and its fireproof coating weighs 50 lb/linear ft, find the concentrated loads on a typical girder.

(c) Show the loading system on the members in **(a)** and **(b)**.

(d) Do part **(a)** and **(b)** using the SI measurement units.

Solution

Typical Beam

In Figure 1.3a, the strip that is 1 ft × 6 ft represents the tributary area for the loading per linear foot on the beam shown in Figure 1.3b. Thus, the uniform load from each load component will be equal to the intensity of that load component times the tributary area. Calculation of the loads on the beam is presented as follows.

Dead Loads

Slab	$1 \times 6 \times 4.5 \times \frac{1}{12} \times 150$	= 337.5 lb/ft
Ceiling load	$1 \times 6 \times 10.0$	= 60.0
Tiling load	$1 \times 6 \times 1$	= 6.0
Subtotal		= **403.5 lb / ft**

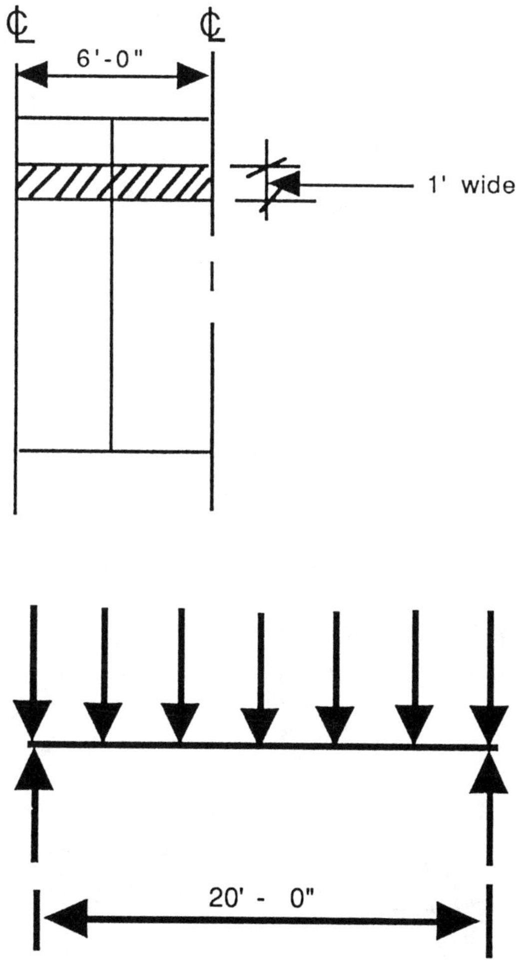

Figure 1.3 Typical beam. (a) Loading strip; (b) simply supported beam.

Live Loads

From Table 1.3, the intensity of the live load is 150 lb/ft^2. Thus, the uniform live load on the beam is calculated as follows:

Live load	$1 \times 6 \times 150$	= 900.0 lb/ft
Total (dead + live)		**= 1303.5 lb / ft**

The intensity of the uniform load on the typical beam shown in Figure 1.4 is therefore found to be 1303.5 lb/ft.

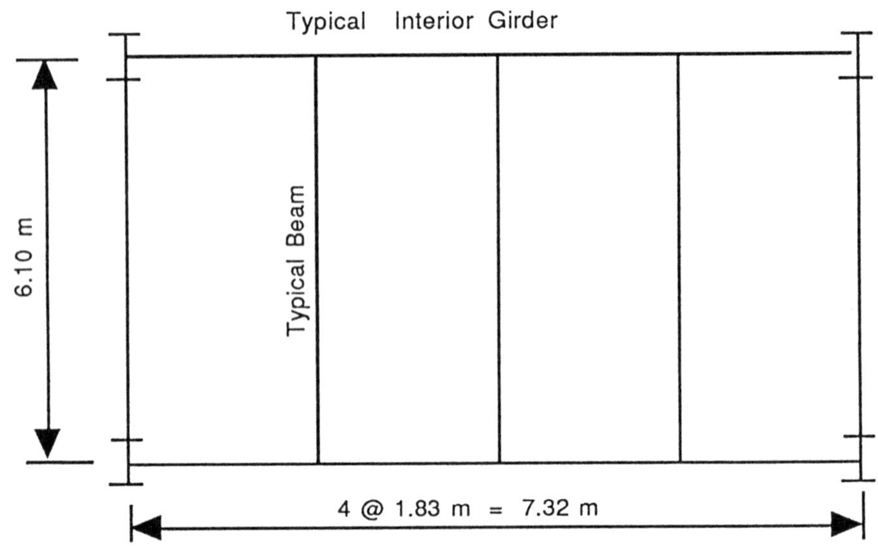

Figure 1.4 Typical interior girder with concentrated loads from typical beams.

Concentrated Loads on Typical Interior Girder

Calculate reaction R at the end of the beam

$$
\begin{aligned}
R &= (DL + LL) \times \frac{L}{2} + \frac{1}{2} \times \text{weight of beam} \\
&= 1303.5 \times \frac{20}{2} + \frac{1}{2} \times 20 \times 50 \\
&= 13{,}035 + 500 \\
&= 13{,}535 \text{ lb}
\end{aligned}
$$

Or expressed in kips, $R = 13.58$ k.

Since an interior girder shown in Figure 1.5 supports typical beams on both sides, the reaction on the girder is double the reaction from one beam. Hence, the concentrated loads on the girder will have the magnitude P of

$$
\begin{aligned}
P &= 13.54 \times 2 \\
&= 27.1 \text{ k}
\end{aligned}
$$

Example 1.2

SI Measurement Units

Repeat the above problem using the SI measurement units showing the floor plan, typical beam, and girder. From Table 1.2, all measurements are

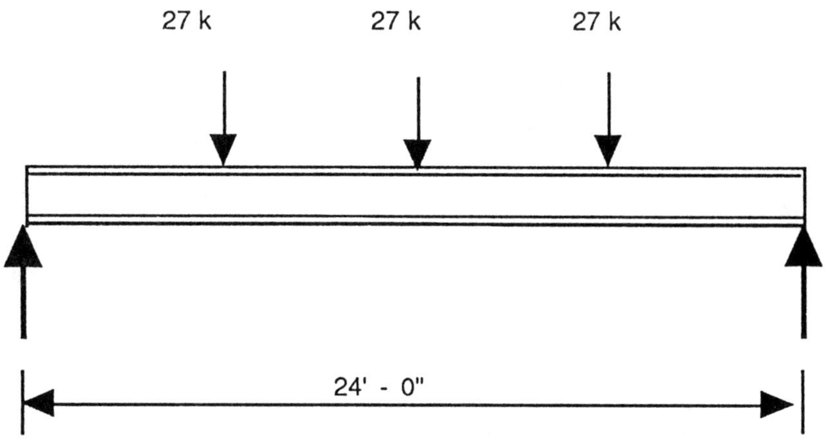

Figure 1.5 Floor plan: typical interior bay in SI units.

converted into SI units. A summary of the results for the above problem is shown below.

Loads	U.S. Units (lb/ft)	Conversion Factor	SI Units (N/m)
		Dead Loads	
Slab	337.5	14.59	4924
Ceiling	60.0		875
Tiling	6.0		88
Subtotal	403.5		5887
		Live Load	
Floor	900.0		13,131
		Weight of Beam	
Beam plus coating	50.0		730
		Total Uniform Load on Beam Including Its Weight	
	1303.5		19,748

Determine the reaction at the end of the beam.

	U.S. Units (lb)	Conversion Factor	SI Units (N)
Reaction at end of beam R	13,535	4.448	60,204
		Concentrated Loads on Girder	
$P = 2R$	27,070		120,408
$P/1000$	27.1 k		120.4 kN

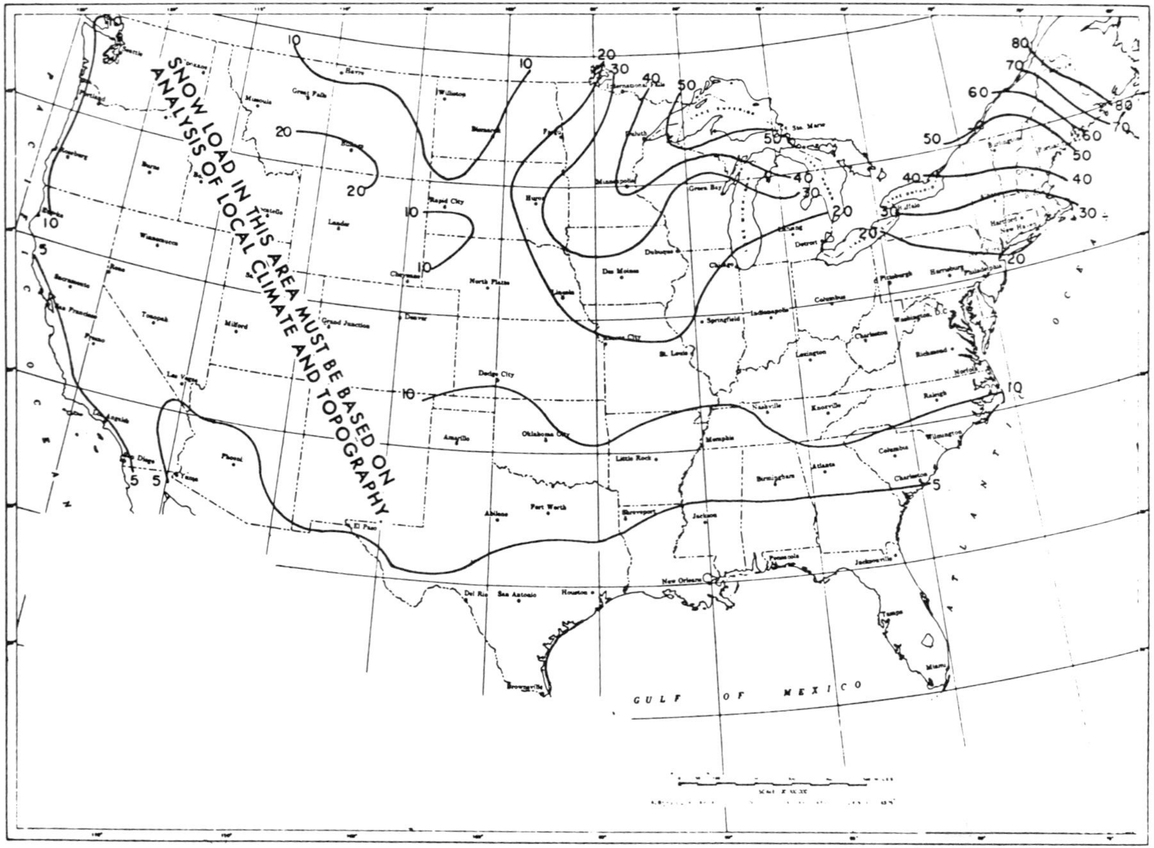

Figure 1.6 Snow load in pound-force per square foot on the ground, 5-year mean recurrence interval.

1.4 LOADS, LOAD FACTORS, AND LOAD COMBINATIONS

In 1986, the AISC introduced the first edition of a manual for steel construction using the load and resistance factor design method (LRFD). It will not be long before designers for steel construction are expected to use this method in preference to the allowable strength method currently in use. LRFD proportions the various members in a given structural system so that no single member or part may exceed the applicable limit state when the structure is subjected to all appropriate factored load combinations. The design strength of each structural member or assemblage must exceed the required strength based on the factored nominal loads.

The following nominal loads are considered in the design of steel structures:

D = dead load due to the weight of the structural element and permanent features on the structure

L = live load due to occupancy and movable equipment

L_r = roof live load

W = wind load

S = snow load (see Figure 1.6 for snow loads in the United States)

E = earthquake load

R = load due to initial rainwater or ice exclusive of the ponding contribution

The required strength of the structure and its elements must be determined from the appropriate critical combination of factored loads listed below:

$$1.4D \tag{1.1}$$

$$1.2D + 1.6L + 0.5(L_r, S, \text{or } R) \tag{1.2}$$

$$1.2D + 1.6(L_r, S, \text{or } R) + (0.5L \text{ or } 0.8W) \tag{1.3}$$

$$1.2D + 1.3W + 0.5L + 0.5(L_r, S, \text{or } R) \tag{1.4}$$

$$1.2D + 1.5E + (0.5L \text{ or } 0.2S) \tag{1.5}$$

$$0.9D - (1.3W \text{ or } 1.5E) \tag{1.6}$$

Exception: The load factor on L in combinations (1.3), (1.4), and (1.5) shall equal 1.0 for garages, areas occupied as places of public assembly, and all areas where the live load is greater than 100 lb/ft^2.

2

Tension Members

2.1 INTRODUCTION

Tension members are simple to design once the forces in these members have been determined. They occur in trusses used for bridges and roofs, towers, bracing systems, cables, and various other applications. A tension member carries only direct axial forces that tend to stretch it. The basic requirement in the selection of a tension member is that it provides enough cross-sectional area so that its maximum carrying capacity is equal to or greater than the designated capacity for a particular design method such as the allowable stress design (ASD) or load and resistance factor design (LRFD).

Figure 2.1 presents two structural systems that include tension members, Structural steel sections of a variety of shapes may be used in the design of tension members. In the past, round bars were used frequently as tension members. Presently, they are not as popular because of the large drift that they cause during a wind storm and the disturbing noise induced by vibrations. Figure 2.2 includes a few examples of the various types of tension members in use in the construction of steel buildings.

2.2 DESIGN CRITERIA

In design, the basic requirement is to provide an assemblage of elements that can resist a given load or combination of loads without exceeding

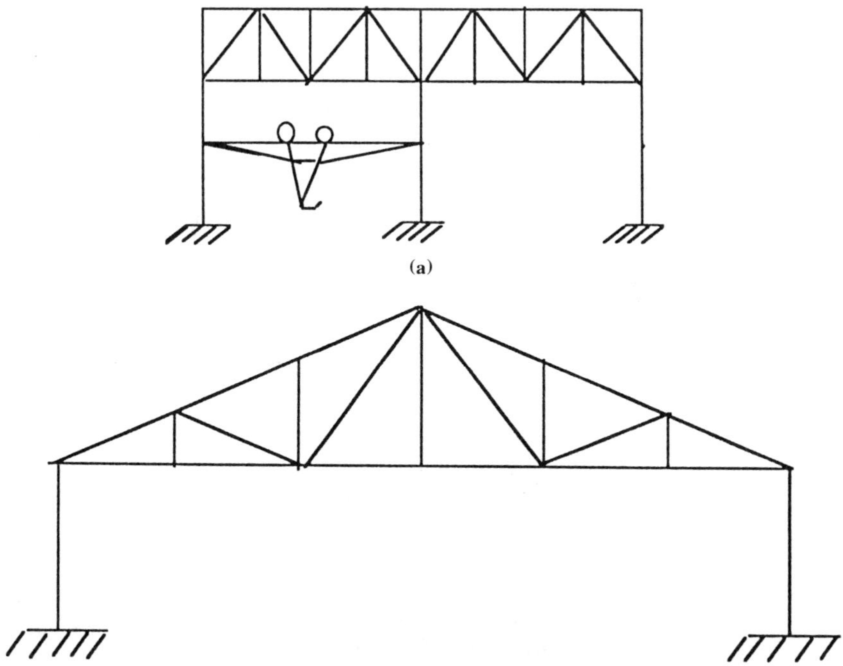

(a)

(b)

Figure 2.1 Typical structures with tension members.

levels of stress or deformation as specified by the building code. The responsibility of the structural designer is to determine the loads that the structure as a whole and with its individual members separately may be subjected to within its life span. As the body of knowledge of the behavior of building materials increases and methods of structural analysis with the availability of sophisticated computer systems are improved, building codes change frequently. Until recently, the AISC had advocated the use of the ASD method. In 1986, AISC introduced the LRFD method as an alternative one. In this text, we will use both methods simultaneously.

2.3 ASD METHOD

The allowable stress F_t shall not exceed $0.60F_y$ on the gross area, nor $0.50F_u$ on the effective net area. For pin-connected members, the allowable stress shall not exceed $0.45F_y$ on the effective net area. The bearing stress on the projected area of a pin shall be equal to or less than $0.90F_y$. In calculating the effective net area of a member in tension, the bolt or

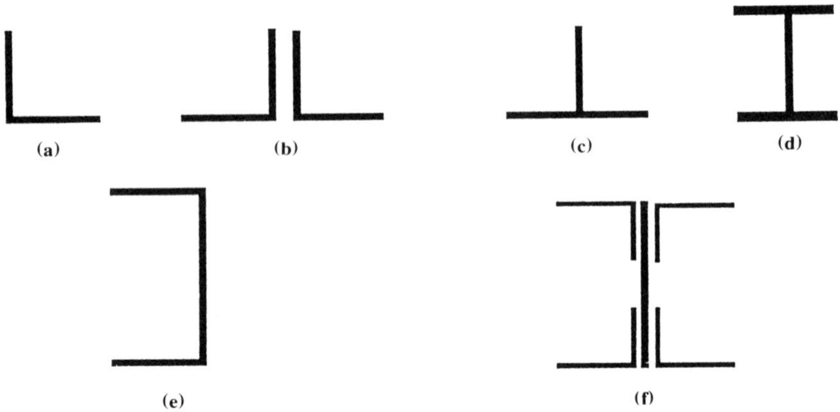

Angle
Double Angles
Structural Tee
W or S Beam
Channel
Built-up Section

Figure 2.2 Structural steel sections used in construction.

rivet hole diameter shall be $\frac{1}{16}$ in. greater than that of the bolt or rivet (AISC, B2). During fabrication, the area around the hole is damaged due to punching. An allowance of $\frac{1}{32}$ in. radially around the hole is made for such damage. Hence, the hole diameter is $\frac{1}{8}$ in. greater than the nominal diameter of the fastener. The determination of the net area of tension members is discussed in another section.

The design strength of a tension member shall be the lower of the two values obtained from Equations (2.1) and (2.2).

For strength based on the gross section

$$P = 0.60F_yA_g \tag{2.1}$$

For strength based on the net section

$$P = 0.50F_uA_e \tag{2.2}$$

where

A_g = the gross sectional area of a tension member
A_e = the effective cross-sectional area of a tension member

When the load is transmitted directly to each of the cross-sectional elements by connectors, the effective area A_e is equal to the net area A_n. If the load is transmitted by bolts or rivets through some but not all the cross-sectional elements of the member, the effective net area A_e is computed by

$$A_e = UA_n$$

where

A_n = net area of the member (in.^{2})
U = reduction coefficient

The following values of U are taken from the AISC specifications, LRFD and ASD, B3:

(a) W, M, or S shapes with flange widths not less than two-thirds the depth and structural tees cut from these shapes, provided the connection is to the flanges. Bolted or riveted connections shall have no fewer than three fasteners per line in the direction of the stress: $U = 0.90$.

(b) W, M, or S shapes not meeting the conditions of the above, structural tees cut from these shapes and all other shapes, including built-up cross sections. Bolted or riveted connections shall have no fewer than three fasteners per line in the direction of the stress: $U = 0.85$.

(c) All members with bolted or riveted connections having only two fasteners per line in the direction of the stress: $U = 0.75$.

For pin-connected members

$$P = 0.45F_yA_e \tag{2.3}$$

The allowable bearing load on the pin

$$P = 0.90F_yDt \tag{2.4}$$

where D and t are the diameter of the pin and thickness of the plate, respectively.

2.4 LRFD METHOD

Structural steel is a ductile material as shown in Figure 2.3. Due to strain hardening, a member without holes and subjected to purely tensile forces

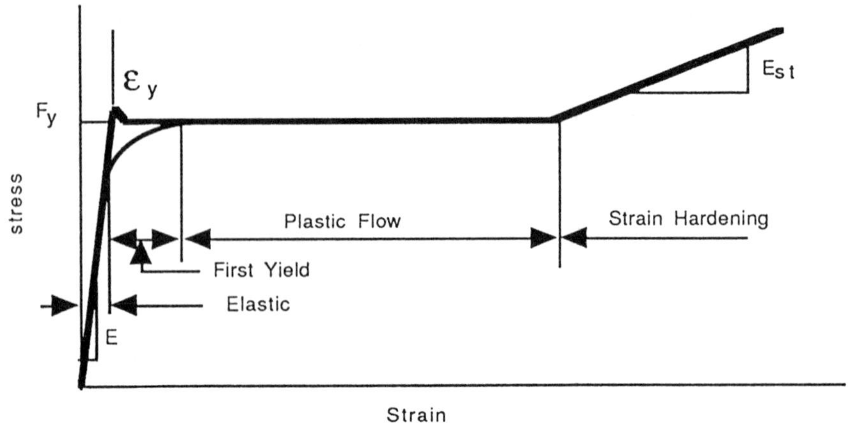

Figure 2.3 Stress vs. strain for structural steel.

can resist a load larger than the product of its cross-sectional area and yield stress. For design purposes, the AISC specifications stipulate that the design strength of a tension member be limited to a value lower than its nominal axial strength P_n. If the ultimate design load is denoted by P_u, then the relationship between the nominal and ultimate design load is expressed by Equation (2.5).

$$P_u = \phi_t P_n \tag{2.5}$$

According to the LRFD specification (D1), the ultimate design load is the lower of the two values obtained from Equations (2.6b) and (2.7b).

For yielding in the gross section

$$\phi_t = 0.90 \tag{2.6a}$$

$$P_n = F_y A_g \tag{2.6b}$$

For fracture in the net section

$$\phi_t = 0.75 \tag{2.7a}$$

$$P_n = F_u A_e \tag{2.7b}$$

where the symbols have the same meaning as in the AISC nomenclature. When the load is transmitted by bolts or rivets through some but not all the cross-sectional elements of the member, the effective net area A_e shall be computed as

$$A_e = UA_n \tag{2.7c}$$

The strength in bearing of milled surfaces, pins in reamed, drilled or bored holes as stated in LRFD specification (J8) is

$$\phi_t = 0.75 \tag{2.8a}$$

$$P_n = 2.0F_y A_e \tag{2.8b}$$

For eyebars and pin-connected members, the design strength shall be the lowest of the following limit states:

(a) Tension on the effective area

$$\phi_t = 0.75 \tag{2.9a}$$

$$P_n = 2tb_{\text{eff}} F_u \tag{2.9b}$$

(b) Shear on the effective area

$$\phi_{sf} = 0.75 \tag{2.9c}$$

$$P_n = A_{sf} F_y \tag{2.9d}$$

(c) Bearing on the projected area

$$\phi = 1.0 \tag{2.9e}$$

$$P_n = A_{pb} F_y \tag{2.9f}$$

where

a = shortest distance from edge of the pin to the edge of the member measured parallel to the direction of the force (in.)

A_{pb} = projected bearing area (in.^{2})

A_{sf} = $2t(a + d/2)$ (in.^{2})

b_{eff} = $2t + 0.63$, but not more than the actual distance from the edge of the hole to the edge of the part measured in the direction normal to the applied force (in.)

d = pin diameter (in.)

t = thickness of the plate (in.)

2.5 EFFECTIVE AREA OF RIVETED AND BOLTED TENSION MEMBERS

Under static loading, the stress around a hole for riveted and bolted members is raised many folds. However, the stress is redistributed due to plastic conditioning before yielding. When a single row of bolts is placed in a tension member as shown in Figure 2.4, the net effective area for load capacity determination is based on the product of the thickness of the plate and the effective width. According to the AISC specification (B3), bolted and riveted splice and gusset plates and other connection fittings subject to tensile forces shall be designed in compliance with the provisions of specification (D1) as listed above. The net effective area shall be taken as the actual net area, except that, for the purpose of design calculations, it shall not exceed 85% of the gross area.

Example 2.1

Determine the net area of the plate shown in Figure 2.4 with the dimensions of $\frac{1}{2}$ in. × 10 in. The diameter of the bolt is $\frac{3}{4}$ in.

Solution

$$\begin{aligned}\text{Effective area} &= A_e\\ &= (\tfrac{1}{2}) \times (10) - (\tfrac{3}{4} + \tfrac{1}{8}) \times \tfrac{1}{2}\\ &= 4.56 \text{ in.}^2\end{aligned}$$

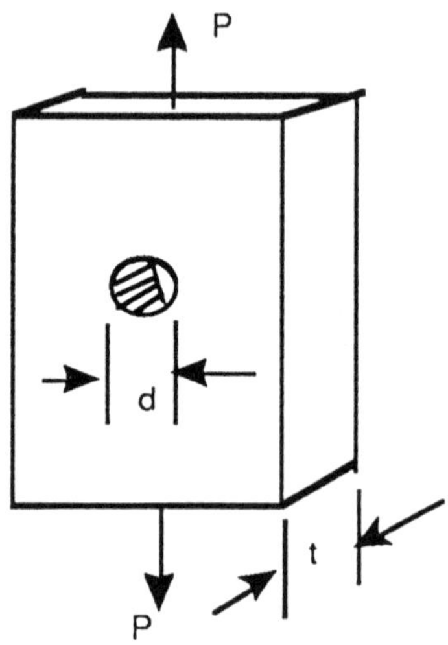

Figure 2.4 Single-bolt connection.

In determining the net area, the diameter of the hole is considered to be $\frac{1}{8}$ in. larger than that of the fastener. This is to allow for the damage in punching of the hole and clearance to facilitate construction. Table 2.1 can be used to obtain the reduction in the area once the thickness of the material and nominal diameter of the hole are known.

Example 2.2

A tension member is made up of a single angle 8 in. × 6 in. × $\frac{5}{8}$ in. with a gross area of 8.36 in.^{2} Two rows of $\frac{3}{4}$-in.-diameter bolt are used. Find the net area. See Figure 2.5.

Solution

Gross area	= 8.36 in.^{2}
Diameter of hole ($\frac{3}{4} + \frac{1}{8}$)	= $\frac{7}{8}$ in.
Thickness of the leg of the angle	= $\frac{5}{8}$ in.
Reduction of area (Table 2.1), 2 at 0.547	= 1.094 in.^{2}
Net area	= 7.266 in.^{2}

2.6 EFFECTIVE AREA FOR STAGGERED HOLES OF TENSION MEMBERS

When there is more than one row of bolts and rivets, it is more efficient to stagger the holes. In the case of no stagger, the net area is determined by subtracting the area taken by the holes from the gross area, as has been shown in Section 2.5. In Figure 2.6a or 2.6b, the net area is the product of the thickness of the plate and the effective width that is obtained by subtracting the sum of the diameter of the holes existing in one column from the gross width of the plate.

For staggered holes as in Figure 2.7, the net width is obtained by deducting from the gross width the sum of the diameters of all holes in the chain, and adding for each gage space in the chain the quantity

$$s^2/4g$$

where

s = longitudinal center-to-center spacing (pitch) of any two consecutive holes (in.)

g = transverse center-to-center spacing (gage) between fastener gage lines (in.)

TABLE 2.1 Reduction of Area for Holes

	Diameter of Hole (in.)							
Thickness (in.)	$\frac{3}{4}$	$\frac{13}{16}$	$\frac{7}{8}$	$\frac{15}{16}$	1	$1\frac{1}{16}$	$1\frac{1}{8}$	$1\frac{13}{16}$
$\frac{3}{16}$	0.141	0.152	0.164	0.176	0.188	0.199	0.211	0.223
$\frac{1}{4}$	0.188	0.203	0.219	0.234	0.250	0.266	0.281	0.297
$\frac{5}{16}$	0.234	0.254	0.273	0.293	0.313	0.332	0.352	0.371
$\frac{3}{8}$	0.281	0.305	0.328	0.352	0.375	0.398	0.422	0.445
$\frac{7}{16}$	0.328	0.355	0.383	0.410	0.438	0.465	0.492	0.520
$\frac{1}{2}$	0.375	0.406	0.438	0.469	0.500	0.531	0.563	0.594
$\frac{9}{16}$	0.422	0.457	0.492	0.527	0.563	0.598	0.633	0.668
$\frac{5}{8}$	0.469	0.508	0.547	0.586	0.625	0.664	0.703	0.742
$\frac{11}{16}$	0.516	0.559	0.602	0.645	0.688	0.730	0.773	0.816
$\frac{3}{4}$	0.563	0.609	0.656	0.703	0.750	0.797	0.844	0.891
$\frac{13}{16}$	0.609	0.660	0.711	0.762	0.813	0.863	0.914	0.965
$\frac{7}{8}$	0.656	0.711	0.766	0.820	0.875	0.930	0.984	1.04
$\frac{15}{16}$	0.703	0.762	0.820	0.879	0.938	0.996	1.05	1.11
1	0.750	0.813	0.875	0.938	1.00	1.06	1.13	1.19
$\frac{1}{16}$	0.797	0.863	0.930	0.996	1.06	1.13	1.20	1.26
$\frac{1}{8}$	0.844	0.914	0.984	1.05	1.13	1.20	1.27	1.34
$\frac{3}{16}$	0.891	0.965	1.04	1.11	1.19	1.26	1.34	1.41
$\frac{1}{4}$	0.938	1.02	1.09	1.17	1.25	1.33	1.41	1.48
$\frac{5}{16}$	0.984	1.07	1.15	1.23	1.31	1.39	1.48	1.56
$\frac{3}{8}$	1.03	1.12	1.20	1.29	1.38	1.46	1.55	1.63
$\frac{7}{16}$	1.08	1.17	1.26	1.35	1.44	1.53	1.62	1.71
$\frac{1}{2}$	1.13	1.22	1.31	1.41	1.50	1.59	1.69	1.78
$\frac{9}{16}$	1.17	1.27	1.37	1.46	1.56	1.66	1.76	1.86
$\frac{5}{8}$	1.22	1.32	1.42	1.52	1.63	1.73	1.83	1.93
$\frac{11}{16}$	1.27	1.37	1.48	1.58	1.69	1.79	1.90	2.00
$\frac{3}{4}$	1.31	1.42	1.53	1.64	1.75	1.86	1.97	2.08
$\frac{13}{16}$			1.59	1.170	1.81	1.93	2.04	2.15
$\frac{7}{8}$			1.64	1.76	1.88	1.99	2.11	2.23
$\frac{15}{16}$			1.70	1.82	1.94	2.06	2.18	2.30
2			1.75	1.88	2.00	2.13	2.25	2.38
$\frac{1}{16}$			1.80	1.93	2.06	2.19	2.32	2.45
$\frac{1}{8}$			1.86	1.99	2.13	2.26	2.39	2.52
$\frac{3}{16}$			1.91	2.05	2.19	2.32	2.46	2.60
$\frac{1}{4}$			1.97	2.11	2.25	2.39	2.53	2.67
$\frac{5}{16}$			2.02	2.17	2.31	2.46	2.60	2.75

Thickness (in.)	*Diameter of Hole (in.)* $\frac{3}{4}$	$\frac{13}{16}$	$\frac{7}{8}$	$\frac{15}{16}$	1	$1\frac{1}{16}$	$1\frac{1}{8}$	$1\frac{13}{16}$
$\frac{3}{8}$			2.08	2.23	2.38	2.52	2.67	2.82
$\frac{7}{16}$			2.13	2.29	2.44	2.52	2.74	2.89
$\frac{1}{2}$			2.19	2.34	2.50	2.66	2.61	2.97
$\frac{5}{8}$			2.30	2.46	2.63	2.79	2.95	3.12
$\frac{3}{4}$			2.41	2.58	2.75	2.92	3.09	3.27
$\frac{7}{8}$			2.52	2.70	2.88	3.05	3.23	3.41
3			2.63	2.81	3.00	3.19	3.38	3.56

Area in square inches = diameter of hole × thickness of material
Source: Courtesy of AISC Steel Manual, Ninth Edition, pp. 4–98.

In calculating the net area for the member as in Figure 2.7, several paths should be considered, and the one with the smallest net area controls.

2.7 TENSION RODS IN DESIGN OF PURLINS

For a steep roof slope, purlins must be supported laterally to account for the component of the vertical load acting in the weak plane of the purlin. Sag rods are installed in the plane of the slope. See Figures 2.8 and 2.9. Usually, there are two sag rods in each bay. They are attached to the ridge purlins that may be constructed of I-beam sections or channels as shown in

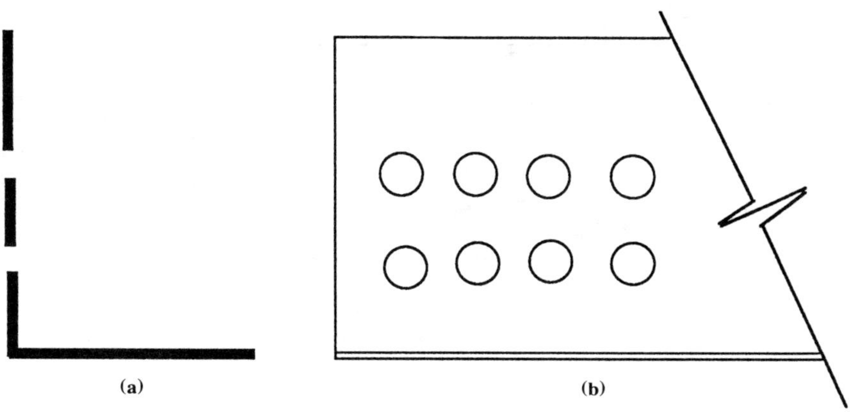

Figure 2.5 Multiple-bolt connection. (a) Cross section. (b) Side view.

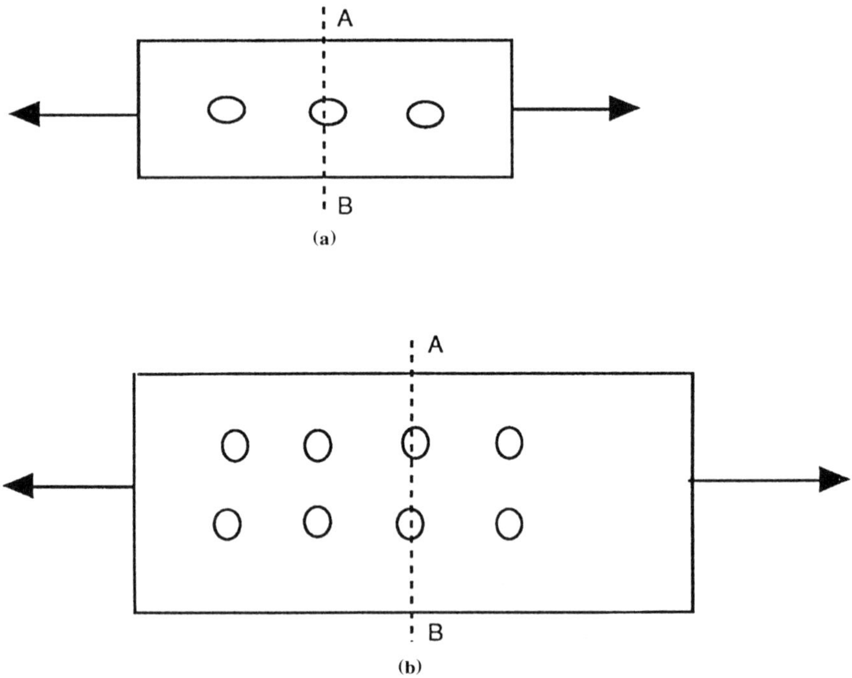

Figure 2.6 Typical connection. (a) Single row. (b) Two nonstaggered rows.

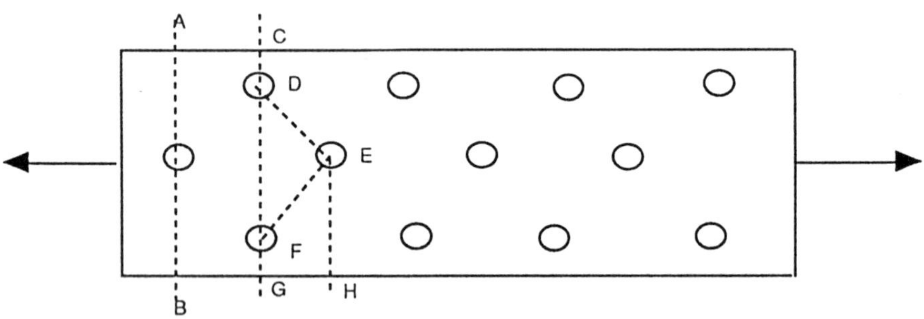

Figure 2.7 Typical staggered bolt connection.

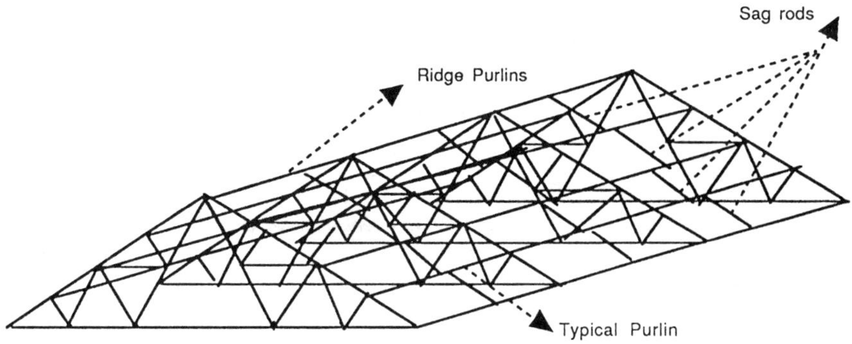

Figure 2.8 Roof truss framing system.

Figure 2.9b. Even though the sag rods cut the span of the purlins in the weak direction, the stress in that direction can be significant and must be accounted for.

2.8 LIMITATION OF LENGTH OF TENSION MEMBERS ON STIFFNESS: SLENDERNESS RATIO

Tension members are not subjected to buckling, and hence the slenderness ratio is not critical in their design. However, to prevent excessive deflection and/or vibration, the AISC specifications impose a limit on the slenderness ratio of 300 (LRFD, First Edition, Part 6, Section B7; ASD, Ninth Edition, Part 5, Section B7). The maximum ratio of 300 is not applicable to rods, which is left to the judgment of the designer.

2.9 APPLICATIONS

Example 2.3

Figure 2.10 shows the plan and a typical truss for the roof system of a warehouse. The warehouse is located in the central United States. The roof is to be constructed of lightweight materials such as corrugated steel sheets. All structural members are to be sprayed with fireproof material good for 1 hr. Determine the loads on the truss, analyze the truss,

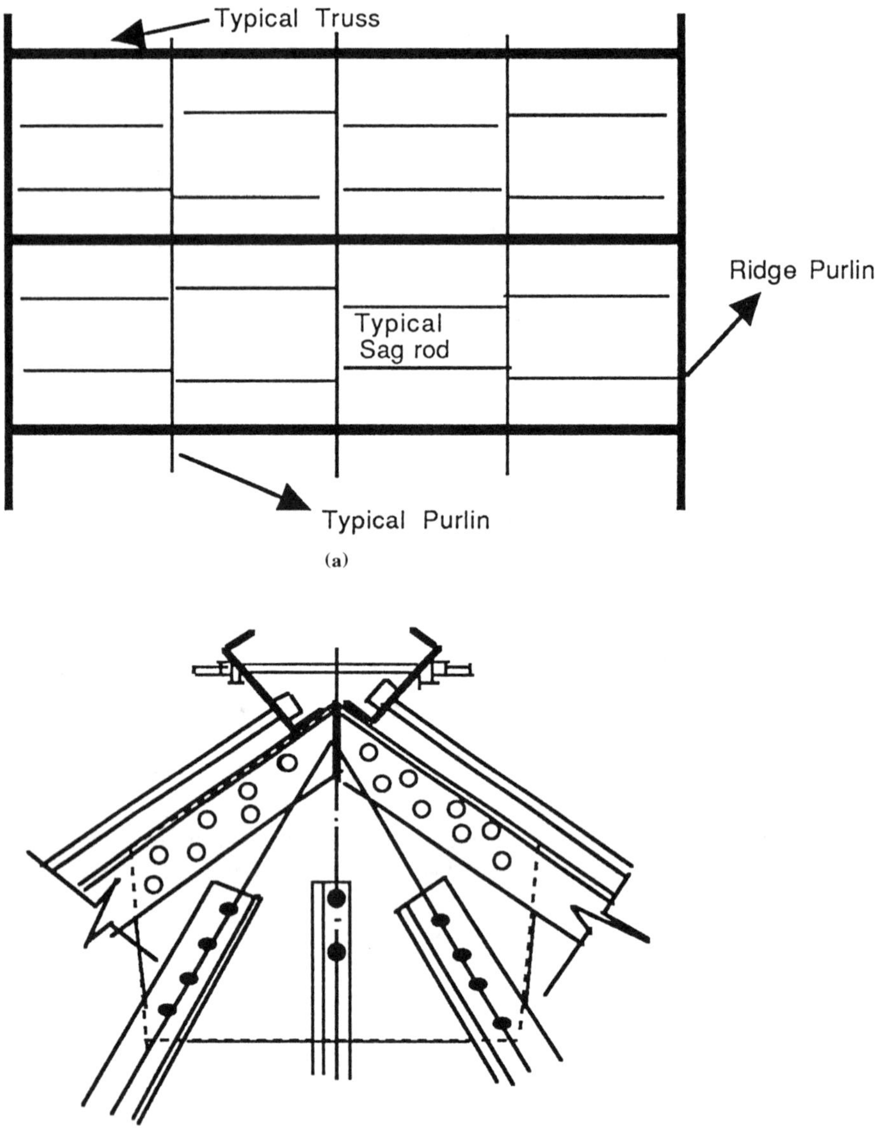

(a)

(b)

Figure 2.9 Roof purlin and tie-rod system. (a) Plan of roof frame. (b) Sag rods for roof purlins.

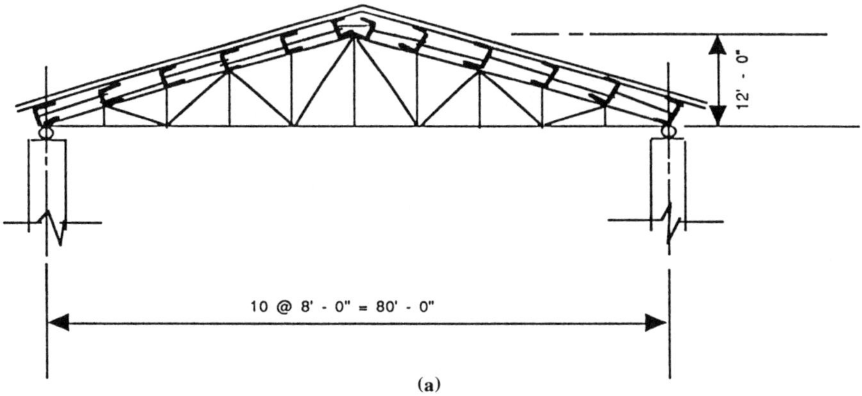

(a)

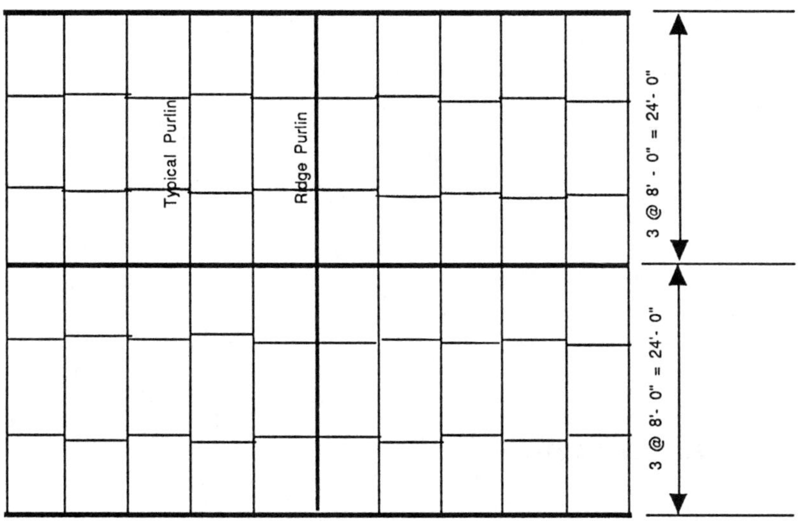

(b)

Figure 2.10 Long-span roof truss system. (a) Typical roof truss. (b) Plan of roof framing system.

determine member forces, and design a typical tension member in each of the following categories: chord, diagonal, and strut.

Solution

First, show the truss members with their geometrical characteristics: length and slope. Second, because of symmetry, only one-half the truss is considered. Third, calculate the roof and ceiling loads and then determine the equivalent node loads. Fourth, determine the reactions at the end of the truss. Fifth, analyze the truss and find the member forces. Then design a typical tension member in each of the following categories: chord, diagonal, and strut.

From Figure 2.10b, the tributary area for each purlin is

$$A_p = 8 \times 24$$
$$= 192 \text{ ft}^2$$

Dead Load Calculations

Weight of steel deck	$= 3.5 \text{ lb/ft}^2$	(see Table 2.2 based on gage 14)
Weight of fire coating	$= 1.5 \text{ lb/ft}^2$	(estimate)
Equivalent weight of purlin on top chord	$= 3 \text{ lb/ft}^2$	(estimate)
Equivalent weight of truss on top chord	$= 1.5 \text{ lb/ft}^2$	(estimate)
Three-ply ready roofing	$= 1 \text{ lb/ft}^2$	
Total dead load on top chord	$= 10.5 \text{ lb/ft}^2$	
Total dead load on top chord joint 192×10.5	$= 2016$ lb, say, 2 k	
Equivalent weight of truss on lower chord	$= 1.5 \text{ lb/ft}^2$	
Total dead load on lower chord 192×1.5	$= 288$ lb, say, $\frac{1}{2}$ k	

Live Load Calculations

Live load on roof	$= 25 \text{ lb/ft}^2$
Total live load on top chord 192×25	$= 4800$ lb, 4.8 k
Ceiling live load	$= 10 \text{ lb/ft}^2$
Total live load on lower chord 192×10	$= 1920$ lb, say, 2 k

Summary of Loads on Top and Bottom Truss Joints

Top (dead + live)	$= 6.8$ k
Bottom (dead + live)	$= 2.5$ k

TABLE 2.2 Properties of $2\frac{1}{2}$ in. × $\frac{1}{2}$ in. Corrugated Steel Sheets

Uncoated						*Galvanized*					
			[b]*Properties per Foot of Corrugated Width*						[b]*Properties per Foot of Corrugated Width*		
U.S. Mfr.'s Gage	Equiv. Thickness (in.)	[a]Weight (lb/ft^2)	*A* (in.^{2})	*I* (in.^{4})	*S* (in.^{3})	Galvanized Sheet Gage	Equiv. Thickness (in.)	[a]Weight (lb/ft^2)	*A* (in.^{2})	*I* (in.^{4})	*S* (in.^{3})
12	0.1046	4.77	1.365	0.0410	0.136	12	0.1084	4.94	1.379	0.0417	0.138
14	0.0747	3.41	0.968	0.0288	0.100	14	0.0785	3.58	0.991	0.295	0.102
16	0.05987	2.73	0.775	0.0229	0.0818	16	0.0635	2.90	0.797	0.0236	0.0839
18	0.0478	2.18	0.620	0.0182	0.0665	18	0.0516	2.35	0.643	0.0189	0.0688
20	0.0359	1.64	0.465	0.0136	0.0509	20	0.0396	1.81	0.487	0.0413	0.0532
22	0.0299	1.36	0.388	0.0113	0.0428	22	0.0336	1.53	0.410	0.0120	0.0451
24	0.0239	1.09	0.310	0.00906	0.0346	24	0.0276	1.26	0.332	0.00971	0.0369
26	0.0179	0.82	0.232	0.00678	0.0262	26	0.0217	0.99	0.255	0.00746	0.0287
28	0.0149	0.68	0.193	0.00564	0.0219	28	0.0187	0.85	0.216	0.00632	0.0245

[a] Weight for roofing style ($27\frac{1}{2}$ in. wide) and no allowance for side or end overlaps.

[b] Steel thickness on which sectional properties were based was obtained by subtracting 0.0020 in. from the galvanized sheet thickness listed. This thickness allowance applies particularly to the 1.25-oz. coating class (commercially).

Source: From Manual of Steel Construction, Seventh Edition, by American Institute of Steel Construction, p. 6-5.

The slopes for the truss members are shown in Figure 2.11. The sum of the total loads on the truss is simply the product of the truss chord nodes and the total load on each one. Since there are nine such nodes, the reaction R at the end of the truss is one-half the total

$$R = \tfrac{1}{2} \times 9 \times (6.8 + 2.5)$$
$$= 41.85, \quad \text{say, } 41.9 \text{ k}$$

The truss loads and reactions are shown acting on the truss in Figure 2.12. The analysis of the truss, being a determinate one, proceeds using the method of joints. Beginning with joint A and summing the forces in the x and y directions to zero yield the required results as shown in Figure 2.13. At joint G, there are two unknown forces. Thus, a two-equation system is written as

$$-0.29F_1 + 0.29F_2 + 32.6 = 0$$
$$-0.96F_1 - 0.96F_2 + 138.7 = 0$$

The solution of these two equations yields

$$F_1 = 128.4 \text{ k}$$
$$F_2 = 16.1 \text{ k}$$

The procedure is continued for each joint to cover the whole truss. At joint K, the horizontal force is found to be 77.0 k. The same result can be obtained by taking moments around joint K. The last step is used as a check for the accuracy of the method. All the member forces are shown in

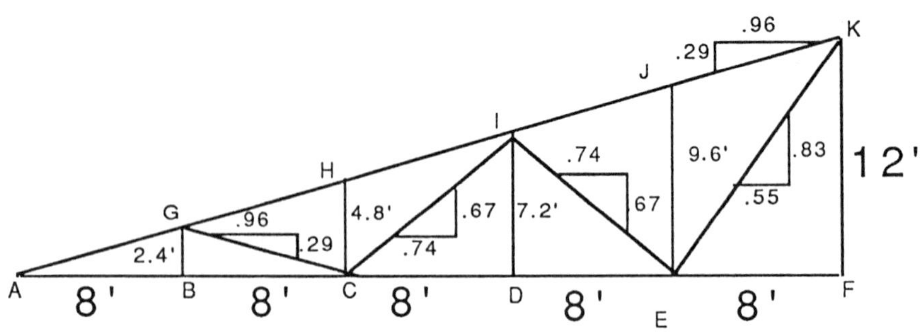

Figure 2.11 Typical loading on roof truss.

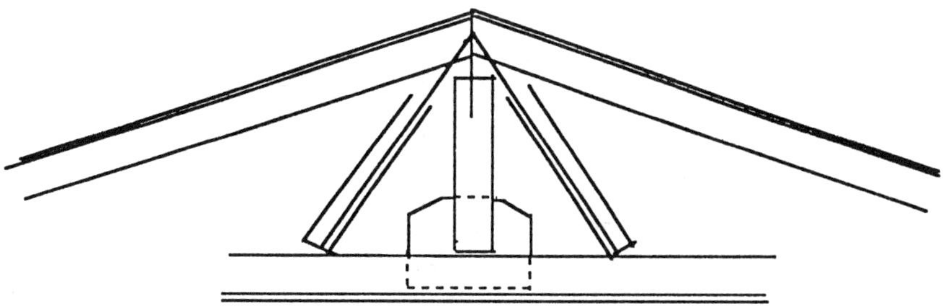

(a)

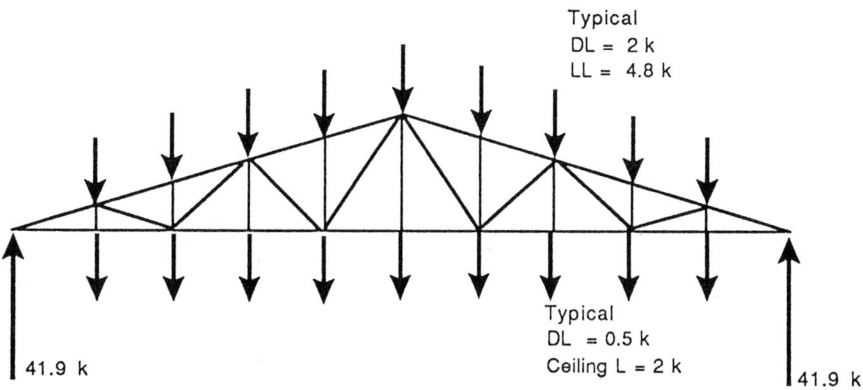

(b)

Figure 2.12 Member direction cosines for truss system. (a) Truss connection. (b) Truss loading.

Figure 2.14. The maximum tensile forces for each category of members are listed as follows:

$$\text{Chord } ABC\ F = 138.7 \text{ k}$$

$$\text{Diagonal } EK\ F = 27.9 \text{ k}$$

$$\text{Strut } BG,\ DI, \text{ and } FK\ F = 2.5 \text{ k}$$

Use a $\frac{3}{4}$-in. Φ bolt, N, STD A-325 steel.

Employ the ASD method. Based on the gross section, from Equation (2.1)

$$P = 0.60 F_y A_g$$

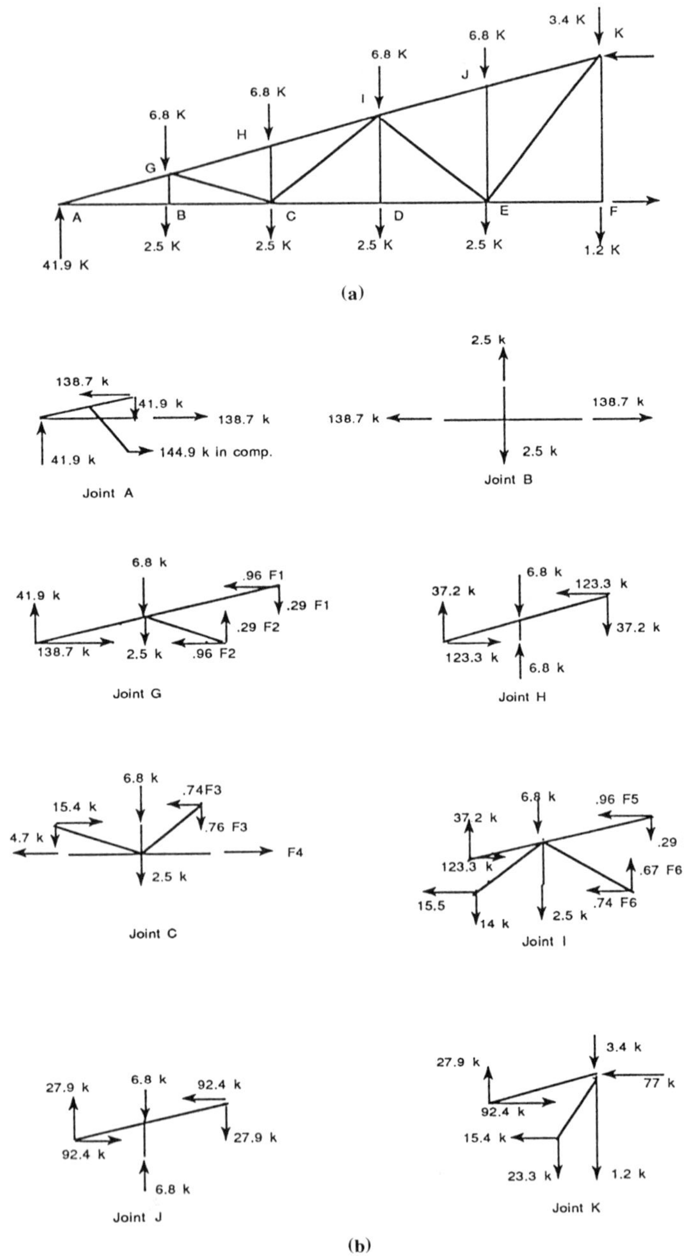

Figure 2.13 Resolution of member forces in the truss.

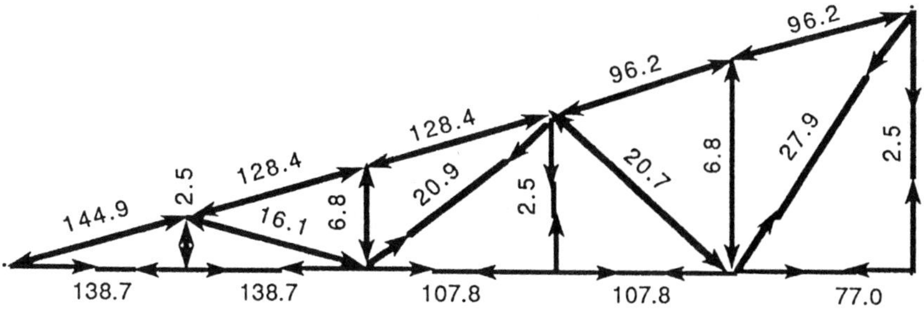

Figure 2.14 Member forces in the truss system.

For the chord member *ABC*, the required gross section is

$$A_g = \frac{P}{0.60F_y}$$

$$= \frac{138.7}{0.60 \times 36}$$

$$= 6.42 \text{ in.}^2$$

Try $2\angle s6 \times 3\frac{1}{2} \times \frac{3}{8}$; the gross sectional area is 6.84 in.^{2}

For the strength based on the net section, it is required to find the net area for a staggered bolt connection as shown in Figure 2.15. For two holes, the net area is obtained from the AISC Manual, 4-97.

$$A_n = 5.53 \text{ in.}^2$$

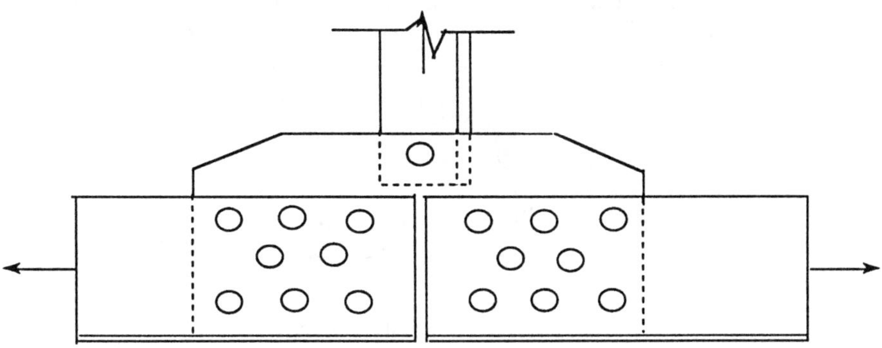

Figure 2.15 Lower chord connection.

From Table 2.1, the reduction in area for one hole is 0.328 in.^{2} There are four holes in a double-angle combination associated with every row of two rivets. Hence, the net area is

$$A_n = 6.84 - 4 \times 0.328$$
$$= 5.53 \text{ in.}^2$$

For the strength based on the net area

$$P = 0.50 F_u A_n$$
$$= 0.50 \times 58 \times 5.53$$
$$= 160.4 \text{ k} > 138.7$$

Check for the staggered condition.

The product of the term $s^2/4g$ and the thickness of the leg is added to the reduced area based on the presence of several holes along a particular path. Note that the above product is multiplied by the number of occurrences that take place along the path of failure

$$A_e = 6.84 - 6 \times 0.305 + 4s^2/4g \times t$$

Using $s = g = 2.5$ in., we find the effective area

$$A_e = 6.84 - (6 \times 0.328) + 2.5 \times \tfrac{3}{8}$$
$$= 5.81 \text{ in.}^2$$

The AISC specifications state that the net area for a connection should not exceed 85% of the gross area

$$0.85 \times 6.84 = 5.81 \text{ in.}^2$$

Notice that the effective area for the path through two holes controls. As a final check, consider the limitation of the slenderness ratio of 300 as required by the AISC specifications for tension members

$$l_x = 96 \text{ in.}$$
$$l_y = 192 \text{ in.}$$

For the double angles of 6 in.$\times$ $3\frac{1}{2}$ in.$\times$ $\frac{3}{8}$ in. and a gusset plate $\frac{3}{8}$ in. thick, the radii of gyrations from the AISC Manual, Ninth Edition, 1-78 are

$$r_x = 1.94 \text{ in.}$$
$$r_y = 1.39 \text{ in.}$$

The slenderness ratio with respect to both axes is

$$l_x/r_x = 96/1.94$$
$$= 49$$
$$l_y/r_y = 192/1.39$$
$$= 138$$

The selection meets all the AISC specification requirements.

Diagonals

The largest force in the diagonals is a tensile force in member *EK*. Since the force is relatively small, the bolt holes will be placed in a single file. Hence, the reduction for the net area is taken from Table 2.1. If we enter the hole size and read vertically down to meet the row that represents the thickness of the plate, a reduction of 0.328 in.^{2} is determined.

Based on the gross section, the required area is

$$A_g = P/0.60F_y$$
$$= 27.9/(0.60 \times 36)$$
$$= 1.29 \text{ in.}^2$$

Try a single angle of $3 \times 2 \times \frac{3}{8}$. The physical properties of the section are

$$A_g = 1.730 \text{ in.}^2$$
$$r_x = 0.940 \text{ in.}$$
$$r_y = 0.559 \text{ in.}$$
$$r_z = 0.430 \text{ in.}$$
$$l_x = l_y = l_z = 173.0 \text{ in.}$$

Note that the slenderness ratio with respect to the z axis exceeds the maximum requirement of 300.

Try another size for a second trial. To minimize the trial-and-error procedure, use the minimum radius of gyration that is admissible

$$r_{\min} = l/300$$
$$= 173/300$$
$$= 0.577 \text{ in.}$$

Using this number, search the table for a single angle of equal legs or unequal legs that has a minimum r value of 0.577 in. The angle $3 \times 3 \times \frac{1}{4}$ seems to fit this requirement

$$A_g = 1.44 \text{ in.}^2$$

$$r_x = 0.930 \text{ in.}$$

$$r_y = 0.930 \text{ in.}$$

$$r_z = 0.592 \text{ in.}$$

The maximum slenderness ratio for this section is

$$l_z/r_z = 173/0.592$$
$$= 292$$

Next, check the strength based on the gross area and net area

$$P = A_g \times 0.60 \times F_y$$
$$= 1.44 \times 0.60 \times 36$$
$$= 31.1 \text{ k}$$

The net area of the section with a single hole is

$$A_n = A_g - 0.328$$
$$= 1.44 - 0.328$$
$$= 1.112 \text{ in.}^2 \text{ controls}$$

Consider 85% of the gross area

$$0.85 \times 1.44 = 1.224 \text{ in.}^2$$
$$P = 0.50F_u \times A_n$$
$$= 0.50 \times 58 \times 1.112$$
$$= 32.2 \text{ k}$$

The capacity of the angle $3 \times 3 \times \frac{1}{4}$ is established to be 31.1 k. At this juncture in the course, the student has not yet been taught how to design connections. Thus, the shear and bearing strength considerations are not included in the solution. These topics will be covered later in this text.

The design of the strut members will be controlled by the slenderness ratio for member *FK*. Its length is 144 in. Using the maximum slenderness ratio of 300 will require a radius of gyration r of 144/300. It means that the minimum acceptable value of r is 0.48 in. A single angle $2\frac{1}{2} \times 2\frac{1}{2} \times \frac{3}{16}$ provides a minimum r of 0.495 in. and an area of 0.902 in.^{2}

Example 2.4

LRFD Method of Design

Figure 2.16a shows the plan of a sports arena. The live and dead loads on the steel joists are given as 194 and 60 lb/ft, respectively. The typical truss is shown in Figure 2.16b. Calculate the truss loads and show then on the upper and lower nodes. Analyze the truss to determine the member forces and then design the tension members in the chord, diagonal, and strut categories. The truss is represented in Figure 2.16b.

Solution

Given Loads

Uniform roof dead load (lightweight roof)	$= 7.5$ lb/ft^2
Uniform roof live load	$= 15.0$ lb/ft^2
Ceiling live loads	$= 10.0$ lb/ft^2
Weight of steel joist (assumed)	$= 20$ lb/ft
Weight of truss (assumed)	$= 60$ lb/ft

Calculate uniform loads on the steel joist.

From Figure 2.16c, the tributary area for the loading on the joist is

$$\begin{aligned} A &= 8 \times 1 \\ &= 9 \text{ ft}^2 \end{aligned}$$

The uniform loads on the steel joist are

Roof dead load	$= 7.5 \times 8$
	$= 60$ lb/ft
Roof live load	$= 15.0 \times 8$
	$= 120$ lb/ft
Ceiling live load	$= 10 \times 8$
	$= 80$ lb/ft
Joist dead load	$= 20$ lb/ft

Calculation of Truss Loads

The loads at the nodes in the truss are equal to the sum of the reactions at the end of the joist for an interior truss and one-half that at an exterior

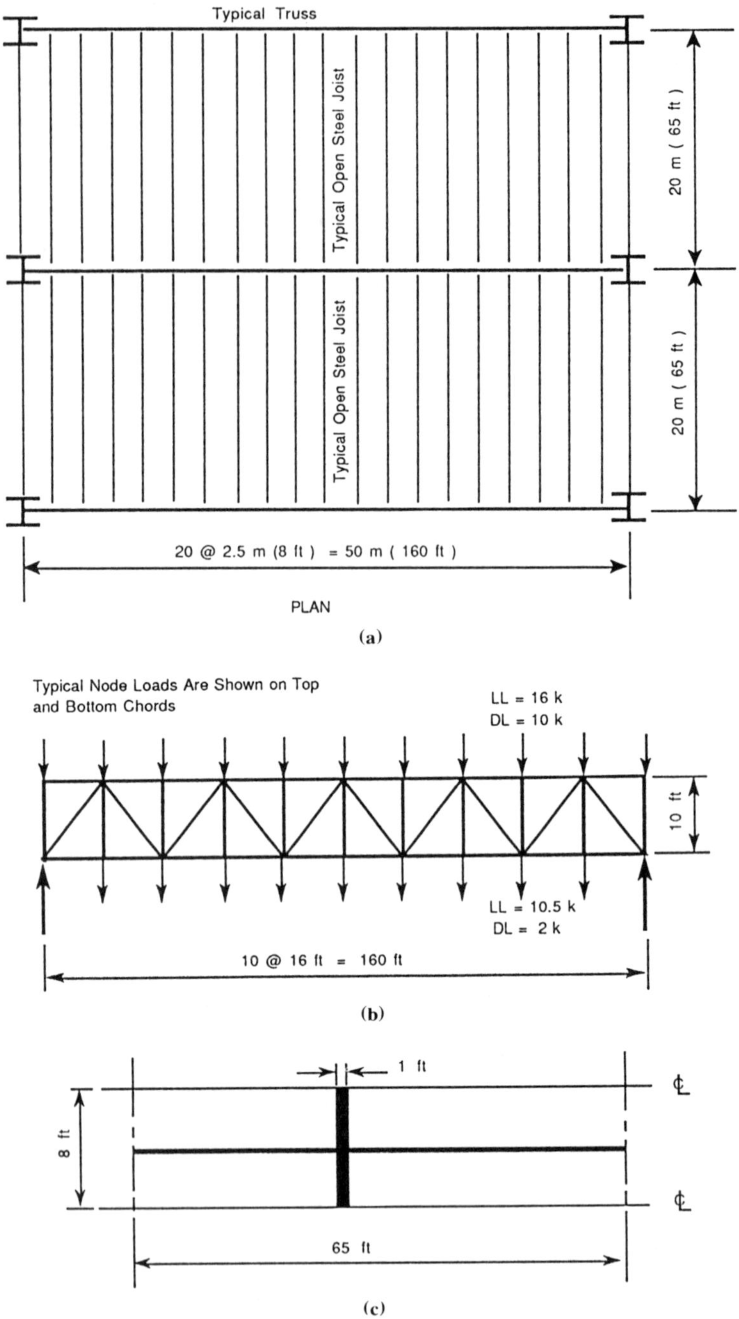

Figure 2.16 (a) Roof plan for a sports arena. (b) Typical node loads on top and bottom chords. (c) Tributary area per foot length of a typical interior truss.

one. Note also that the dead load from the weight of the joist and the truss is divided equally between the top and bottom chord nodes in the truss. This is a simplification that is reasonable.

Top Chord Loads

Total roof dead load on the joist 65 × 60	= 3900 lb
One-half the weight of the joist $\frac{1}{2} \times 65 \times 20$	= 650 lb
Subtotal	= 4550 lb
Two joists for every truss panel 2 × 4550	= 9100 lb
Load from truss weight $2 \times 8 \times \frac{1}{2} \times 60$	= 480 lb
Total dead load on top truss node	= 9580 lb, say, 10 k
Live load from one joist 65 × 120	= 7800 lb
Two joists for every truss panel 2 × 7800	= 15,600 lb, say, 16 k
Total live load on top truss node	= 16 k

Bottom Chord Loads

Dead load from joist $\frac{1}{2} \times 65 \times 20$	= 650 lb
Two joists per truss panel 2 × 650	= 1300 lb
Load from truss weight $16 \times \frac{1}{2} \times 60$	= 480 lb
Subtotal	= 1780 lb, say, 2 k
Total dead load on bottom truss node	= 2 k
Live load on bottom chord of the joist 80 × 65	= 5200 lb
Two joists per panel 2 × 5200	= 10,400 lb, say, 10.5 k
Total live load on bottom truss node	= 10.5 k

The factored loads for the top and lower nodes are based on the LRFD specifications (A4-2)

$$P_u = 1.2D + 1.6L$$

where D and L represent the service dead and live loads, respectively.

The factored loads for the nodes are

Top node

$$P_u = 1.2 \times 10 + 1.6 \times 16$$
$$= 37.6 \text{ k}$$

Lower node

$$P_u = 1.2 \times 2 + 1.6 \times 10.5$$
$$= 19.2 \text{ k}$$

Place these loads on the truss and start analyzing for member forces. See Figure 2.16b. From the given loads on the truss, the reactions can easily be found as seen in Figure 2.17a.The method of joints provides a

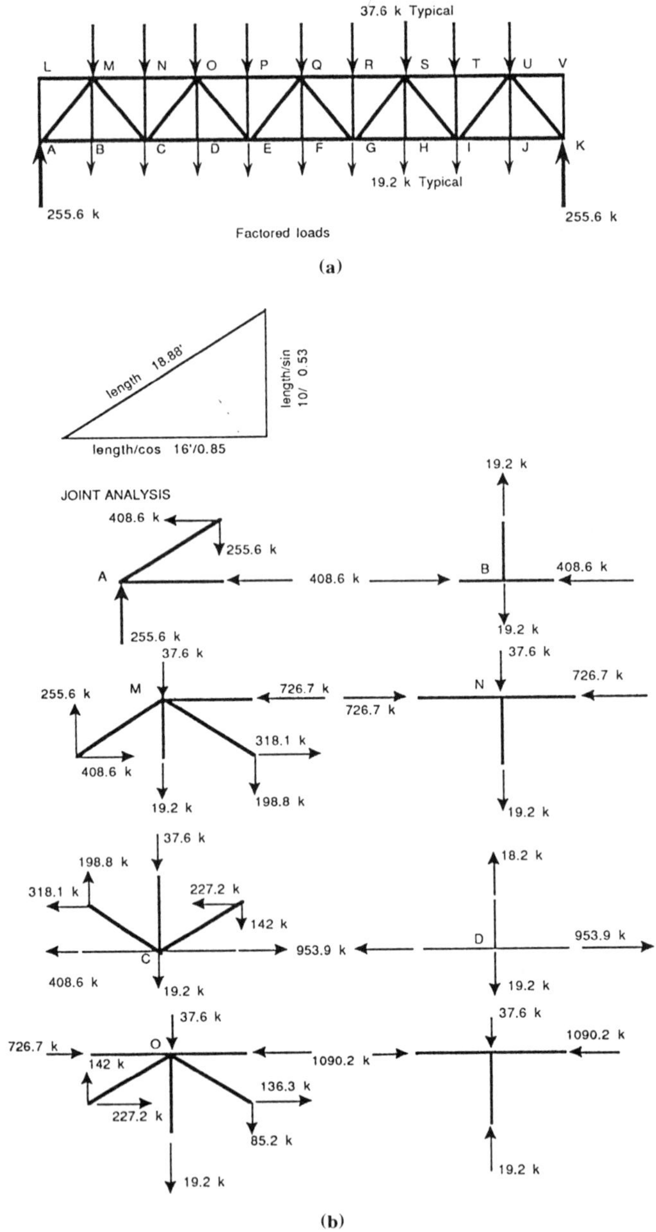

Figure 2.17 (a) Factored load on the truss system. (b) Resolution of member forces for the truss system. (c) Member Forces in the truss.

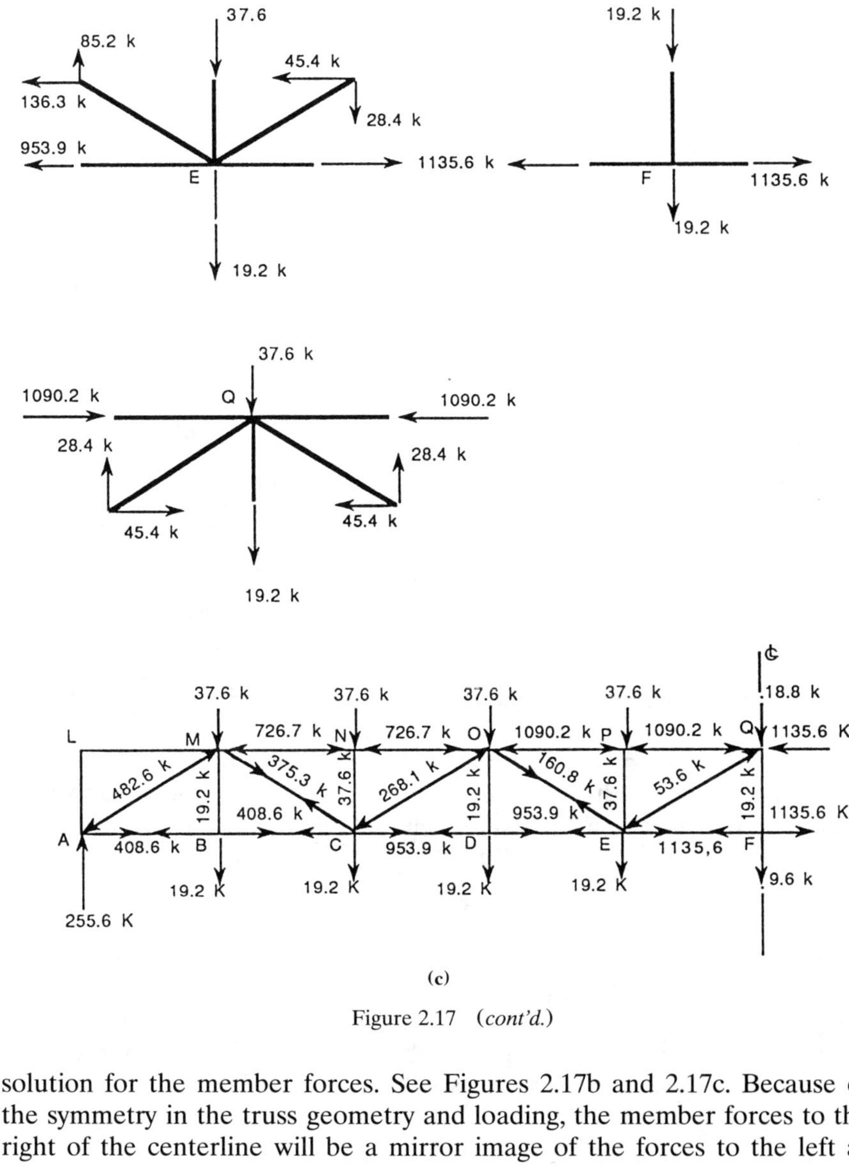

(c)

Figure 2.17 (*cont'd.*)

solution for the member forces. See Figures 2.17b and 2.17c. Because of the symmetry in the truss geometry and loading, the member forces to the right of the centerline will be a mirror image of the forces to the left as shown in Figure 2.17c. The lower chord will be designed in five sections. The middle section will stretch from F to G. The other sections will include the members AC, CE, and their counterparts on the right. The lower chord is subjected to tension and bending stresses. Their design will be left to a later chapter where bending members are discussed. The remaining tension members are the two diagonals and three struts as shown in Figure 2.17c.

Let us consider the design of member *CM*. The tension in this member is 375.3 k. The factored load-induced force in the member is

$$P_u = 375.3 \text{ k}$$

From Equation (2.5),

$$P_u = \Phi_t \times P_n$$

where

$$\Phi_t = 0.90$$

and

$$P_n = F_y \times A_g$$

from which the expression for the gross area is expressed as

$$A_g = \frac{P_u}{\Phi_t F_y} = \frac{375.3}{0.90 \times 36} = 11.58 \text{ in.}^2$$

The strength based on the net section is given by Equations (2.7a) and (2.7b). To find the effective net area, the three paths of the fracture must be discussed. Path one goes through a single hole. Path two goes through a pair of holes, and path three through three staggered holes as shown in Figure 2.18. It is evident that path one is ruled out. For path two, the reduction in the area is

$$A_r = 2 \times (2 \times t \times d_h)$$

For the third path,

$$A_r = 2 \times [3 \times t \times d_h - 2 \times t \times (s^2/4g)]$$

where t and d_h are the thickness of the plate and diameter of the hole, respectively. The other terms are as defined before. The larger of the two values will be subtracted from the gross area to provide the equivalent area.

Using a $\frac{3}{4}$-in.-diameter bolt, A325-N, STD steel in double shear yields the following result. Use a thickness of $\frac{3}{4}$ in. for double angles

$$A_r = 2 \times [2 \times 0.75 \times (\tfrac{3}{4} + \tfrac{1}{8})] = 2.625 \text{ in.}^2 \text{ controls}$$

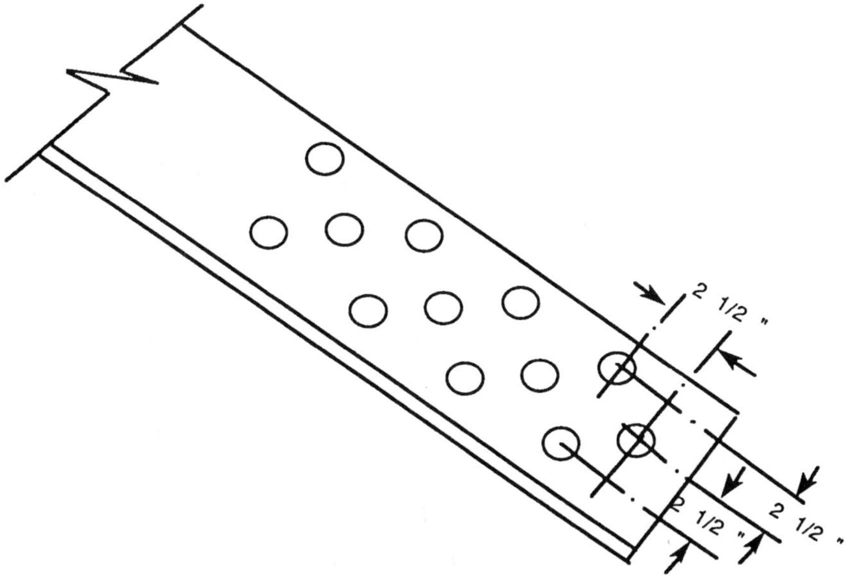

Figure 2.18 Diagonal tension member in the truss system.

and

$$A_r = [3 \times (\tfrac{3}{4} + \tfrac{1}{8}) - \times 2 \times (2.5^2/4 \times 2.5)] \times 2 \times 0.75$$
$$= 2.0625 \text{ in.}^2$$

Thus, the effective area is

$$A_n = A_g - 2.625$$
$$= 11.58 - 2.625$$
$$= 8.955 \text{ in.}^2$$

Based on the gross area, the capacity is

$$P_u = 0.90 \times 36 \times 11.58$$
$$= 375.2 \text{ k}$$

Using the fracture criterion as a basis for strength (LRFD B3-1), we obtain

$$P_u = 0.85 \times 0.75 \times 58 \times 9.5175$$
$$= 351.9 \text{ k}$$

The first of the two values controls.

Based on the above results, a WT12 × 42 with a gross cross-sectional area of 12.4 in.^{2} will suffice.

Example 2.5

SI LRFD

The factored load on the diagonal member in Figure 2.19 is given as 1112 kN. Select a section that can carry this load.

Solution

From Equations (2.5) through (2.7) and based on the tensile strength in the gross section,

$$A_g = \frac{P_u}{\Phi_t F_y}$$

$$= \frac{1112 \times 10^3}{0.90 \times 300 \text{ MPa}}$$

$$= \frac{1.112}{270} \text{ m}^2$$

$$= 4118.5 \text{ mm}^2$$

Choosing from the Canadian Institute of Steel Construction (CISC), Third Edition, try 2∠s100 × 75 × 13 with an area of 4210 mm^2. The radii

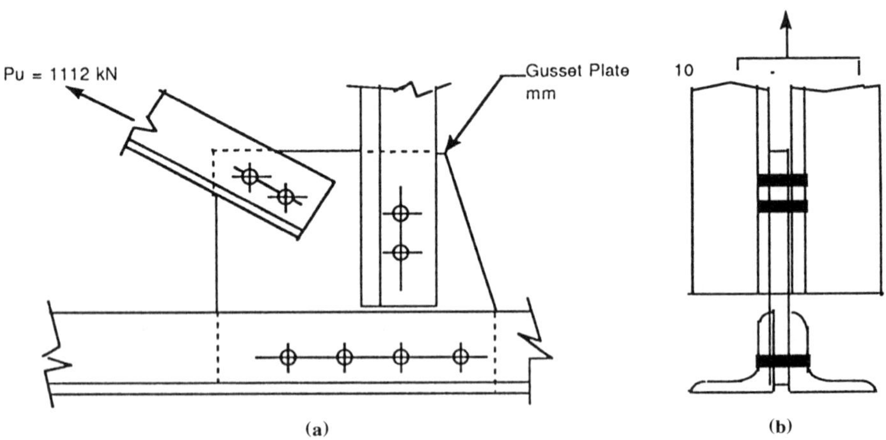

Figure 2.19 Tension member using the SI system of measurement.

of gyration with respect to the x and y axes are 31.1 and 33.6 mm, respectively.

Based on the net section,

$$P_u = 0.85\Phi_t A_n F_u$$
$$= 0.85 \times 0.75 \times 450 \text{ MPa } A_n$$

The required net area is

$$A_n = 3867.5 \text{ mm}^2$$

If we use bolts 20 mm in diameter, the net area is

$$A_n = 4210 - [(20 + 2) \times 10] \times 2$$
$$= 3770 \text{ mm}^2$$

Since this is smaller than required, the needed gross area can be obtained from

$$A_g = 3867.5 + [(20 + 2) \times 10] \times 2$$
$$= 4307.5 \text{ mm}^2$$

Select $2\angle s90 \times 90 \times 13$, $A_g = 4340 \text{ mm}^2$, and the radii of gyration are $r_x = 27.2$ mm and $r_y = 42.2$ mm.

3

Compression Members

3.1 INTRODUCTION

Compression members differ behaviorally from those in tension under load. Whereas tension members remain straight under all levels of loading until they fail, members in compression tend to fail at levels lower than their yield capacity. The inability of compression members to reach yield is attributed to their slenderness. Under compressive loads, a member deflects in a direction perpendicular to that of the load. The deflection occurs along the weaker of the two axes of the section. There are several types of compression members: column, strut, post, stanchion, and top chords of trusses.

It is well established from the basic mechanics of materials that only very short members can reach their yield capacity under compressive loading. Usually, buckling due to instability occurs before the material reaches its full strength. In the middle of the eighteenth century, a Swiss mathematician by the name of Leonhard Euler wrote a significant paper concerning the buckling of compression members. The failure load for a compression member came to be known as the Euler buckling load. Euler's paper marked the beginning of the theoretical and experimental investigation of columns. Research in the theory of column behavior is extensive, and many formulas have been introduced for predicting column behavior under compression. The Euler formula is derived in the next section.

3.2 DERIVATION OF EULER'S FORMULA

Euler's formula applies to a straight, concentrically loaded, homogeneous, long slender member with simply supported ends as shown in Figure 3.1. The x and y axes are oriented as shown in Figure 3.1. The bending moment in the column about the z axis at any point y is given by

$$M_y = -Px \tag{3.1}$$

It is known from the theory of beams that

$$\frac{d^2x}{dy^2} = \frac{M_y}{EI} \tag{3.2}$$

Substituting Equation (3.1) into (3.2) and rearranging the terms yield

$$\frac{d^2x}{dy^2} + \frac{Px}{EI} = 0 \tag{3.3}$$

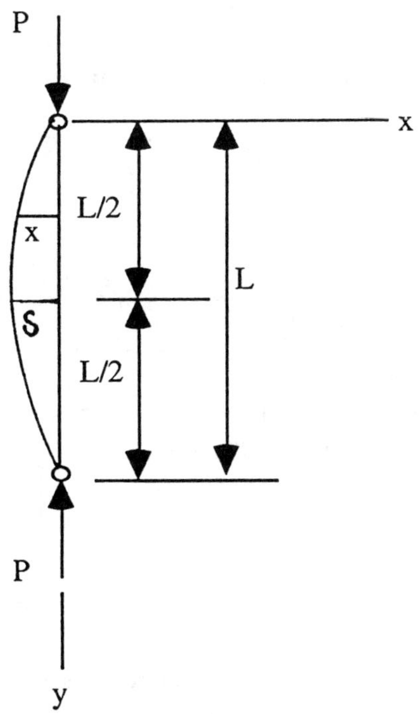

Figure 3.1 Critical axial column load.

Equation (3.3) is an ordinary second-order homogeneous linear differential equation, the solution of which may be expressed as

$$x = A \sin ky + B \cos ky \tag{3.4}$$

where

$$k^2 = \frac{P}{EI}$$

Applying the end conditions of zero deflection at $x = 0$ and $x = L$ leads to three conclusions:

1. $B = 0$
2. $A = \delta$
3. $kL = N\pi$

where N represents the number of buckling mode. Thus,

$$k^2 = \left(\frac{N\pi}{L}\right)^2 \tag{3.5}$$

Using the first buckling load ($N = 1$), we obtain

$$P_{cr} = \frac{\pi^2 EI}{L^2} \tag{3.6}$$

The average compression stress is

$$F_{cr} = \frac{P_{cr}}{A_g} = \frac{\pi^2 EI}{L^2 A_g} \tag{3.7}$$

Using $I = A_g r^2$ where A_g and r represent the gross area of the section and its radius of gyration, respectively, yields the buckling compression stress

$$F_{cr} = \frac{\pi^2 E}{\left(\frac{L}{r}\right)^2} \tag{3.8}$$

The term (L/r) is known as the slenderness ratio that plays a significant role in the design of structural members both in tension and compression. The slenderness ratio must be multiplied by a factor of K that reflects the support condition as seen in Figure 3.2. Thus, Equation (3.8) is modified as follows:

$$F_{cr} = \frac{\pi^2 E}{\left(\dfrac{Kl}{r}\right)^2} \tag{3.9}$$

Early investigators of column behavior observed the discrepancy between theoretical results based on the Euler formula and those obtained from test data. The Euler approach ignores a number of factors that affect the behavior of a column under a compression load. These factors are listed as follows:

1. The stress–strain properties do not remain constant throughout the section.
2. Residual stresses due to cooling after rolling the steel section and those imposed by welding during construction exist in the section before loading.
3. The column may not be perfectly straight as the load is applied to it.
4. Due to construction details, the load is not perfectly concentric.
5. End conditions vary from case to case.
6. Secondary stresses due to bending are developed in the section due to a small deflection in the column.
7. Twisting may occur during loading.

3.3 DESIGN CRITERIA FOR COMPRESSION MEMBERS UNDER CONCENTRIC LOAD: ASD METHOD

According to AISC specifications, the allowable stress on the gross section of an axially loaded member is

$$F_a = \frac{(1 - \frac{1}{2}\lambda^2)F_y}{\frac{5}{3} + \frac{3}{8}\lambda - \frac{1}{8}\lambda^3} \tag{3.10}$$

where

$$\lambda = \frac{(Kl/r)}{C_c}$$

and

$$C_c = \left(2\pi^2 E/F_y\right)^{1/2}$$

	(a)	(b)	(c)	(d)	(e)	(f)
Buckled shape of columns shown by dashed line	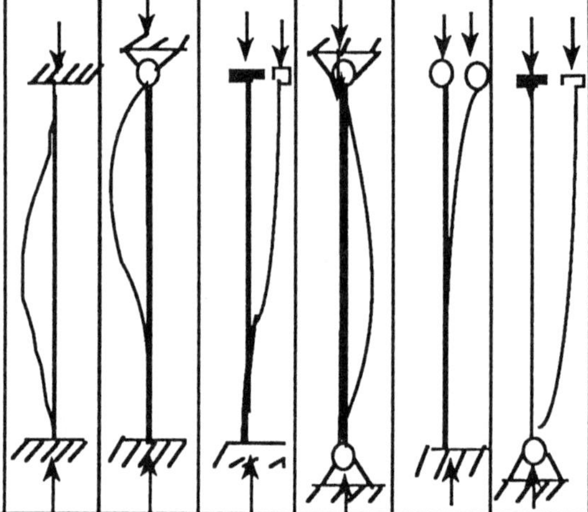					
Theoretical K value	0.5	0.7	1.0	1.0	2.0	2.0
Recommended design valuewhen ideal conditions are approximated	0.65	0.80	1.20	1.0	2.10	2.0

End condition code		Rotation fixed and translation fixed Rotation free and translation fixed Rotation fixed and translation free Rotation free and tanslation free

Figure 3.2 Column effective length. (Figure C-C2-1 from Manual of Steel Construction, ASD, 9th Edition, 1989 by American Institute of Steel Construction.)

for λ equal or less than 1,

$$F_a = \frac{12\pi^2 E}{23(Kl/r)^2} \tag{3.11}$$

for λ greater than 1.

Equations (3.10) and (3.11) are used to prepare Table 3.1 (AISC, Ninth Edition, Table C-36).

3.4 EFFECTIVE LENGTH AND SLENDERNESS RATIO

The design of columns and other compression members conforms with the latest AISC specifications, Section E1 through E6. The effective length factor is determined from Section C2 and can be obtained from Figure 3.2. In determining the slenderness ratio of a member axially loaded, the length is taken as its effective length Kl and r as the corresponding radius of gyration. Compression members consisting of two or more rolled shapes separated by intermittent fillers need to be connected to these fillers at intervals such that the slenderness ratio Kl/r of either element, between the fasteners, does not exceed $\frac{3}{4}$ times the governing slenderness ratio of the built-up member. The smallest radius of gyration r is used in computing the slenderness ratio of each component part. A minimum of two intermediate connectors is recommended to be used along the length of the built-up member. In truss systems, the AISC ASD commentary recommends that $K = 1$ for truss members to be used in determining the effective length of a compression member, and that the slenderness ratio of compression members be limited to 200.

Example 3.1

In Example 2.3, the member *AG* has a compression force of 144.9 k. Assuming a purely concentric action and using A-36 steel, select an adequate member to carry this load. Use double angles. The length of member *AG* is 8.35 ft. See Figure 2.11.

Solution

From the double-angle column load tables, AISC, 3-68, try $2\angle s\ 6 \times 4 \times \frac{1}{2}$.

$$Kl_x = 8.35 \text{ ft}$$

$$P_{cr} = 172 \text{ k}$$

$$Kl_y = 8.35 \text{ ft}$$

$$P_{cr} = 155 \text{ k}$$

TABLE 3.1 Allowable Stress for Compression Members A-36 Yield Stress Steel: ASD Method

Kl/r	F_a (k/in.^{2})	Kl/r	F_a (k/in.^{2})	Kl/r	F_a (k/in.^{2})	Kl/r	F_a (k/in.^{2})	Kl/r	F_a (k/in.^{2})
1	21.56	41	19.11	81	15.24	121	10.14	161	5.76
2	21.52	42	19.03	82	15.13	122	9.99	162	5.69
3	21.48	43	18.95	83	15.02	123	9.85	163	5.62
4	21.44	44	18.86	84	14.90	124	9.70	164	5.55
5	21.39	45	18.78	85	14.79	125	9.55	165	5.49
6	21.35	46	18.70	86	14.67	126	9.41	166	5.42
7	21.30	47	18.61	87	14.56	127	9.26	167	5.35
8	21.25	48	18.53	88	14.44	128	9.11	168	5.29
9	21.21	49	18.44	89	14.32	129	8.97	169	5.23
10	21.16	50	18.35	90	14.20	130	8.84	170	5.17
11	21.10	51	18.26	91	14.09	131	8.70	171	5.11
12	21.05	52	18.17	92	13.97	132	8.57	172	5.05
13	21.00	53	18.08	93	13.84	133	8.44	173	4.99
14	20.95	54	17.99	94	13.72	134	8.32	174	4.93
15	20.89	55	17.90	95	13.60	135	8.19	175	4.88
16	20.83	56	17.81	96	13.48	136	8.07	176	4.82
17	20.78	57	17.71	97	13.35	137	7.96	177	4.77
18	20.72	58	17.62	98	13.23	138	7.84	178	4.71
19	20.66	59	17.53	99	13.10	139	7.73	179	4.66
20	20.60	60	17.43	100	12.98	140	7.62	180	4.61
21	20.54	61	17.33	101	12.85	141	7.51	181	4.56
22	20.48	62	17.24	102	12.72	142	7.41	182	4.51
23	20.41	63	17.14	103	12.59	143	7.30	183	4.46
24	20.35	64	17.04	104	12.47	144	7.20	184	4.41
25	20.28	65	16.94	105	12.33	145	7.10	185	4.36
26	20.22	66	16.84	106	12.20	146	7.01	186	4.32
27	20.15	67	16.74	107	12.07	147	6.91	187	4.27
28	20.08	68	16.64	108	11.94	148	6.82	188	4.23
29	20.01	69	16.53	109	11.81	149	6.73	189	4.18
30	19.94	70	16.43	110	11.67	150	6.64	190	4.14
31	19.87	71	16.33	111	11.54	151	6.55	191	4.09
32	19.80	72	16.22	112	11.40	152	6.46	192	4.05
33	19.73	73	16.12	113	11.26	153	6.38	193	4.01
34	19.65	74	16.01	114	11.13	154	6.30	194	3.97
35	19.58	75	15.90	115	10.99	155	6.22	195	3.93
36	19.50	76	15.79	116	10.85	156	6.14	196	3.89
37	19.42	77	15.69	117	10.71	157	6.06	197	3.85
38	19.35	78	15.58	118	10.57	158	5.98	198	3.81
39	19.27	79	15.47	119	10.43	159	5.91	199	3.77
40	19.19	80	15.36	120	10.28	160	5.83	200	3.73

When an element's width-to-thickness ratio exceeds the noncompact section limits of Section B5.1, see Appendix B5.

Note: C_c = 126.1.

Source: AISC Manual, Ninth Edition, Table C-36.

The section properties are

$$\text{Area} = 9.50 \text{ in.}^2$$
$$r_x = 1.91 \text{ in.}$$
$$r_y = 1.64 \text{ in.}$$

The section properties of a single $\angle 6 \times 4 \times \frac{1}{2}$ are

$$r_x = 1.86 \text{ in.}$$
$$r_y = 1.86 \text{ in.}$$
$$r_z = 1.18 \text{ in.}$$

Check the slenderness ratio of the built-up section

$$\frac{Kl_x}{r_x} = 8.35 \times 12/1.91$$
$$= 52$$
$$\frac{Kl_y}{r_y} = 8.35 \times 12/1.86$$
$$= 54 \text{ controls}$$

From Table 3.1, the allowable compression stress is found to be equal to 17.99 k/in.^{2} Thus, the capacity of the member is calculated as follows:

$$P = 9.50 \times 17.99$$
$$= 170.9 \text{ k}$$

The maximum slenderness of the individual angle is

$$\frac{a}{r_z} = 0.75 \times 54$$
$$= 41$$

where

a = the distance between two connectors to the pair of angles
r_z = the radius of gyration round the z axis

The above slenderness ratio represents the maximum allowable one for an individual angle. Use two intermediate connectors spaced equally. The slenderness ratio of the individual angle is

$$\frac{8.35 \times 12}{(3 \times 1.180)} = 28$$

This value is less than $\frac{3}{4}$ times the slenderness ratio of the built-up section. The selection satisfies the design requirements.

3.5 DESIGN CRITERIA FOR COMPRESSION MEMBERS UNDER CONCENTRIC LOAD: LRFD METHOD

The design strength of compression members is a function of the material and geometrical properties of the section. For members whose elements have a width-thickness ratio less than λ_r, the design strength is $\phi_c P_n$, where

$$\lambda_r = \frac{76}{(F_y)^{1/2}} \tag{3.12}$$

and

$$\phi_c = 0.85 \tag{3.13}$$

For A-36 yield steel, λ_r is 12.7. The factored load P_u is given by

$$P_u = \phi_c P_n \tag{3.14}$$

Since

$$P_n = A_g F_{cr} \tag{3.15}$$

then

$$A_g = \frac{P_u}{\phi_c F_{cr}} \tag{3.16}$$

where

A_g = gross area of member (in.^{2})
F_{cr} = critical stress (k/in.^{2})

The critical stress is obtain from

$$F_{cr} = (0.658 \lambda_c^2) F_y \tag{3.17}$$

for $\lambda_c \leq 1.5$ at $Kl/r = 134$ and

$$F_{cr} = \left(\frac{0.877}{\lambda_c^2} \right) F_y \tag{3.18}$$

for $\lambda_c > 1.5$,

where

$$\lambda_c = \frac{Kl}{r\pi}\sqrt{\frac{F_y}{E}}$$

$$= 0.0112 \times (Kl/r)$$

Equations (3.17) and (3.18) are expressed in Table 3.2 for A-36. A similar table can be prepared from the above equations for any yield-stress steel. Compression members composed of two or more components need to have connectors so that the l/r ratio for either component, between connectors, does not exceed the governing slenderness ratio of the built-up member. For buckling about the x-x axis, both angles move parallel to each other so that the connectors do not have any effect on the strength of the built-up member. On the other hand, for buckling about the y-y axis, the presence of intermediate connectors reduces the slenderness ratio and thus increases the strength of the built-up member. Tabulated loads assume a gusset plate of $\frac{3}{8}$ in. between the angles.

To ensure that shear forces in the connectors between individual shapes remain under control, the modified column slenderness ratio $(Kl/r)_m$ needs to be checked in accordance with AISC, E4.

(a) For snug-tight bolted connectors

$$\left(\frac{Kl}{r}\right)_m = \sqrt{\left(\frac{Kl}{r}\right)^2 + \left(\frac{a}{r_i}\right)^2} \tag{3.19}$$

(b) For welded connectors and fully bolted connectors as required for slip-critical joints

With $a/r_i > 50$

$$\left(\frac{Kl}{r}\right)_m = \sqrt{\left(\frac{kl}{r}\right)^2 + \left(\frac{a}{r_i} - 50\right)^2} \tag{3.20}$$

With $a/r_i \leq 50$

$$\left(\frac{Kl}{r}\right)_m = \left(\frac{Kl}{r}\right)_0 \tag{3.21}$$

TABLE 3.2 Design Stress for Compression Members of A-36 Steel: LRFD Method

Kl/r	$\phi_c F_{cr}$ (k/in.^{2})	Kl/r	$\phi_c F_{cr}$ (k/in.^{2})	Kl/r	$\phi_c F_{cr}$ (k/in.^{2})	Kl/r	$\phi_c F_{cr}$ (k/in.^{2})	Kl/r	$\phi_c F_{cr}$ (k/in.^{2})
1	30.60	41	28.01	81	21.66	121	14.16	161	8.23
2	30.59	42	27.89	82	21.48	122	13.98	162	8.13
3	30.59	43	27.76	83	21.29	123	13.80	163	8.08
4	30.57	44	27.64	84	21.11	124	13.62	164	7.93
5	30.56	45	27.51	85	20.92	125	13.44	165	7.84
6	30.54	46	27.37	86	20.73	126	13.27	166	7.74
7	30.52	47	27.24	87	20.54	127	13.09	167	7.65
8	30.50	48	27.11	88	20.36	128	12.92	168	7.56
9	30.47	49	26.97	89	20.17	129	12.74	169	7.47
10	30.44	50	26.83	90	19.98	130	12.57	170	7.38
11	30.41	51	26.68	91	19.79	131	12.40	171	7.30
12	30.37	52	26.54	92	19.60	132	12.23	172	7.21
13	30.33	53	26.39	93	19.41	133	12.06	173	7.13
14	30.29	54	26.25	94	19.22	134	11.88	174	7.05
15	30.24	55	26.10	95	19.03	135	11.71	175	6.97
16	30.19	56	25.94	96	18.84	136	11.54	176	6.89
17	30.14	57	25.79	97	18.65	137	11.37	177	6.81
18	30.08	58	25.63	98	18.46	138	11.20	178	6.73
19	30.02	59	25.48	99	18.27	139	11.04	179	6.66
20	29.96	60	25.32	100	18.08	140	10.89	180	6.59
21	29.90	61	25.16	101	17.89	141	10.73	181	6.51
22	29.83	62	24.99	102	17.70	142	10.58	182	6.44
23	29.76	63	24.83	103	17.51	143	10.43	183	6.37
24	29.69	64	24.67	104	17.32	144	10.29	184	6.30
25	29.61	65	24.50	105	17.13	145	10.15	185	6.23
26	29.53	66	24.33	106	16.94	146	10.q1	186	6.17
27	29.45	67	24.16	107	16.75	147	9.87	187	6.10
28	29.36	68	23.99	108	16.56	148	9.74	188	6.04
29	29.28	69	23.82	109	16.37	149	9.61	189	5.97
30	29.18	70	23.64	110	16.19	150	9.48	190	5.91
31	29.09	71	23.47	111	16.00	151	9.36	191	5.85
32	28.99	72	23.29	112	15.81	152	9.23	192	5.79
33	28.90	73	23.12	113	15.63	153	9.11	193	5.73
34	28.79	74	22.94	114	15.44	154	9.00	194	5.67
35	28.69	75	22.76	115	15.26	155	8.88	195	5.61
36	28.58	76	22.58	116	15.07	156	8.77	196	5.55
37	28.47	77	22.40	117	14.89	157	8.66	197	5.50
38	28.36	78	22.22	118	14.70	158	8.55	198	5.44
39	28.25	79	22.03	119	14.52	159	8.44	199	5.39
40	28.13	80	21.85	120	14.34	160	8.33	200	5.33

where

$(Kl/r)_0$	= column slenderness ratio of built-up member acting as a unit
a/r_i	= largest column slenderness of individual components
$(Kl/r)_m$	= modified column slenderness of built-up members
a	= distance between connectors
r_i	= minimum radius of gyration of individual component

Example 3.2

The site for the building shown in Figure 3.3 is in the northern region of the United States. Roof dead load is estimated at 10 lb/ft^2. The dead and live load at the ceiling are estimated at 2 and 10 lb/ft^2, respectively. Calculate the loads on the top and lower nodes for the truss shown in Figure 3.3b; analyze the truss and design member GH using the LRFD method and A-36 yield steel.

Solution

The snow load for the northern region of the United States is 30 lb/ft^2. The loads at nodes G and B for the typical interior truss are calculated below.

Top node G

$$\begin{aligned} \text{Live load} &= 10 \times 10 \times 30 \\ &= 3000 \text{ lb} \\ \text{Dead load} &= 10 \times 10 \times 10 \\ &= 1000 \text{ lb} \end{aligned}$$

The factored load

$$\begin{aligned} P_u &= 1.2 \times 1000 + 1.6 \times 3000 \\ &= 6000 \text{ lb} \\ &= 6 \text{ k} \end{aligned}$$

Lower node B

$$\begin{aligned} \text{Live load} &= 10 \times 10 \times 10 \\ &= 1000 \text{ lb} \\ \text{Dead load} &= 10 \times 10 \times 2 \\ &= 200 \text{ lb} \end{aligned}$$

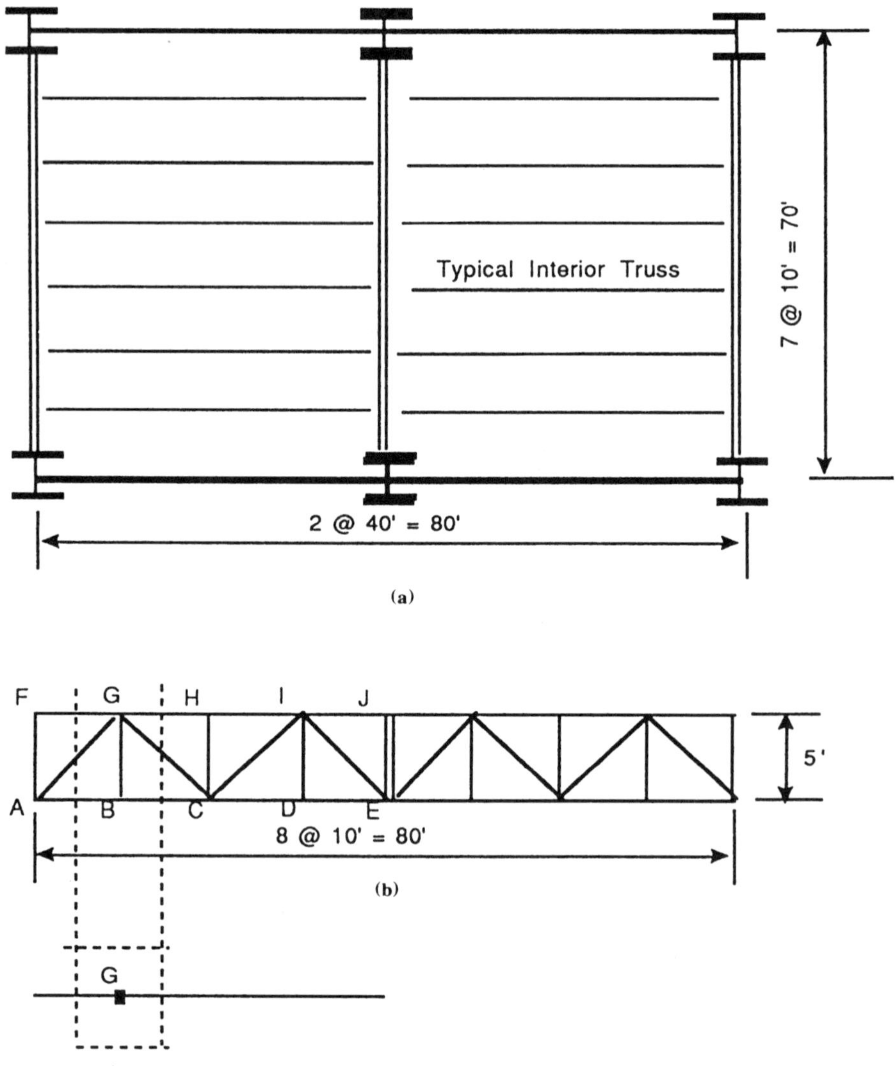

Figure 3.3 Building floor plan.

The factored load

$$P_u = 1.2 \times 200 + 1.6 \times 1000$$
$$= 1840 \text{ lb}$$
$$= 1.84 \text{ k}$$

Referring to Figure 3.4a and 3.4b and taking moments around H yield

$$F_{GH} = 31.36 \text{ k/in.}^2$$

Try a Kl/r of 120. From Table 3.2, the design compression stress is 14.34 k/in.^{2} Using Equation (3.16) yields a gross area of 2.19 in.^{2}. Select $2\angle s\ 3 \times 2\frac{1}{2} \times \frac{1}{4}$. The section properties are

$$A = 2.63 \text{ in.}^2$$
$$r_x = 0.945 \text{ in.}$$
$$r_y = 1.12 \text{ in.}$$

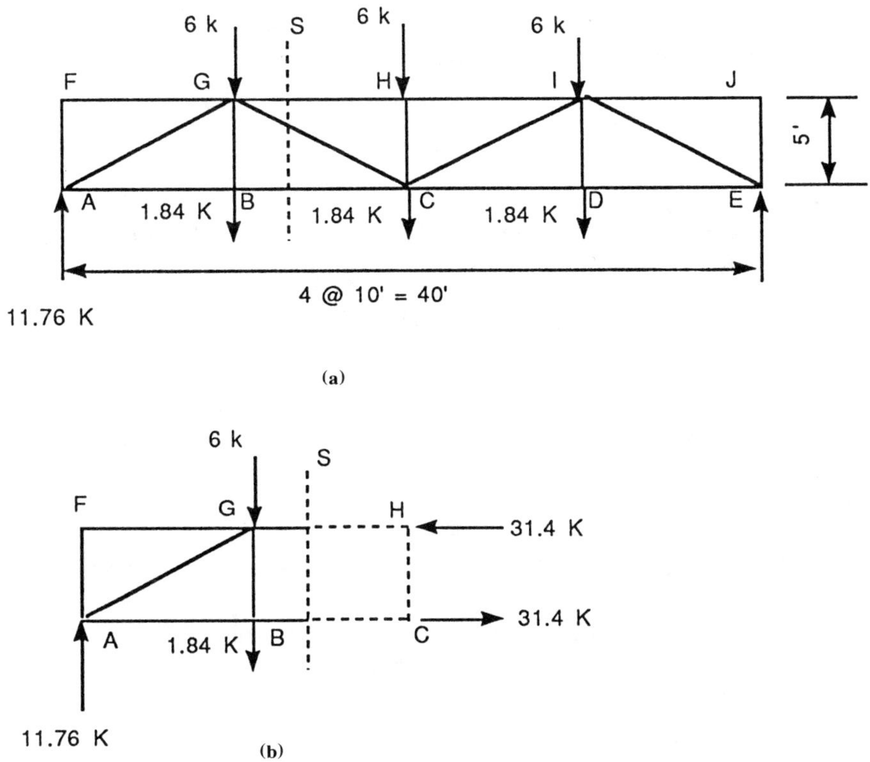

Figure 3.4 Typical truss load system.

and for the single angle

$$r_z = 0.528 \text{ in.}$$

Check the slenderness ratio for this selection. The length l is 10 ft.

$$\frac{l_x}{r_x} = \frac{10 \times 12}{0.945}$$

$$= 127 \text{ controls}$$

$$\frac{l_y}{r_y} = \frac{10 \times 12}{1.12} = 107$$

The compression strength taken from Table 3.2 is

$$F_{cr} = 13.09 \text{ k/in.}^2$$

$$P_u = 2.63 \times 13.09$$

$$= 34.4 \text{ k} > 31.6 \text{ k}$$

Using a connector at midpoint between the supports for the above selection yields a slenderness ratio of

$$\frac{l_z}{r_z} = \frac{\frac{1}{2} \times 10 \times 12}{0.528}$$

$$= 114 < 127$$

Check that the modified slenderness ratio does not govern. Since $l_z/r_z >$ 50, use Equation (3.20)

$$\left(\frac{K_x l_x}{r_x}\right)_m = \sqrt{(107)^2 + (114 - 50)^2}$$

$$= 125 < 127$$

The selection of $2\angle s\ 3 \times 2\frac{1}{2} \times \frac{1}{4}$ satisfies the LRFD specification requirements.

3.6 SI LRFD DESIGN CRITERIA (AXIAL COMPRESSION)

The factored axial compressive strength for rolled structural steel shapes manufactured in Canada is obtained from the following

(a) $0 < \dfrac{Kl}{r} < 13$

$$P_{ucr} = \phi_c A F_y$$

(b) $13 < \frac{Kl}{r} < 89$

$$P_{ucr} = \phi_c A F_y (1.035 - 0.202\lambda_c - 0.222\lambda_c^2) \qquad (3.22)$$

(c) $89 < \frac{Kl}{r} < 178$

$$P_{ucr} = \phi_c A F_y (-0.111 + 0.636\lambda_c^{-1} + 0.087\lambda_c^{-2}) \qquad (3.23)$$

(d) $178 < \frac{Kl}{r} < 320$

$$P_{ucr} = \phi_c A F_y (0.009 + 0.877\lambda_c^{-2}) \qquad (3.24)$$

(e) $320 < \frac{Kl}{r}$

$$P_{ucr} = \phi_c A \left[\frac{1,970,000}{\left(\frac{Kl}{r}\right)^2} \right] \qquad (3.25)$$

where for A-36 steel

$$\lambda_c = 0.0112 \frac{Kl}{r}$$

Table 3.3 is based on the Canadian code. It is calculated in accordance with Clause 13.3.1 of CAN3-S16.1-M78. $\phi_c = 0.90$.

Example 3.3

Find the factored compressive resistance of a W14 × 257 column of ASTM A-36 steel. The column has an unsupported length in the x and y directions of 11 ft, 7 in. Use the SI system.

TABLE 3.3 Design Stress for Compression Members F_y = 248 MPa

Kl/r	F_{cr}	Kl/r	F_{cr}	Kl/r	F_{cr}	Kl/r	E_{cr}	Kl/r	E_{cr}
1	223	41	200	81	149	121	90.4	161	59.8
2	223	42	199	82	148	122	89.4	162	59.3
3	223	43	198	83	146	123	88.4	163	58.7
4	223	44	197	84	145	124	87.4	164	58.2
5	223	45	196	85	143	125	86.4	165	57.7
6	223	46	195	86	142	126	85.5	166	57.1
7	223	47	194	87	140	127	84.5	167	56.6
8	223	48	192	88	138	128	83.6	168	56.1
9	223	49	191	89	137	129	82.70	169	55.6
10	223	50	190	90	135	130	81.8	170	55.1
11	223	51	189	91	133	131	80.9	171	54.6
12	223	52	188	92	131	132	80.0	172	54.1
13	223	53	187	93	129	133	79.2	173	53.6
14	223	54	186	94	127	134	78.3	174	53.1
15	222	55	184	95	126	135	77.5	175	52.6
16	221	56	183	96	124	136	76.7	176	52.2
17	221	57	182	97	122	137	75.9	177	51.7
18	220	58	181	98	121	138	75.1	178	51.3
19	219	59	180	99	119	139	74.3	179	50.6
20	218	60	178	100	117	140	73.6	180	50.1
21	218	61	177	101	116	141	72.8	181	49.6
22	217	62	176	102	114	142	73.0	182	49.0
23	216	63	174	103	113	143	71.3	183	48.5
24	215	64	173	104	111	144	70.6	184	48.0
25	214	65	172	105	110	145	70.8	185	47.5
26	214	66	171	106	108	146	69.2	186	47.0
27	213	67	169	107	107	147	68.5	187	46.6
28	212	68	168	108	106	148	67.9	188	46.1
29	211	69	167	109	104	149	67.2	189	45.6
30	210	70	165	110	103	150	66.5	190	45.2
31	209	71	164	111	102	151	65.9	191	44.7
32	208	72	162	112	101	152	65.2	192	44.3
33	208	73	161	113	99.4	153	64.6	193	43.8
34	207	74	160	114	98.2	154	64.0	194	43.4
35	206	75	158	115	97.0	155	63.4	195	43.0
36	205	76	157	116	95.9	156	62.8	196	42.6
37	204	77	155	117	94.8	157	62.2	197	42.2
38	203	78	154	118	93.7	158	61.6	198	41.8
39	202	79	152	119	92.6	159	61.0	199	41.4
40	201	80	151	120	91.5	160	60.4	200	41.0

Source: Adapted from Canadian Handbook of Steel Construction, CAN3-S16.1-M78, clause 13.3.1.

Solution

The section properties for both the U.S. and SI systems are listed below.

	U.S. System	SI System
A	75.6 in.^{2}	48,774 mm^2
I_x	3400 in.^{4}	1415×10^6 mm^4
I_y	1290 in.^{4}	537
r_y	4.13 in.	105 mm
r_x/r_y	1.62	1.62
l_x, l_y	139 in.	3531 mm

The slenderness ratio of the column (using $K = 1$) is controlled by the y-axis characteristic

$$Kl/r_y = 34$$

$$F_{cr} = 223 \text{ MPa}$$

$$P_{ucr} \text{ (compressive load capacity)} = 10{,}876.6 \text{ kN} \quad (2445 \text{ k})$$

3.7 COMPRESSION MEMBERS IN BRACED FRAMES: ASD METHOD

So far, the discussion of compression members has dealt with truss systems. Although trusses are quite prevalent in the construction industry, columns play a more significant role. Steel columns in buildings receive their loads from beams and girders that frame into them. Forces and moments are transmitted to columns through the beams and girders at each floor or roof level in the structure. In braced frames with standard connections, the beams and girders are assumed to be simply supported. Hence, they have zero moments at their ends. However, the connection between beam and column dictates that the reactions transmitted from the beams and girders act outside the centroidal axes of the column. For the purpose of illustration, see Figure 3.5. If the reactions on both sides are equal, the moment M will be identical to zero. An imbalanced condition of the moments in the column at the connection develops when the live load is removed from one side and not from the other when both spans are equal. Another imbalanced moment condition arises when the spans of the girders or beams are unequal, or the live loads on both sides of the column differ by more than 25%.

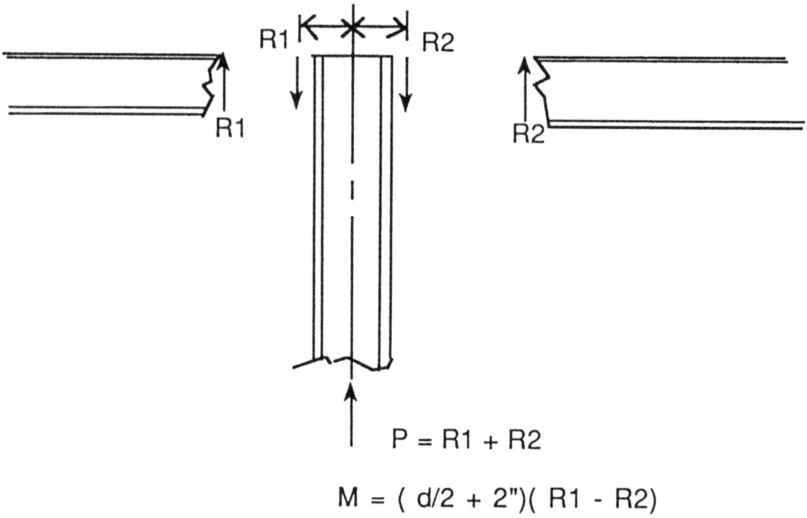

Figure 3.5 Beam/girder column reactions.

3.8 AXIAL COMPRESSION AND BENDING: ASD METHOD

When the column in a braced system that uses standard connections is exposed to bending moments in addition to axial loads, it must be designed to satisfy the following AISC requirements:

$$\frac{f_a}{F_a} + \frac{C_{mx} f_{bx}}{\left(1 - \frac{f_a}{F'_{ex}}\right) F_{bx}} + \frac{C_{my} f_{by}}{\left(1 - \frac{f_a}{F'_{ey}}\right) F_{by}} \leq 1.0 \tag{3.26}$$

$$\frac{f_a}{0.60 F_y} + \frac{f_{bx}}{F_{bx}} + \frac{f_{by}}{F_{by}} \leq 1.0 \tag{3.27}$$

When $f_a/F_a \leq 0.15$, Equation (3.28) is permitted instead of Equations (3.26) and (3.27)

$$\frac{f_a}{F_a} + \frac{f_{bx}}{F_{bx}} + \frac{f_{by}}{F_{by}} \leq 1.0 \tag{3.28}$$

where

F_a = axial compressive stress that would be permitted if axial force alone existed (k/in.^2)

F_b = compressive bending stress that would be permitted if bending moment alone existed (k/in.^2)

F'_e = $12\pi^2 E/23(Kl_b/r_b)^2$
= Euler stress divided by a factor of safety (k/in.^2). (In the expression for F'_e, l_b is the actual unbraced length in the plane of bending and r_b the corresponding radius of gyration. F'_e may be increased by $\frac{1}{3}$ for wind and earthquake forces.)

f_a = computed axial stress (k/in.^2)

f_b = computed compressive bending stress at the point under consideration (k/in.^2)

C_m = a coefficient whose value is taken from the following:

(a) For compression members in frames subject to translation (sideways), $C_m = 0.85$.

(b) For rotationally restrained compression members in frames braced against joint translation and not subject to transverse loading between their supports in the plane of bending, $C_m = 0.6 - 0.4(M_1/M_2)$, where M_1/M_2 is the ratio of the smaller to larger moments at the ends of that portion of the member unbraced in the plane of bending under consideration. M_1/M_2 is positive when the member is bent in reverse curvature, negative when bent in single curvature.

(c) For compression members in frames braced against joint translation in the plane of loading and subjected to transverse loading between their supports, the value of C_m may be determined by an analysis. However, in lieu of such analysis, the following values are permitted:

(i) For members whose ends are restrained against rotation in the plane of bending, $C_m = 0.85$.

(ii) For members whose ends are unrestrained against rotation in the plane of bending, $C_m = 1.0$.

3.9 REDUCTION IN LIVE LOADS

Live load reduction applied to members with an influence area of 400 ft^2 or more is designed for reduced live load determined by the following equation:

$$L = L_0\left(0.25 + \frac{15}{\sqrt{A_I}}\right) \tag{3.29}$$

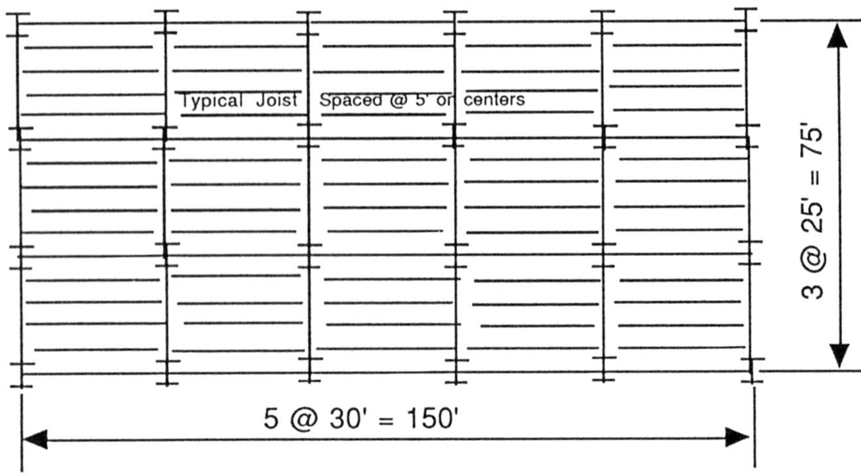

Figure 3.6 Seven-story building.

where

L = reduced design live load per square foot of area supported by the member
L_0 = unreduced design live load per square foot of area supported by the member (see Table 1.3)
A_I = influence area (ft^2)

The influence area A_I is four times the tributary area for a column, two times the tributary area for a beam, and equal to the panel area for a two-way slab. The reduced live load cannot be less than 50% of the unit live load L_0 for members supporting one floor nor less than 40% of the unit live load L_0 otherwise.

Example 3.4

The drawings in Figure 3.6 and 3.7 show the plan and elevations of a seven-story office building. The building is located in the northern region of the United States. Determine the roof and typical floor loads on an interior column. See the sketch in Figure 3.8 for the orientation of the open steel joists.

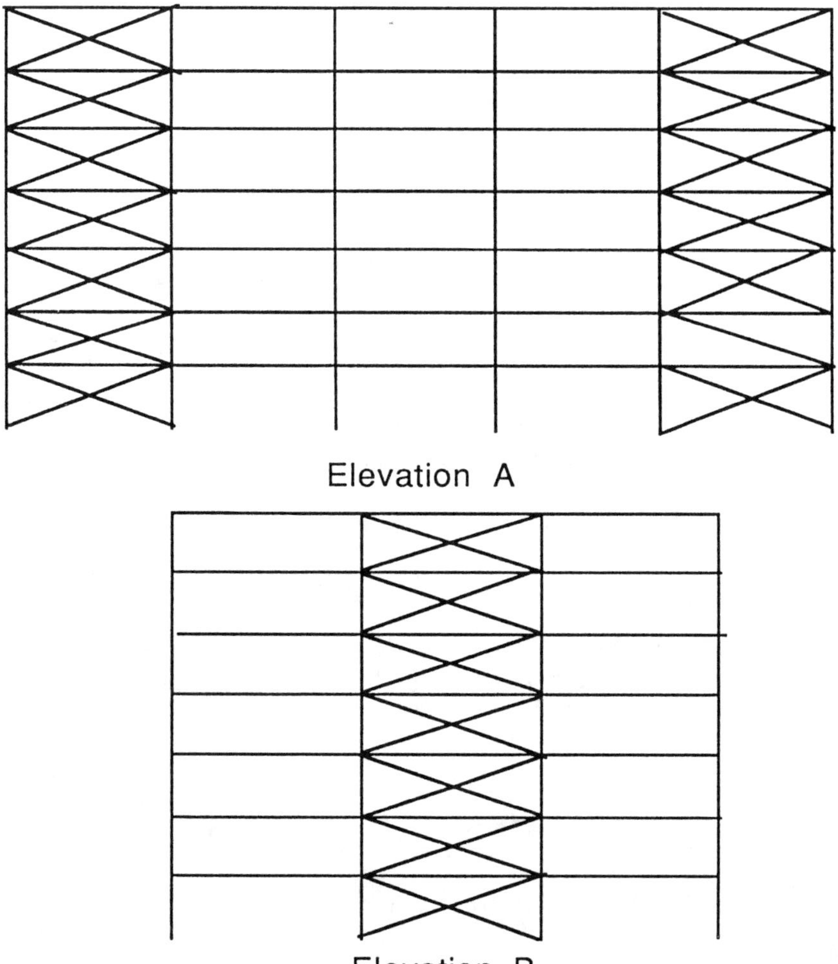

Figure 3.7 Elevation of seven-story building.

Solution

Referring to Figure 3.8, we see that the column with a dotted rectangle around it represents the typical interior column. The dotted lines include the influence loading area for this type of column.

Since the building is located in the northern region of the United States, the snow load is 30 lb/ft^2. The live load is 80 lb/ft^2, which includes partition loads (see Table 1.3). There will be no reduction in the live load

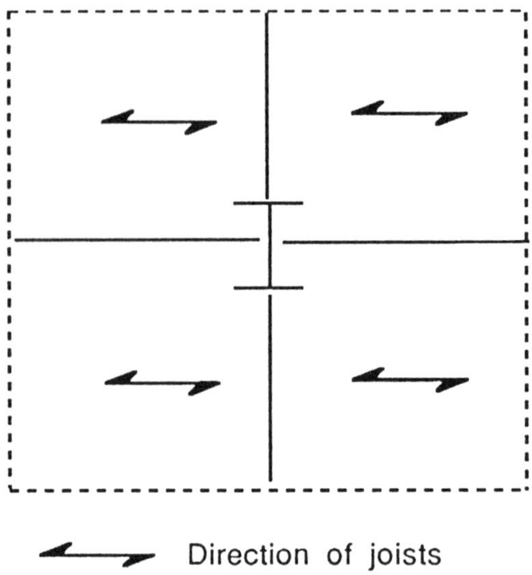

Figure 3.8 Direction of steel joists.

on the roof. The live load on an interior column at the first floor must be reduced in accordance to Equation (3.29). The tributary area for one floor is 25 ft × 30 ft = 750 ft^2. Then the reduced live load is obtained as follows:

$$L = L_0\left(0.25 + \frac{15}{\sqrt{25 \times 30 \times 4 \times 6}}\right)$$
$$= 0.362L_0 < 0.40$$

Use a live load reduction of 0.4

$$L = 80 \times 0.4$$
$$= 32 \text{ lb/ft}^2$$

Roof Loads

Live load (snow) = 30 lb/ft^2
Dead loads (assumed) = 15 lb/ft^2

Floor Loads

Reduced live load	$= 32\ \text{lb/ft}^2$
$3\frac{1}{2}$-in. reinforced concrete slab on a metal deck	$= 45\ \text{lb/ft}^2$

Calculation of the Loads on a Typical Joist

Roof

$$\text{Live load} = 1 \times 5 \times 30/1000$$
$$= 0.150\ \text{k/ft}$$
$$\text{Dead load} = 1 \times 5 \times 15/1000$$
$$= 0.075\ \text{k/ft}$$

Floor

$$\text{Live load} = 1 \times 5 \times 32/1000$$
$$= 0.160\ \text{k/ft}$$
$$\text{Dead load} = 1 \times 5 \times 45/1000$$
$$= 0.225\ \text{k/ft}$$

The reaction at each end of the joist is
Roof

$$\text{Live load} = 0.15 \times 30/2$$
$$= 2.25\ \text{k}$$
$$\text{Dead load} = 0.075 \times 30/2$$
$$= 1.125\ \text{k}$$

Floor

$$\text{Live load} = 0.16 \times 30/2$$
$$= 2.4\ \text{k}$$
$$\text{Dead load} = 0.225 \times 30/2$$
$$= 3.375\ \text{k}$$

The above reactions are placed on the joists for both roof and floor loads as shown in Figure 3.9. For the reactions on the interior column from the roof and a typical floor, see Figure 3.10.

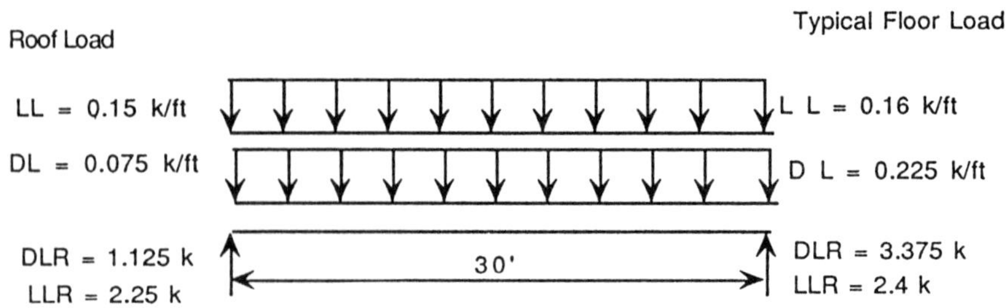

Figure 3.9 Loads on a typical roof or floor joist.

The joist reactions are placed on the column as shown in Figure 3.10. Thus, the load on a typical interior column at the first floor is the sum of the roof loads and the six levels below it.

$$\text{Total roof loads} = 33.75 \text{ k}$$

$$\text{Total floor loads } 6 \times 57.75 = 346.5 \text{ k}$$

$$\text{The grand total} = 380.2 \text{ k}$$

If we consider an unsupported length of the column to be 10 ft, the selection of a column for the first floor or any succeeding one would require load analysis to determine the worst loading condition. Assume that the live loads in one quadrant have been removed from every floor except the roof. The reduction in the axial load is 6.0 k for every floor. The loading condition for this case is summarized below

$$P = 380.2 - 6 \times 6$$
$$= 344.2 \text{ k}$$

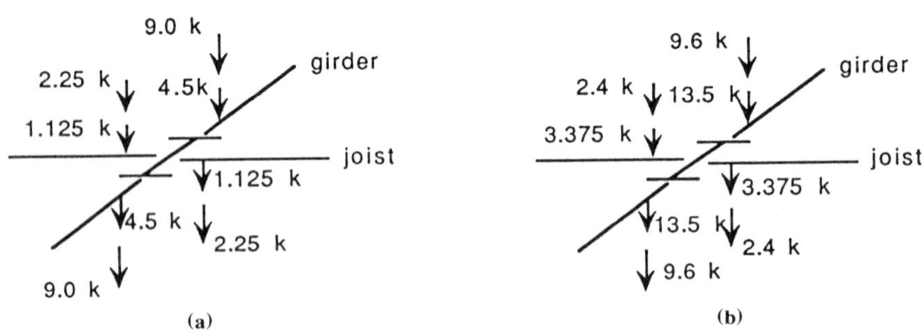

Figure 3.10 Loads on a typical interior column. (a) Roof loads. (b) Floor loads.

and

$$M_x = 6 \times 4.8\left(\frac{d}{2} + 2\right) \text{ in. k}$$

$$M_y = 6 \times 1.2 \times 2 \text{ in. k}$$

Two inches are assumed to be the distance from the back of the column flange or its web to the position of the reaction at the end of the beam or girder.

A third loading pattern that has to be checked is the case of no live loads in two adjacent quadrants. The loads resulting from this condition are

$$P = 380.2 - 12 \times 6$$

$$= 308 \text{ k}$$

$$M_x = 6 \times 9.6\left(\frac{d}{2} + 2\right) \text{ in. k}$$

$$M_y = 0$$

Having determined the loads for each of the three cases, we see that a column design analysis provides the optimum section.

For an unsupported length of 10 ft, try W12 × 79 (AISC Manual, Section 3–28). P_a is the capacity of the column and is 466 k. The ratio $(P/P_a) = (f_a/F_a)$.

$$\frac{f_a}{F_a} = \frac{380.2}{446}$$

$$= 0.852 > 0.15$$

Check Equations (3.26) and (3.27) (AISC, H1-1, H1-2).

$$\frac{f_a}{F_a} + \frac{C_{mx} f_{bx}}{\left(1 - \dfrac{f_a}{F'_{ex}}\right) F_{bx}} + \frac{C_{my} f_{by}}{\left(1 - \dfrac{F_a}{F'_{ey}}\right) F_{by}} \leq 1.0$$

From paragraph H1, C_{mx} and C_{my} are taken to be 1.0 for the purpose of simplification. The properties of this section are listed in the AISC

Manual as follows:

$$P_a = 446 \text{ k}$$

$$A = 23.2 \text{ in.}^2$$

$$l_y = 3.05$$

$$\frac{r_x}{r_y} = 1.75$$

$$\frac{K_x l_x}{r_x} = 23$$

$$\frac{k_y l_y}{r_y} = 40$$

$$F_a = 19.19 \text{ k/in.}^2$$

$$F'_{ex} = 282.29 \text{ k/in.}^2 \quad \text{(AISC Manual, Table 8)}$$

$$F'_{ey} = 93.33 \text{ k/in.}^2$$

$$S_x = 107 \text{ in.}^3$$

$$S_y = 35.8 \text{ in.}^3$$

$$d = 12.38 \text{ in.}$$

From condition 2, the loads are

$$P = 344.2 \text{ k}$$

$$M_x = 6 \times 4.8\left(\frac{d}{2} + 2\right)$$

$$= 6 \times 4.8\left(\frac{12.38}{2} + 2\right)$$

$$= 236.0 \text{ in. k}$$

$$M_y = 6 \times 1.2 \times 2$$

$$= 14.4 \text{ in. k}$$

$$f_{bx} = \frac{M_x}{S_x} = \frac{236}{107} = 2.206 \text{ k/in.}^2$$

$$f_{by} = \frac{M_y}{S_y} = \frac{14.4}{35.8} = 0.402 \text{ k/in.}^2$$

$$F_{bx} = 0.66F_y = 24 \text{ k/in.}^2$$

$$F_{by} = 0.75F_y = 27 \text{ k/in.}^2$$

$$f_a = \frac{P}{A} = \frac{344.4}{23.2} = 14.836 \text{ k/in.}^2$$

Applying these values in Equation (3.26) yields the following:

$$\frac{14.836}{19.19} + \frac{1.0 \times 2.206}{\left(1 - \frac{14.836}{282.29}\right)24} + \frac{1.0 \times 0.402}{\left(1 - \frac{14.836}{93.33}\right)27} = 0.88 < 1.0$$

Check the loading for condition 3

$$P = 308 \text{ k}$$

$$M_x = 6 \times 9.6 \times \left(\frac{12.38}{2} + 2\right)$$

$$\times 472 \text{ in. k}$$

$$\frac{308}{446} + \frac{1 \times \frac{472}{107}}{\left(1 - \frac{308}{282.29 \times 23.2}\right)24} = 0.883 < 1$$

The column section satisfies the AISC specification. Use W12 × 79.

3.10 COLUMNS SUBJECT TO BENDING AND AXIAL FORCE IN A BRACED SYSTEM: LRFD METHOD

In a braced frame, columns may be subjected to bending about one axis or both due to the unequal loads on the sides of the column. This condition arises from the removal of the live load from one side of the column and not the other. The interaction of axial and flexural stresses in symmetric shapes is given by the following equations:

For $P_u/\phi_c P_n \geq 0.2$

$$\frac{P_u}{\phi_c P_n} + \frac{8}{9}\left(\frac{M_{ux}}{\phi_b M_{nx}} + \frac{M_{uy}}{\phi_b M_{ny}}\right) \leq 1.0 \tag{3.30}$$

For $P_u/\phi_c P_n < 0.2$

$$\frac{P_u}{2\phi_c P_n} + \left(\frac{M_{ux}}{\phi_b M_{nx}} + \frac{M_{uy}}{\phi_b M_{ny}}\right) \leq 1.0 \tag{3.31}$$

where

P_u = required compressive strength (k)
P_n = nominal compressive strength determined from Equation (3.15) or column selection table (k)
M_u = required flexural strength determined (k-in.)
M_n = nominal flexural strength (k-in.), where M_n is obtained from Equation (4.23b)
ϕ_c = resistance factor for compression, 0.85
ϕ_b = resistance factor for flexure, 0.90

The design of columns under the combined action of bending and axial forces is a trial-and-error procedure. A trial section is checked for compliance with formulas H1-a and H1-b. A simplified and fast method is to assume an effective axial load on the basis of which the trial section will carry the bending and axial forces adequately

$$P_{u\text{eff}} = P_u + mM_{ux} + mUM_{uy} \tag{3.32}$$

where

P_u = actual factored load (k)
M_{ux} = actual factored moment about the strong axis (k-ft)
M_{uy} = actual factored moment about the minor axis (k-ft)
m = factor taken from Table B, LRFD, p. 2–10
U = factor taken from column load table

The procedure is summarized in the following steps:

1. With the known KL (effective unbraced length), select values for m and U.
2. Solve for $P_{u\text{eff}}$.
3. Make the appropriate selection from the column load table.
4. Based on the selection of Step 3, select the appropriate values for m and U.
5. With these values, solve for $P_{u\text{eff}}$.
6. Repeat Steps 3 and 4 until the values of m and U converge.

On the basis of elastic analysis, M_u is determined from

$$M_u = B_1 M_{nt} \tag{3.33}$$

where

$$B_1 = \frac{C_m}{(1 - P_u/P_e)} \tag{3.34}$$

$C_m = 0.85$ for members whose ends are restrained

$C_m = 1.0$ for members whose ends are unrestrained

$P_e = \dfrac{A_g F_y}{\lambda_c^2}$ or from Table 9, LRFD

Example 3.5

Figure 3.11 shows part of a building frame. The building has 11 stories. The first story is 15 ft, 0 in. high. All others are 10 ft, 6 in. high. The loads are as follows:

$$\text{Snow load} = 20 \text{ lb/ft}^2$$

$$\text{Roof dead load} = 45 \text{ lb/ft}^2$$

Typical floor loads

$$\text{Live load} = 50 \text{ lb/ft}^2$$

$$\text{Partition load} = 25 \text{ lb/ft}^2$$

$$\text{Ceiling load} = 5 \text{ lb/ft}^2$$

$$\text{Floor dead load} = 55 \text{ lb/ft}^2$$

Use A-36 steel to choose the column size. Splice columns every third floor. Use the LRFD method.

Solution

Load Calculations

Roof Loads

$$\text{Dead load on typical beam} = 45 \times 7.5 \text{ ft} \times \frac{1}{1000}$$

$$= 0.338 \text{ k/ft}$$

$$\text{Live load} = 20 \times 7.5 \text{ ft} \times \frac{1}{1000}$$

$$= 0.150 \text{ k/ft}$$

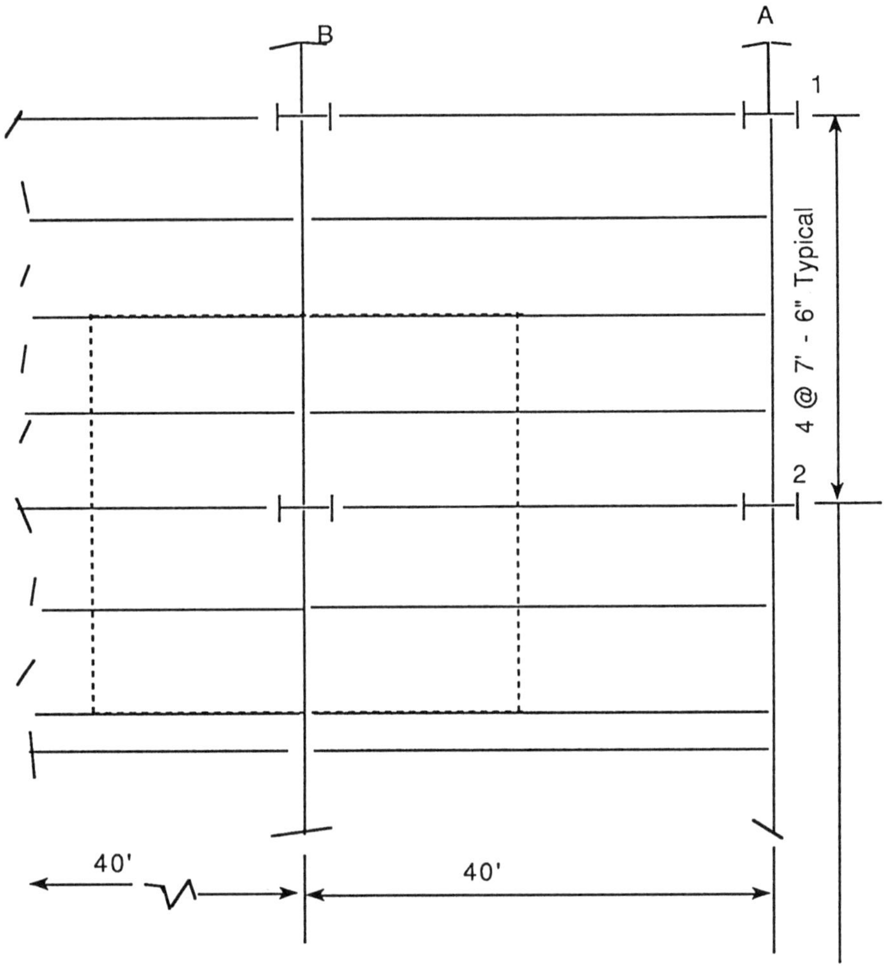

Figure 3.11 Typical interior column tributary area.

Typical floor load

$$\text{Dead load + ceiling load} = 60 \times 7.5 \text{ ft} \times \frac{1}{1000}$$
$$= 0.45 \text{ k/ft}$$

$$\text{Live + partition load} = 75 \times 7.5 \text{ ft} \times \frac{1}{1000}$$
$$= 0.562 \text{ k/ft}$$

The tributary area for a typical interior column is 30 ft $\times$ 40 ft = 1200 ft^2. Live load reduction applies to a member with an influence area that exceeds 400 ft^2. There will be no reduction in the roof live load. The tributary area of the lightest (the uppermost column) column section is three times the tributary area for that column per floor

$$A_l = 3 \times 4 \times 1200$$

$$= 14{,}400 \text{ ft}^2$$

$$L = 75 \times \left(0.25 + \frac{15}{\sqrt{14{,}400}}\right)$$

$$= 28.1 \text{ lb/ft}^2$$

Check the minimum reduction allowed in the live load. The reduced live load should not be less than 40% of the actual live load. Thus, the total live load on the column from a group of three columns is 30 lb/ft^2.

Beam reactions

Roof

$$1.2 \times \text{dead load} = 1.2 \times 0.338 \times \frac{40}{2}$$

$$= 8.1 \text{ k}$$

$$1.6 \times \text{live load} = 1.6 \times 0.150 \times \frac{40}{2}$$

$$= 4.8 \text{ k}$$

Typical floor

$$1.2 \times \text{dead load} = 1.2 \times 0.450 \times \frac{40}{2}$$

$$= 10.8 \text{ k}$$

$$1.6 \times \text{live load} = 1.6 \times 0.225 \times \frac{40}{2}$$

$$= 7.2 \text{ k}$$

To understand the action of the loads on an interior column, refer to Figures 3.11 and 3.12. The loads are transmitted to the column at four different points as shown in Figures 3.12c and 3.12d.

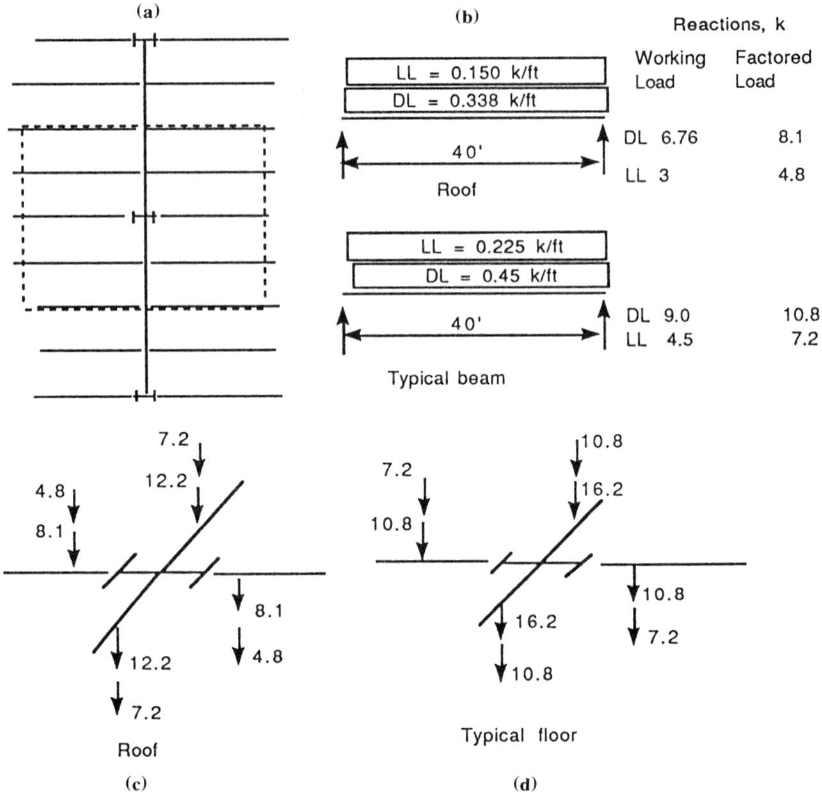

Figure 3.12 Typical load calculation on an interior column.

Design of the Column on the First Floor

Three loading conditions will be considered. First, the live load exists in all quadrants. Second, the live load is removed from the first quadrant. Third, the live load is removed from two adjacent quadrants.

Condition 1

Live load in all quadrants

$$
\begin{aligned}
P_u &= \text{Roof load} + 10 \times \text{typical floor load} \\
&= 64.6 + 10 \times 90 \\
&= 964.6 \text{ k} \\
&= 15 \text{ ft}
\end{aligned}
$$

From the column selection table, try W14 × 159

$$\phi_c P_n = 1280 \text{ k} > 964.6 \text{ k}$$

Condition 2

The live load is removed from quadrant *I*.

Load removed from each floor = 9 k

$$\text{For 10 floors} = 90 \text{ k}$$
$$\sum P_u = 964.6 - 90 = 874.6 \text{ k}$$

Load removed along the x axis = 5.4 × 10 = 54 k

$$\begin{aligned} M_{uy} &= P_{ux} \times 2 \text{ in.} \\ &= 54 \times 2 \\ &= 108 \text{ k-in.} \end{aligned}$$

The reaction is assumed to be applied 2 in. from the extreme edge of the column to where the connection reaction is supposed to take place.

Load removed along the y axis = 3.6 × 10 = 36 k

$$\begin{aligned} M_{ux} &= P_{uy}\left(\frac{d}{2} + 2 \text{ in.}\right) \\ &= 36\left(\frac{14.98}{2} + 2\right) \\ &= 341.4 \text{ k-in.} \end{aligned}$$

If we consider no intermediate stiffners, $L_b = 15$ ft. Also, it is assumed that there is no torsion.

List the column load requirements:

$$\begin{aligned} P_u &= 874.6 \text{ k} \\ M_{ux} &= 341.4 \text{ k-in.} \\ &= 28.5 \text{ k-ft} \\ M_{uy} &= 108 \text{ k-in.} \\ &= 9 \text{ k-ft} \\ l_b &= 15 \text{ ft} \\ C_{mx} &= 0.85 \\ C_{my} &= 0.85 \end{aligned}$$

Step 1. For $Kl = 15$ ft, from Table B (LRFD, p. 2–10) and the column selection table as a first trial

$$m = 2.6$$
$$U = 1.26$$

Step 2. From Equation (3.32)

$$P_{\text{eff}} = 874.6 + 2 \times 28.5 + 2 \times 1.26 \times 9$$
$$= 954.3 \text{ k}$$

Step 3. From the column load tables, select W14 × 120 ($\phi_c P_n = 956$).

Step 4. Adopt W14 × 120. With $m = 2$ and $U = 1.35$,

$$P_{\text{eff}} = 874.6 + 2 \times 28.5 + 2 \times 1.35 \times 9$$
$$= 956 \text{ k}$$
$$A = 35.3 \text{ in.}^2$$
$$r_x = 6.25 \text{ in.}$$
$$r_y = 3.74 \text{ in.}$$
$$\frac{Kl}{r_y} = \frac{15 \times 12}{3.74}$$
$$= 48$$
$$\frac{Kl}{r_x} = \frac{15 \times 12}{6.25}$$
$$= 29$$
$$\lambda_{cx} = \frac{29}{\pi}\sqrt{\frac{36}{29{,}000}}$$
$$= 0.325$$
$$\lambda_{cy} = \frac{48}{\pi}\sqrt{\frac{36}{29{,}000}}$$
$$= 0.538$$
$$P_{ex} = \frac{35.3 \times E_y}{\lambda_{cx}^2}$$
$$= \frac{35.3 \times 36}{0.325^2}$$
$$= 12{,}031 \text{ k}$$

$$P_{ey} = \frac{35.3 \times F_y}{\lambda_{cy}^2}$$

$$= \frac{35.3 \times 36}{0.538^2}$$

$$= 4390 \text{ k}$$

$$B_x = \frac{C_m}{1 - \dfrac{P_u}{P_{ex}}}$$

$$= \frac{0.85}{1 - \dfrac{874.6}{12{,}031}}$$

$$= 0.917 \quad (\text{use } B_x = 1.0)$$

$$B_y = \frac{C_m}{1 - \dfrac{P_u}{P_{ey}}}$$

$$= \frac{0.85}{1 - \dfrac{874.6}{4390}}$$

$$= 1.061$$

$$M_{ux'} = B_x \times M_{ux}$$

$$= 1.0 \times 28.5$$

$$= 28.5 \text{ k-ft}$$

$$M_{uy'} = B_y \times M_{uy}$$

$$= 1.061 \times 9$$

$$= 9.6 \text{ k-ft}$$

Apply the above values to Equation (3.30).

$$\frac{874.6}{956} + \frac{8}{9}\left(\frac{28.5}{636} + \frac{9.6}{306}\right) = 0.983 < 1.0$$

$$L_p = \frac{300 r_y}{\sqrt{F_y}} = \frac{300 \times 4.0}{\sqrt{36}} = 200 \text{ in.}$$

$$= 16.67 \text{ ft}$$

$$L_b < L_p$$

No intermediate stiffners are needed in the column.

3.11 DESIGN OF COLUMNS FOR BRACED AND UNBRACED FRAMES: ASD METHOD

In a braced frame, lateral stability is provided by diagonal bracing as shown in Figure 3.13. When the beams or girders are rigidly connected to the column, the use of an alignment chart is used to simplify the sizing of the column. These charts were developed for the elastic column. However, by multiplying the G values from the alignment chart by the proper reduction factor, the charts may be used for the inelastic column. The procedure is outlined as follows:

1. Determine the axial load on the column.
2. Select a column size and calculate f_a.

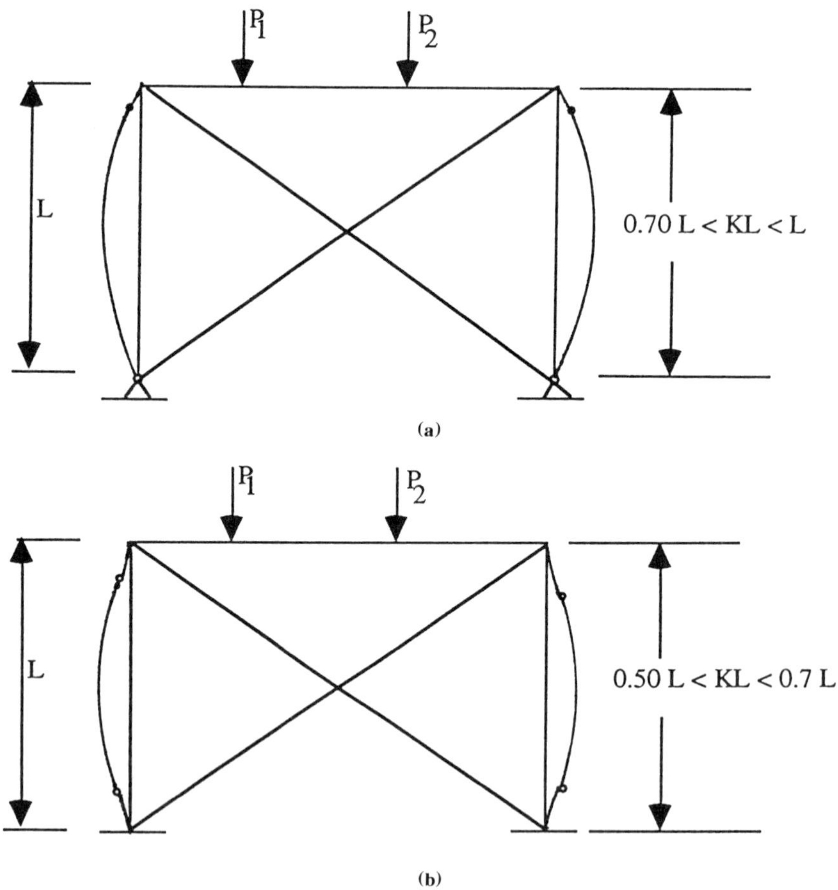

Figure 3.13 Braced frames. (a) Hinged support. (b) Fixed support.

3. Enter the value of f_a as determined in Step 2 in Table A, which is included in the AISC Manual, and find the reduction factor.
4. Determine the G value at the bottom and top of the column

$$G = \frac{\sum I_c/I_c}{\sum I_g/I_g}$$

$$= 10 \text{ for a simply supported column to the footing}$$

$$= 1.0 \text{ for a fixed column to the footing}$$

5. Using the values established in Step 4, enter the alignment chart (Figure 3.14) for the particular stability condition (sidesway or no sidesway) and find K.
6. Multiply the k factor from Step 5 by the reduction factor from Step 2 to obtain the correct K factor for the inelastic column.
7. To determine the effective length for the column, multiply the unsupported length of the column by the k factor calculated in Step 6.

Example 3.6

The frame shown in Figure 3.15 is at the lower level of a multistory building. Bracing is used to prevent sidesway. Select a column size to carry the given axial load for an interior column.

Solution

Try a section for the column of W18 × 35

$$I_g = 510 \text{ in}^4$$

$$I_c = 2400 \text{ in.}^4$$

$$P_a = 1069 \text{ k}$$

$$A = 56.8 \text{ in.}^2$$

$$G_{\text{top}} = \frac{2400/15}{510/30}$$

$$= 9.41 \quad \text{(for the elastic column)}$$

$$G_{\text{bottom}} = 10 \quad \text{(for a simple support at the base)}$$

$$f_a = \frac{970}{56.8}$$

$$= 17.1 \text{ k/in.}^2$$

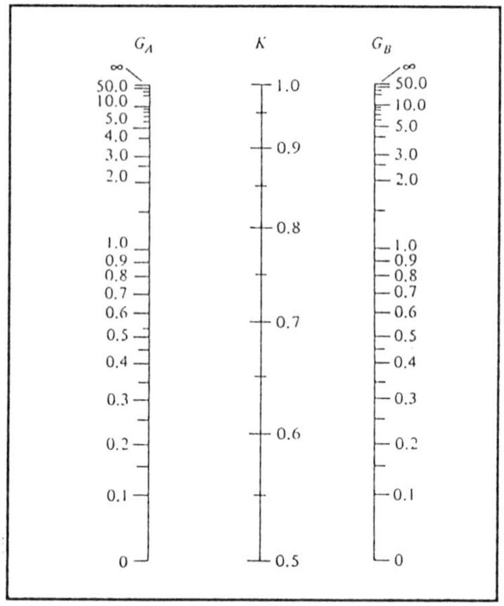

(a)

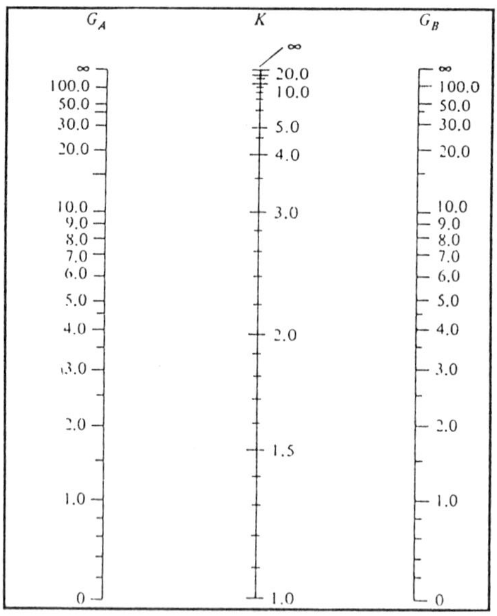

(b)

Figure 3.14 Chart for the K column factor. (a) Sidesway inhibited. (b) Sidesway uninhibited.

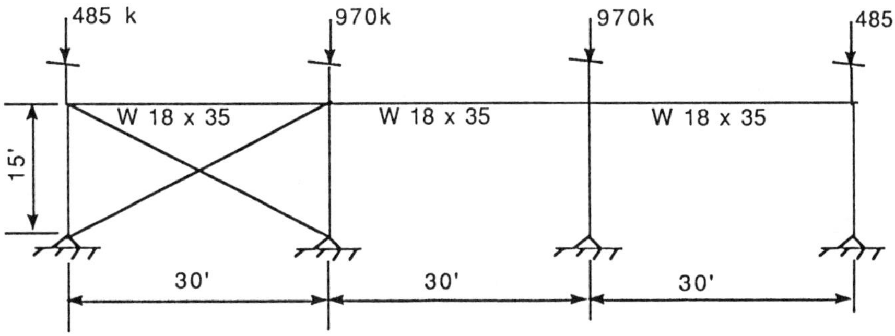

Figure 3.15 Braced frame of a multistory building.

From Table A in the AISC Manual, the reduction factor is found to be 0.460

$$G_{\text{inelastic}} = 0.460 \times 9.41$$

$$= 4.3$$

Use the values of 4.3 and 10 for the G values at the top and bottom of the column, respectively, in the alignment chart in Figure 3.14 for no sidesway to obtain a K value of 0.95. Hence, the effective slenderness ratio of the column is

$$\frac{Kl_x}{r_x} = \frac{0.95 \times 15 \times 12}{6.5}$$

$$= 27$$

Enter this value in Table 3.1 to obtain a value of $F_a = 20.15\ \text{k/in.}^2$ that is higher than required. The selection meets the load requirements and those of the AISC specifications.

Example 3.7

Consider the sidesway in the frame shown in Figure 3.15 to be uninhibited. See Figure 3.16. Check the frame as designed in Example 3.6.

$$G_B = 10$$

$$G_{A\ \text{inelastic}} = 4.3$$

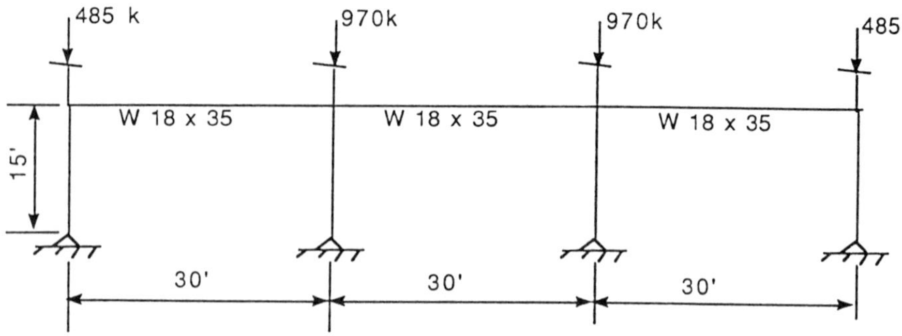

Figure 3.16 Unbraced frame of a multistory building.

Solution

Use these values with the alignment chart in Figure 3.14 for an uninhibited sidesway. The K factor is found to be 2.5. The slenderness ratio for the column is

$$\frac{K_x l_x}{r_x} = \frac{2.5 \times 15 \times 12}{6.5}$$

$$= 69$$

The allowable compression stress corresponding to this slenderness ratio is $F_a = 16.53 \text{ k/in.}^2 < 17.1 \text{ k/in.}^2$ (required). Try a stronger section, W14 $\times$ 211.

$$A = 62 \text{ in.}^2$$

$$I_x = 2660 \text{ in.}^4$$

$$I_y = 1030 \text{ in.}^4$$

$$r_y = 4.07 \text{ in.}$$

$$\frac{r_x}{r_y} = 1.61$$

$$d = 15.72 \text{ in.}$$

The compression stress is

$$f_a = \frac{970}{62}$$

$$= 15.64 \text{ k/in.}^2$$

The corresponding reduction factor is obtained from Table A, AISC Manual, p. 3–8 and found to be equal to 0.626 by interpolation. Then the inelastic g factor for the top of the column is determined

$$G_{A\ \text{inelastic}} = 0.626 \times G_{A\ \text{elastic}}$$

$$G_{A\ \text{elastic}} = \frac{2660/15}{510/30}$$

$$= 10.43$$

$$G_{A\ \text{inelastic}} = 0.626 \times 10.43$$

$$= 6.5$$

With the modified value for the g-factor at the top of the column, enter the alignment chart in Figure 3.14 for the uninhibited sidesway to obtain a k factor of 2.7. The new slenderness ratio is

$$\frac{K_x l_x}{r_x} = \frac{2.7 \times 15 \times 12}{4.07 \times 1.61}$$

$$= 74$$

$$F_a = 16.01\ \text{k/in.}^2 > 15.64\ \text{k/in.}^2$$

Use W14 × 211.

Example 3.8

The building shown in Figure 3.17 is eight stories high. The frames are braced against lateral movement as shown in the figure. Design an interior column for the first floor. The tributary area is indicated by dashed lines in the diagram.

Solution

Use the ASD method for A-36 steel.

$$A_t = 35 \times 25$$

$$= 875\ \text{ft}^2$$

The influence area A_I for an interior column is

$$A_I = 4 \times A_t$$

$$= 4 \times 875$$

$$= 3500\ \text{ft}^2 > 400\ \text{ft}^2$$

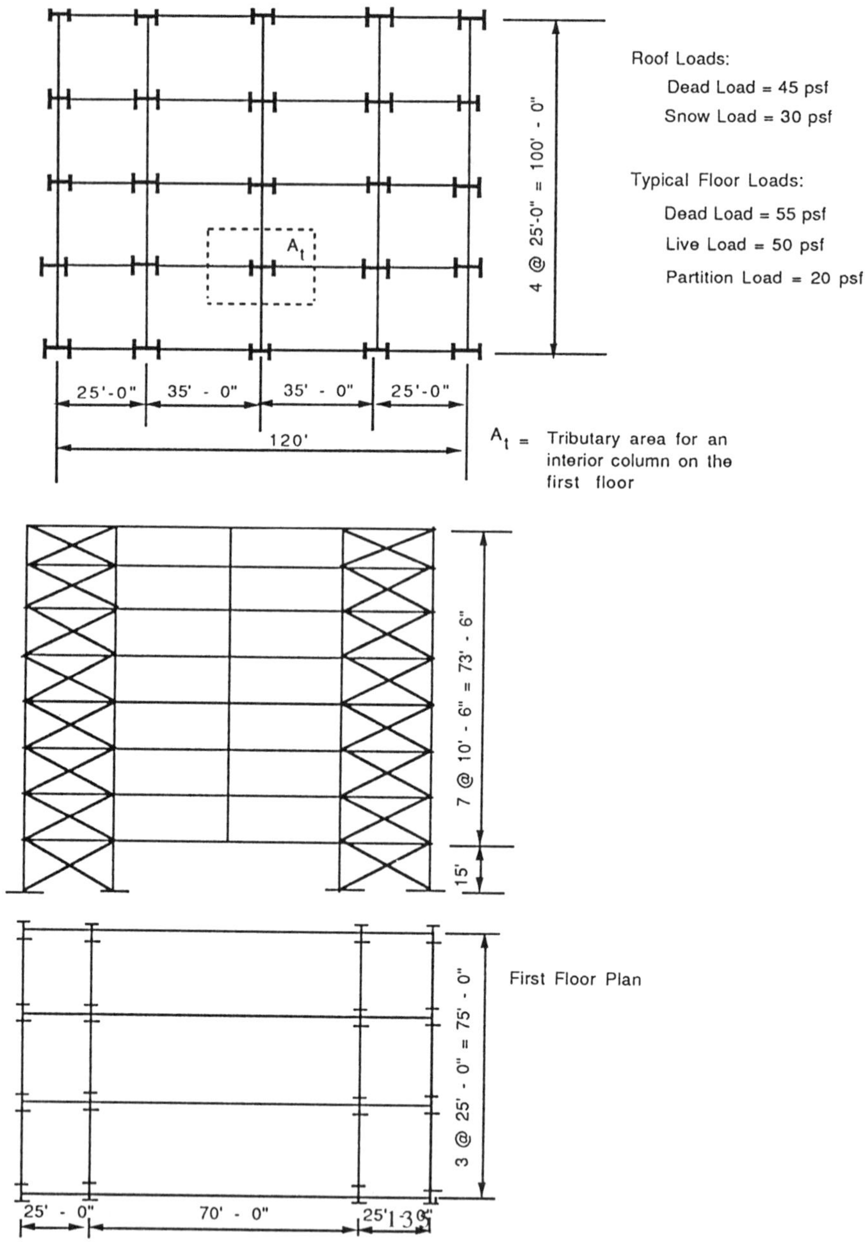

Figure 3.17 Floor and elevation of a seven-story building.

Reduction in the live load for column loading must be used in the design. The total influence area for an interior column on the first floor is $3500 \times 7 = 24{,}500\ \text{ft}^2$. The reduced live load on the floors is computed as follows:

$$L = 70\left(0.25 + \frac{15}{\sqrt{24{,}500}}\right)$$

Note that the partition load is considered a live load

$$\begin{aligned} L &= 0.346 \times 70 \\ &= 24.2\ \text{lb/ft}^2 \end{aligned}$$

The minimum live load reduction for an interior column is limited to 40%.

$$\begin{aligned} L &= 0.40 \times 70 \\ &= 28\ \text{lb/ft}^2 \text{ controls} \end{aligned}$$

The joist loads are shown in Figures 3.18a and 3.18b. The joists run in the short direction at 5 ft on center. The calculations of the loads carried by the joists are included below.

Roof loads

$$\begin{aligned} \text{Snow load} &= 0.03 \times 5 \\ &= 0.15\ \text{k/ft} \\ \text{Dead load} &= 0.045 \times 5 \\ &= 0.225\ \text{k/ft} \end{aligned}$$

Typical floor loads

$$\begin{aligned} \text{Live load} &= 0.028 \times 5 \\ &= 0.14\ \text{k/ft} \\ \text{Dead load} &= 0.055 \times 5 \\ &= 0.275\ \text{k/ft} \end{aligned}$$

Determine the reactions at the end of the joists:

Roof joists (total DL and LL)

$$\begin{aligned} R &= (0.15 + 0.225) \times 25/2 \\ &= 4.7\ \text{k} \end{aligned}$$

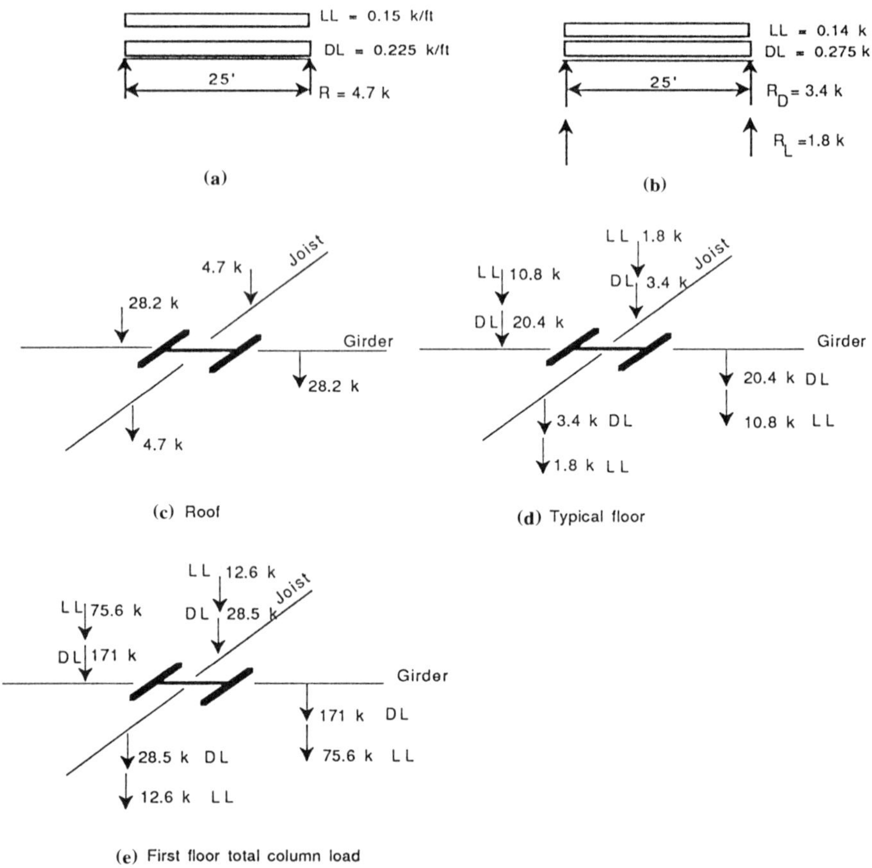

Figure 3.18 Interior column load calculations for the first floor. (a) Typical roof joist loads. (b) Typical floor joist loads. (c). Roof. (d) Typical floor. (e) First floor total column load.

Typical floor joist (dead and live loads are separate):

$$R_{DL} = 0.275 \times 25/2$$
$$= 3.4 \text{ k}$$
$$R_{LL} = 0.140 \times 25/2$$
$$= 1.8 \text{ k}$$

Impose these reactions on the interior column. See Figures 3.18c and 3.18d. Note that the girder supports reactions from a set of three pairs of joists. Hence, the load on the girder is three times that of one joist. The total load components acting on an interior column on the first floor are shown in Figure 3.18e.

Load Calculations for the First Floor Interior Column

From the girders

Floor

$$DL = 7 \times 20.4$$
$$= 142.8 \text{ k}$$
$$LL = 7 \times 10.8$$
$$= 75.6 \text{ k}$$

Roof

$$DL + LL = 28.2 \text{ k}$$

Total roof and floor loads

$$DL = 142.8 + 28.2$$
$$= 171 \text{ k}$$
$$LL = 75.6 \text{ k}$$

From the joists

Floor

$$DL = 7 \times 3.4$$
$$= 23.8 \text{ k}$$
$$LL = 7 \times 1.8$$
$$= 12.6 \text{ k}$$

Roof

$$DL + LL = 4.7 \text{ k}$$

Total roof and floor loads

$$DL = 23.8 + 4.7$$
$$= 28.5 \text{ k}$$
$$LL = 12.6 \text{ k}$$

The column can be subjected to any of several loading conditions in its life span. As discussed before, these conditions are listed as follows:

1. Live load exists everywhere.
2. Live load is removed uniformly from one quadrant from every floor.
3. Live load is removed from two adjacent quadrants.

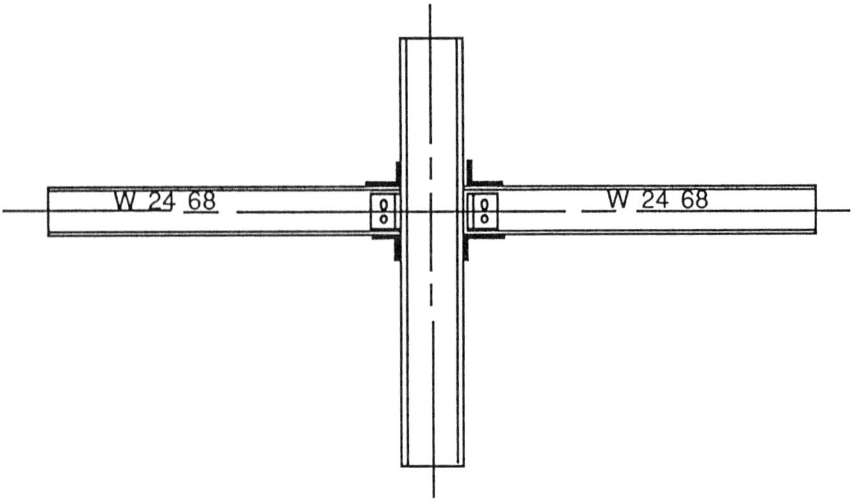

Figure 3.19 Interior column-girder connection.

For condition 1, the load is concentric. The column can be selected on that basis and later checked for the other two conditions.

Condition 1

$$P = 575.4 \text{ k}$$
$$M_x = 0$$
$$M_y = 0$$

The girders are rigidly connected to the columns as shown in Figure 3.19. Because of the bracing, the frame is inhibited from moving laterally. As a first attempt, try W14 × 132.

Column Properties

$$I_c = 1530 \text{ in.}^4$$
$$A = 38.8 \text{ in.}^2$$
$$r_y = 3.76 \text{ in.}$$
$$\frac{r_x}{r_y} = 1.67$$
$$d = 14.66 \text{ in.}$$
$$P_a = 719 \text{ k}$$

Since the column is simply supported to the foundation,

$$G_B = 10$$

The elastic G factor at the top of the column is obtained from the following expression:

$$G_A = \frac{1530/15}{1830/35}$$
$$= 1.95$$

The compression stress in the column is

$$f_a = \frac{575.4}{38.8}$$
$$= 14.83 \text{ k/in.}^2$$

The reduction factor to multiply the elastic G_A in order to obtain the equivalent inelastic factor is determined from Table A in the AISC Manual, ASD. Thus, the reduction factor is found by interpolation to be 0.71

$$G_A = 0.71 \times 1.95$$
$$= 1.38$$

Enter the G_B and modified G_A values in Figure 3.14a to obtain the K factor

$$K = 0.88$$

The slenderness ratio for this column is

$$\frac{K_x l_x}{r_x} = \frac{15 \times 12 \times 0.88}{3.76 \times 1.67}$$
$$= 25$$

From Table 3.1,

$$F_a = 20.28 \text{ k/in.}^2 > 14.83 \text{ k/in.}^2$$

Condition 2

List the load requirements and stress intensities for this condition. The change in the load is equal to one-half of each of the live load components on the joist and girder adjacent to the quadrant from which the live load has been removed. The load change is as follows:

From the joist

$$\Delta P = \tfrac{1}{2} \times 12.6$$
$$= 6.3 \text{ k}$$

From the girder

$$\Delta P = \tfrac{1}{2} \times 75.6$$
$$= 37.8 \text{ k}$$

The total change in the load on the column is

$$\sum \Delta P = 6.3 + 37.8$$
$$= 44.1 \text{ k}$$

The magnitude of the load for condition 2 is

$$P = 575.4 - 44.1$$
$$= 531.3 \text{ k}$$

Refer to Figure 3.21e to determine the moments about the x and y axes, respectively. Note that the y axis is parallel to the web and the x axis perpendicular to it.

$$M_x = 37.8 \times \left(\frac{14.66}{2} + 2\right)$$
$$= 353 \text{ in. k}$$

Similarly, the moment about the y axis is

$$M_y = 6.3 \times 2$$
$$= 12.6 \text{ in. k}$$
$$F_a = 20.28 \text{ k/in.}^2$$
$$f_a = \frac{531.3}{38.8}$$
$$= 13.69 \text{ k/in.}^2$$
$$S_x = 209 \text{ in.}^3$$
$$S_y = 74.5 \text{ in.}^3$$
$$f_{bx} = \frac{353}{209}$$
$$= 1.689 \text{ k/in.}^2$$
$$f_{by} = \frac{12.6}{74.5}$$
$$= 0.169 \text{ k/in.}^2$$

Check Equations (3.26) and (3.27).

$$F_{bx} = 24 \text{ k/in.}^2$$

$$F_{by} = 27 \text{ k/in.}^2$$

$$\frac{f_a}{F_a} + \frac{C_{mx} f_{bx}}{\left(1 - \dfrac{f_a}{F_{ex'}}\right) F_{bx}} + \frac{C_{my} f_{by}}{\left(1 - \dfrac{f_a}{F_{ey'}}\right) F_{by}} < 1.0$$

$$C_{mx} = 1$$

$$C_{my} = 1$$

$$K_y = K_x$$

$$= 0.88$$

$$\frac{k_x l_x}{r_x} = 25$$

$$F_{ex'} = 238.8 \text{ k/in.}^2 \quad \text{(from Table 8, AISC ASD)}$$

$$\frac{k_y l_y}{r_y} = \frac{15 \times 12 \times 0.88}{1.67 \times 3.76}$$

$$= 42$$

$$F_{ey'} = 84.65 \text{ k/in.}^2 \quad \text{(from Table 8, AISC ASD)}$$

$$\frac{13.69}{20.28} + \frac{1.689}{\left(1 - \dfrac{13.69}{238.93}\right) 24} + \frac{0.169}{\left(1 - \dfrac{13.69}{84.65}\right) 27} = 0.757 < 1.0$$

and

$$\frac{f_a}{0.60 F_y} + \frac{f_{bx}}{F_{bx}} + \frac{f_{by}}{F_{by}} < 1.0$$

$$\frac{13.69}{0.60 \times 36} + \frac{1.689}{24} + \frac{0.169}{27} = 0.71 < 1.0$$

The column seems to be a little oversized. For a more economical design, the process can be repeated with W14 × 120 or W14 × 109.

Example 3.9

Design an interior column adjacent to the exterior bay as indicated in Figure 3.20 on the first floor.

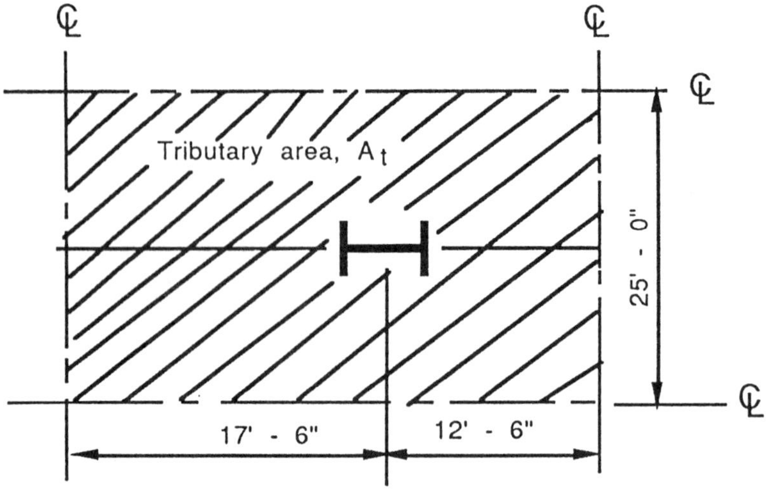

(a)

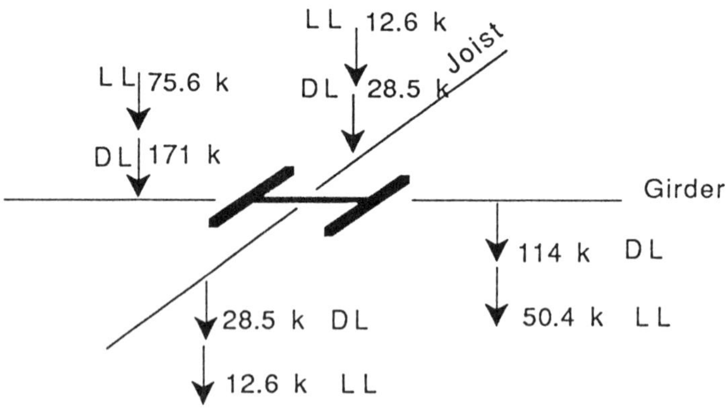

(b)

Figure 3.20 Exterior column load calculations. (a) Typical interior column adjacent to exterior bay. (b) Loads on a typical interior column.

Solution

Calculate the tributary area

$$A_t = 25 \times \frac{35 + 25}{2}$$

$$= 750 \text{ ft}^2$$

$$A_I = 4 \times 750$$

$$= 3000 \text{ ft}^2 > 400 \text{ ft}^2 \text{ per floor}$$

Live load reduction must be utilized in determining the size of the column. For the first floor column, the tributary area is

$$A_I = 7 \times 3000$$
$$= 21{,}000 \text{ ft}^2$$

The reduced live load is

$$L = 70\left(0.25 + \frac{15}{\sqrt{21{,}000}}\right)$$
$$= 24.7 \text{ lb/ft}^2$$
$$0.4L_0 = 70 \times 0.4$$
$$= 28 \text{ lb/ft}^2$$

The loads on this column differ from the previous example only by the difference on the right girder. The reactions on the right girder are two-thirds what they were in the previous problem. Examine the loads as shown in Figure 3.20b. Two loading conditions will be considered.

Condition 1

Live load covers the entire space around the column

$$P = 493.1 \text{ k}$$
$$M_x = 82.2\left(\frac{d}{2} + 2\right)$$
$$= 740 \text{ in. k} \quad (\text{for } d \text{ approximately 14 in.})$$
$$M_y = 0$$

Select W14 × 109 for the first trial. The section properties are listed as follows:

$$P_a = 592 \text{ k}$$
$$I_x = 1240 \text{ in.}^4$$
$$r_y = 3.73 \text{ in.}$$
$$\frac{r_x}{r_y} = 1.67$$
$$A = 32 \text{ in.}^2$$
$$d = 14.32 \text{ in.}$$
$$S_x = 173 \text{ in.}^3$$

The girders on the left and right have the sections W24 × 68 and W21 × 44, respectively. From these sizes, the G factors are obtained in the usual procedure as developed before

$$G_B = 10$$

$$G_{A\ \text{elastic}} = \frac{1240 \times 2/15}{(1830/35 + 843/25)}$$

$$= 1.92$$

$$f_a = \frac{493.1}{32}$$

$$= 15.41\ \text{k/in.}^2$$

The stiffness reduction factor is found to be equal to 0.653. The equivalent $G_{A\ \text{inelastic}}$ factor is

$$G_{A\ \text{inelastic}} = 0.653 \times 1.92$$

$$= 1.25$$

The k factor is obtained from the alignment chart

$$K = 0.88$$

$$\frac{Kl_x}{r_x} = 26$$

$$F_a = 20.22\ \text{k/in.}^2 > 15.41$$

Next, check the column interaction equation

$$f_a = \frac{493.1}{32}$$

$$= 15.41\ \text{k/in.}^2$$

$$f_{bx} = \frac{740}{173}$$

$$= 4.58\ \text{k/in.}^2$$

$$F_{ex'} = 220.9\ \text{k/in.}^2 \quad \text{(Table 8, AISC Manual)}$$

$$\frac{f_a}{F_a} + \frac{C_{mx} f_{bx}}{\left(1 - \dfrac{f_a}{F_{ex'}}\right) F_{bx}} = 0.967 < 1.0$$

Condition 2

Live load is removed from the first quadrant. The total load is reduced by 6.3 k from the joist side and 25.2 k from the left girder side. Thus, the load for the second condition is

$$P = 461.6\ \text{k}$$

$$M_x = [(171 + 75.6) - (114 + 25.2)]\left(\frac{14.32}{2} + 2\right)$$

$$= 984\ \text{k/in.}$$

$$f_{bx} = \frac{984}{173}$$

$$= 5.69\ \text{k/in.}^2$$

$$M_y = 6.3 \times 2$$

$$= 12.6\ \text{in. k}$$

$$f_{by} = \frac{12.6}{61.2}$$

$$= 0.20\ \text{k/in.}^2$$

$$F_{ey'} = 80.76\ \text{k/in.}^2$$

The column interaction equation yields

$$\frac{14.42}{20.22} + \frac{5.69}{\left(1 - \dfrac{14.42}{220.92}\right)24} + \frac{0.2}{\left(1 - \dfrac{14.42}{80.76}\right)27} = 0.976 < 1.0$$

Example 3.10

Design the first floor corner column for the frame shown in Figure 3.21. The building is a ten-story building.

Solution

The loads are as follows:
Roof

$$\text{Dead load} = 45\ \text{lb/ft}^2$$

$$\text{Live load} = 35\ \text{lb/ft}^2$$

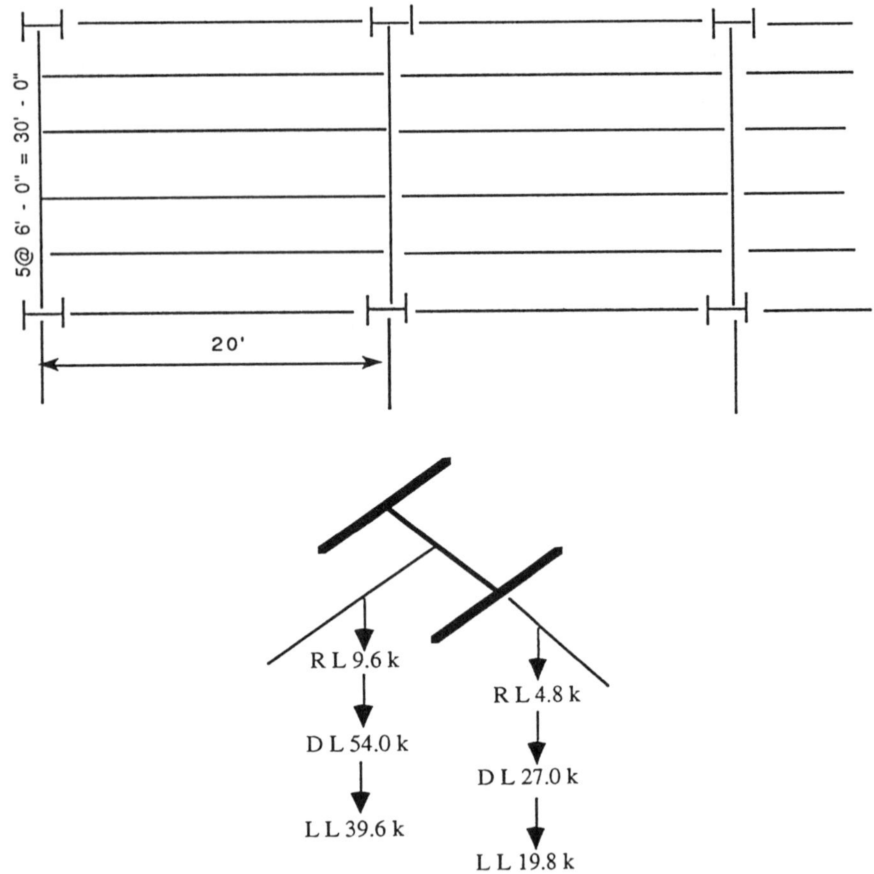

Figure 3.21 Corner column load calculations.

Floor

$$\text{Dead load} = 50 \text{ lb/ft}^2$$
$$\text{Live load} = 80 \text{ lb/ft}^2$$

Use A-36 steel.

The tributary are for the corner column is

$$A_t = 10 \times 15$$
$$= 150 \text{ ft}^2$$

The influence area is four times the tributary area for a column

$$A_I = 600 \text{ ft}^2 > 400$$

Use reduction in the live load on the floors. The reduced live load is

$$
\begin{aligned}
L &= L_0\left(0.25 + \frac{15}{\sqrt{4 \times A_t}}\right) \\
&= 80\left(0.25 + \frac{15}{\sqrt{4 \times 150 \times 9}}\right) \\
&= 36.3\ \text{lb/ft}^2
\end{aligned}
$$

Roof Loads

When we enter the flange of the column, the load is

$$
\begin{aligned}
&= \frac{(35 + 45)}{1000} \times 6 \times 10 \\
&= 4.8\ \text{k}
\end{aligned}
$$

When we enter the web of the column, the load is

$$
\begin{aligned}
&= 2 \times \left[\frac{(35 + 40)}{1000} \times 6 \times 10\right] \\
&= 9.6\ \text{k}
\end{aligned}
$$

Dead Floor Loads

When we enter the flange of the column, the load is

$$
\begin{aligned}
&= \frac{50}{1000} \times 6 \times 10 \times 9 \\
&= 27\ \text{k}
\end{aligned}
$$

When we enter the web of the column, the load is

$$
\begin{aligned}
&= 2 \times \frac{50}{1000} \times 6 \times 10 \times 9 \\
&= 54\ \text{k}
\end{aligned}
$$

Live Floor Loads

When we enter the flange of the column, the load is

$$
\begin{aligned}
&= \frac{36.3}{1000} \times 6 \times 10 \times 9 \\
&= 19.8\ \text{k}
\end{aligned}
$$

When we enter the web of the column, the load is

$$= 2 \times \frac{36.3}{1000} \times 6 \times 10 \times 9$$

$$= 39.6 \text{ k}$$

Show the placement of the loads on the column on the first floor. See Figure 3.21b. Total load on the column is

$$\sum_{i=1}^{9} P_i = 145.2 \text{ k}$$

$$M_y = 93.6 \times 2$$

$$= 187.2 \text{ in. k}$$

Refer to Figure 3.22.

$$M_x = 51.6 \times (d/2 + 2)$$

For an unsupported length of 11.5 with respect to the y axis and from the selection table for column capacity, try W8 $\times$ 58.

$$P_a = 288 \text{ k}$$

$$A = 17.1 \text{ in.}^2$$

$$f_a = \frac{145.2}{17.1}$$

$$= 8.50 \text{ k/in.}^2$$

$$r_y = 2.10 \text{ in.}$$

$$r_x = 3.65 \text{ in.}$$

$$\frac{l_x}{r_x} = \frac{11.5 \times 12}{3.65}$$

$$= 39$$

$$F_{ex'} = 98.2 \text{ k/in.}^2$$

$$\frac{l_y}{r_y} = \frac{11.5 \times 12}{2.10}$$

$$= 66$$

$$F_{ey'} = 34.3 \text{ k/in.}^2$$

$$S_x = 52.0 \text{ in.}^3$$

$$S_y = 18.3 \text{ in.}^3$$

$$M_x = 51.6 \times 6.375$$

$$= 329 \text{ in. k}$$

$$f_{bx} = \frac{329}{52}$$

$$= 6.326 \text{ k/in.}^2$$

$$f_{by} = \frac{187.2}{18.3}$$

$$= 10.23 \text{ k/in.}^2$$

$$\frac{f_a}{F_a} = \frac{P}{P_a}$$

$$= \frac{146.2}{288}$$

$$= 0.508 > 0.15$$

Check only Equations H1-1 and H1-2, AISC Manual, p. 5–54. Examine H1-2 first.

$$\frac{f_a}{0.6F_y} + \frac{f_{bx}}{F_{bx}} + \frac{f_{by}}{F_{by}} = \frac{8.50}{21.6} + \frac{6.326}{24} + \frac{10.23}{27}$$

$$= 1.04 > 1.00$$

The results are not satisfactory. Try a larger section, W8 × 67. The section properties and requirements are as follows:

$$P_a = 334 \text{ k}$$

$$A = 19.7 \text{ in.}^2$$

$$r_y = 2.12 \text{ in.}$$

$$r_x = 1.75 \times 2.12 \text{ in.}$$

$$\frac{l_x}{r_x} = \frac{11.5 \times 12}{2.12 \times 1.75}$$

$$= 37$$

$$\frac{l_y}{r_y} = 37 \times 1.75$$

$$= 65$$

$$F_{ex'} = 109.1 \text{ k/in.}^2$$

$$F_{ey'} = 35.34 \text{ k/in.}^2$$

$$S_x = 60.4 \text{ in.}^3$$

$$S_y = 21.4 \text{ in.}^3$$

$$d = 9.0 \text{ in.}$$

$$M_y = 187.2 \text{ in. k}$$

$$M_x = 51.6 \times 6.5$$

$$= 335 \text{ in. k}$$

$$f_a = \frac{45.2}{19.7}$$

$$f_{bx} = \frac{335}{60.4}$$

$$f_{by} = \frac{187.2}{21.4}$$

Check Equation H1-2 first.

$$\frac{7.421}{21.6} + \frac{5.546}{24} + \frac{8.748}{27} = 0.90 < 1.0$$

Examine H1-1 next.

$$\frac{f_a}{F_a} + \frac{C_{mx}}{\left(1 - \dfrac{f_a}{F_{ex'}}\right)} \times \frac{f_{bx}}{F_{bx}} + \frac{C_{my}}{\left(\dfrac{1 - f_a}{F_{ey'}}\right)} \times \frac{f_{by}}{F_{by}} \leq 1.0$$

Substituting the proper values for the terms in H1-1 yields

$$\frac{7.421}{16.94} + \frac{0.85}{\left(1 - \dfrac{7.421}{109.8}\right)} \times \frac{5.546}{24} + \frac{0.85}{\left(\dfrac{1 - 7.421}{35.34}\right)} \times \frac{8.745}{27} = 0.997$$

Use W8 × 67.

Example 3.11

Given the column shown in Figure 3.22, select an adequate size using the ASD method with A-36 steel. The unsupported length of the column with respect to the x and y axes is 10.5 ft.

The solution is left to the student.

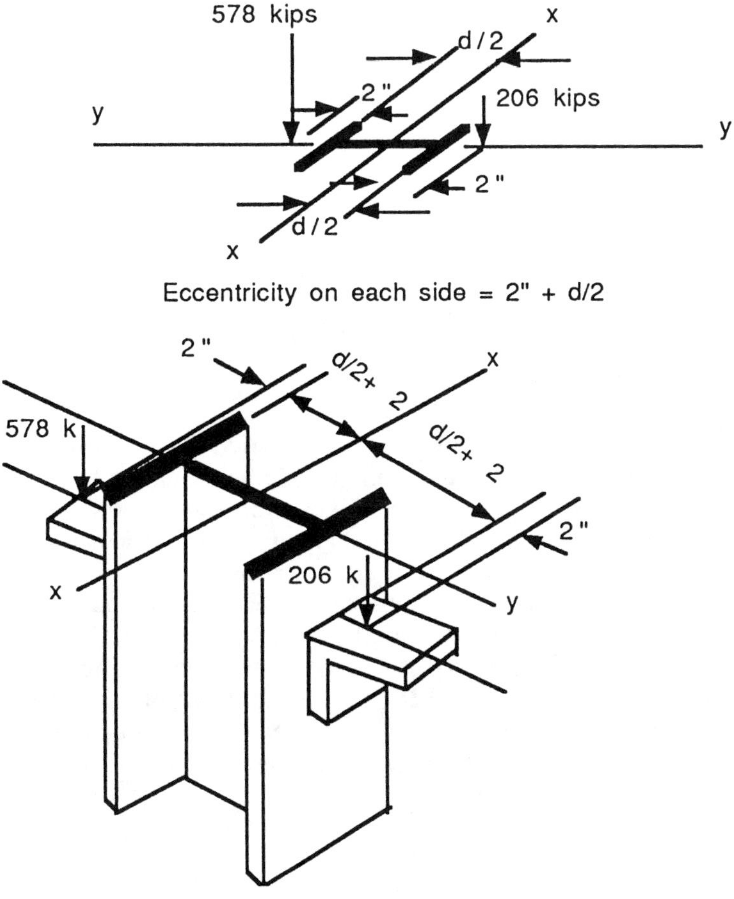

Figure 3.22 Determination of eccentricity of a typical column beam connection.

4

Design of Bending Members

4.1 INTRODUCTION

In buildings, loads of any kind have to be carried down to the foundation. Dead loads (the weight of the building elements and other permanent fixtures in the building) and live loads (furniture, snow, people, and movable equipment) are usually called gravity loads. Wind and earthquake forces acting against buildings are known as lateral loads. Structural members subjected to loads acting perpendicularly to their longitudinal axis are known as flexural (bending) members. In construction, they are referred to as beams, girders, joists, purlins, etc. Such members constitute the major and minor components of almost all types of structures.

Beams and girders are used interchangeably. A girder may suggest a more important role in terms of load-carrying capacity. However, both are subjected to tension and compression stresses due to bending. See Figure 4.1.

4.2 SIMPLE BENDING

The theory of the bending of beams is based on the principle that a plane section before bending remains very nearly plane after bending (Figure 4.2) and that within the elastic range stress is proportional to strain. When a beam is loaded, it deflects as shown in Figure 4.1a. At each point along the length of the beam, its axis has a definite radius of curvature ρ that

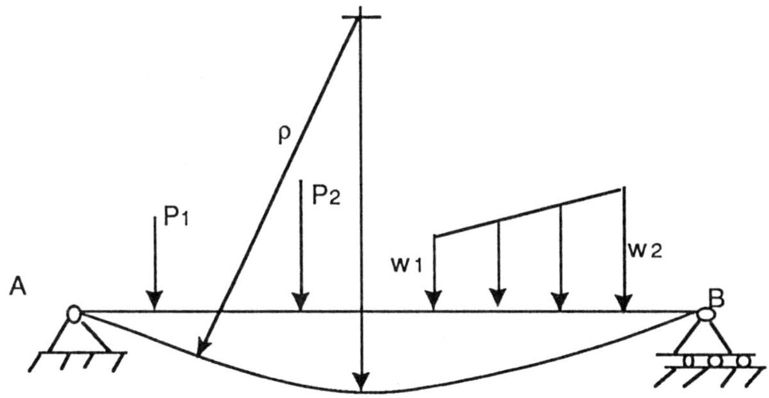

(a)

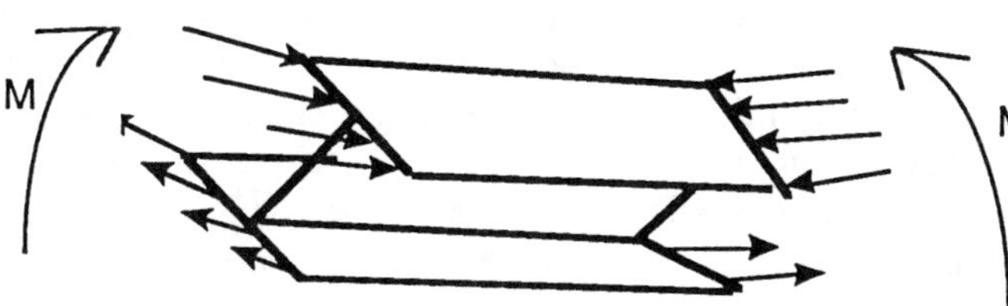

(b)

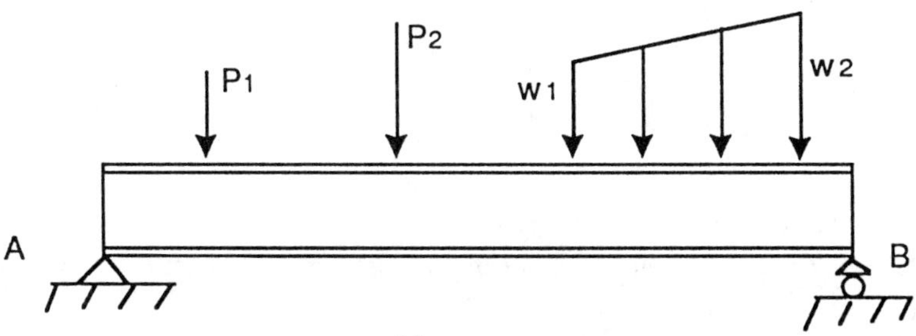

(c)

Figure 4.1 Beam deflection due to bending.

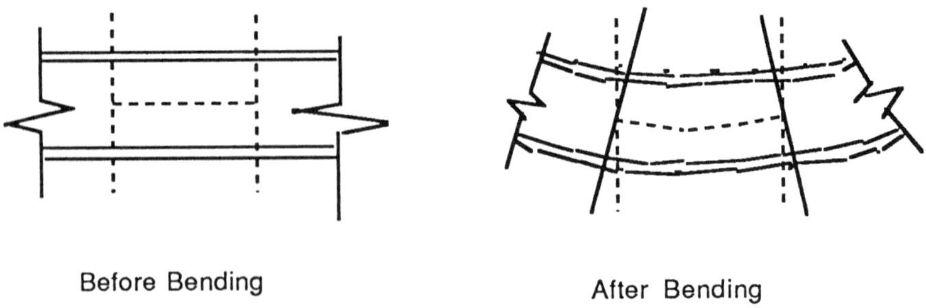

Figure 4.2 Beam section before and after bending.

varies from point to point. Considering a small part of the beam before and after loading as seen in Figure 4.2, we notice that the fibers across the section of the beam are subjected to compression in the upper fibers and tension in the lower ones. For a symmetrical section, the neutral axis falls in the middle of the section and the compression and tension stresses are equally distributed. By using the equations of equilibrium, simple equations for normal and shearing stresses in the beam (Figure 4.3) are obtained when bending takes place about a principal axis in the absence of twisting of the section

$$f_b = \frac{M_x y}{I_x} \tag{4.1}$$

$$f_v = \frac{V_y Q_x}{I_x b} \tag{4.2}$$

where

f_b = the fiber stress in bending
M_x = the bending moment about the x axis along the length of the beam
y = the fiber position along the y axis
I_x = the moment of inertia about the x axis
f_v = shear stress at a distance y from the neutral axis
V_y = the total shear at any point along the longitudinal axis of the beam
Q_x = the first moment of area at y about the x axis
b = the width of the section at the point under consideration

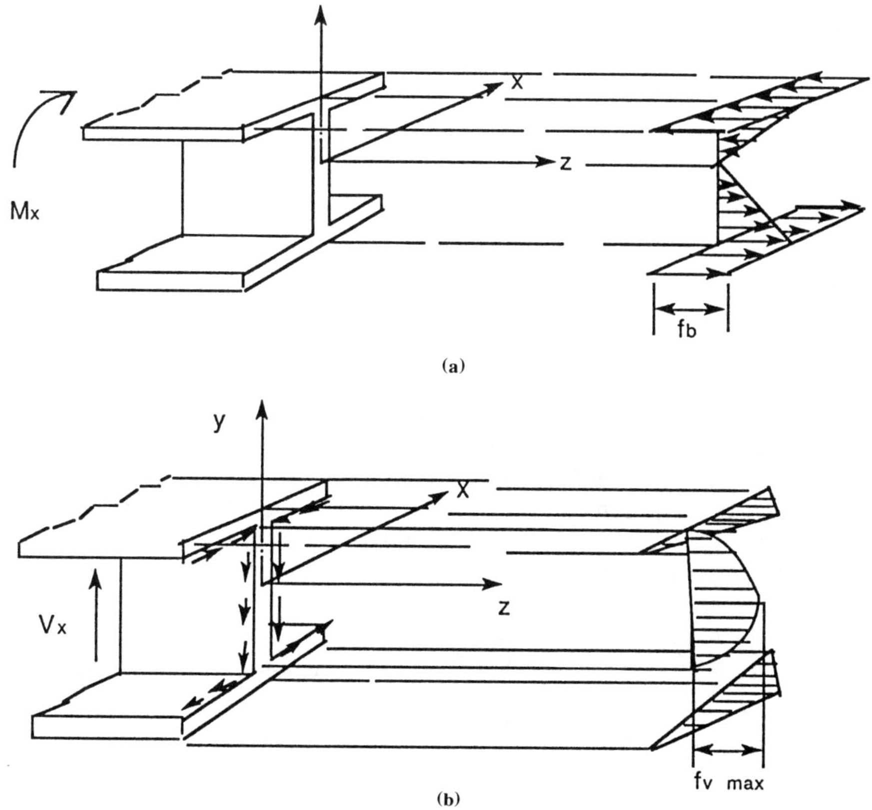

Figure 4.3 (a) Normal bending stress. (b) Sheer bending stress.

4.3 DESIGN OF BEAMS AND OTHER FLEXURAL MEMBERS: ASD ALLOWABLE BENDING STRESS

Beams are differentiated from plate girders by the fact that the ratio of h/t_w is lower than $760/\sqrt{F_b}$, where F_b is the allowable bending stress. The allowable shear stresses and stiffener requirements are given in Chapter F of the ASIC specifications. It applies to singly or doubly symmetric beams, including hybrid beams and girders loaded in the plane of symmetry.

The allowable stress for compact members loaded in the plane of their minor axis and symmetrical about the same is

$$F_b = 0.66F_y \tag{4.3}$$

provided the flanges are connected continuously to the web and the unsupported length of the compression flange L_b does not exceed L_c, where

$$L_c = \frac{76b_f}{\sqrt{F_y}} \tag{4.4a}$$

or

$$L_c = \frac{20{,}000}{\dfrac{d}{A_f}F_y} \tag{4.4b}$$

In addition to the above restrictions, the width-thickness ratios of its compression elements must not exceed the applicable limiting width-thickness ratios from Table 4.1 (AISC Manual, Table B5.1). For members that meet the requirements of F1.1 except that their flanges are noncompact (excluding built-up members and those with yield points greater than 65 k/in.^{2}), the allowable yield stress is

$$F_b = F_y\left(0.79 - 0.002\frac{b_f}{2t_f}\sqrt{F_y}\right) \tag{4.5}$$

For the members described above (in addition, their webs are noncompact), the allowable stress is

$$F_b = F_y\left(0.79 - 0.002\frac{b_f}{2t_f}\sqrt{\frac{F_y}{k_c}}\right) \tag{4.6}$$

where

$$k_c = \frac{4.05}{\left(\dfrac{h}{t_w}\right)^{0.46}} \qquad \text{if} \qquad \frac{h}{t_w} > 70 \tag{4.7}$$

Otherwise, $k_c = 1.0$.

For noncompact sections not covered above and loaded through the shear center and braced laterally in the region of compression stress at intervals not exceeding the limit set in Equation (4.4), the allowable bending stress is

$$F_b = 0.60F_y \tag{4.8}$$

TABLE 4.1 Limiting Width-Thickness Ratios for Compression Elements

Description of Elements	Width-Thickness Ratio	*Limiting Width-Thickness Ratios* Compact	Noncompact[c]
Flanges of I-shaped rolled beams and channels in flexure[a]	b/t	$65/\sqrt{F_y}$	$95/\sqrt{F_y}$
Flanges of I-shaped welded beams in flexure	b/t	$65/\sqrt{F_y}$	$95/\sqrt{\frac{F_y f_e}{k_c}}$
Outstanding legs of pairs of angles in continuous contact; angles or plates projecting from rolled beams or columns; stiffeners on plate girders	b/t	NA	$95/\sqrt{F_y}$
Angles or plates projecting from girders, built-up columns, or other compression members; compression flanges of plate girders	b/t	NA	$95/\sqrt{F_y/k_c}$
Stems of tees	d/t	NA	$127/\sqrt{F_y}$
Unstiffened elements simply supported along one edge, such as legs of single-angle struts, legs of double-angle struts with separators, and star-shaped sections	b/t	NA	$76/\sqrt{F_y}$
Flanges of square and rectangular box and hollow structural selections of uniform thickness subject to bending or compression[b] flange cover plates and diaphram plates between lines of fasteners or welds	b/t	$190/\sqrt{F_y}$	$238/\sqrt{F_y}$
Unsupported width of cover plates perforated with a succession of access holes[d]	b/t	NA	$317/\sqrt{F_y}$
All other uniformly compressed stiffened elements, i.e., supported along two edges	b/t h/t_w	NA	$253/\sqrt{F_y}$

(cont'd.)

TABLE 4.1 (*cont'd.*)

Description of Elements	Width-Thickness Ratio	*Limiting Width-Thickness Ratios*	
		Compact	Noncompact[c]
Webs in flexural compression[a]	d/t	$640/\sqrt{F_y}$	—
	h/t_w	—	$760/\sqrt{F_b}$
Webs in combined flexural and axial compression	d/t_w	For $(f_a/F_y < 0.16)$ $\times \frac{640}{\sqrt{F_y}}(1 - 3.74\frac{f_a}{F_y})$ For $f_a/F_y > 0.16$ $257/\sqrt{F_y}$	
	h/t_w		$760/\sqrt{F_b}$
Circular hollow sections in axial compression in flexure	D/t D/t	$3300/F_y$ $3300/F_y$	

[a] For hybrid beams, use the yield strength of the flange F_{yt} instead of F_y.

[b] See also Section F3.1, AISC Manual, Ninth Edition.

[c] For the design of slender sections that exceed the noncompact limits, see Appendix B5, AISC Manual, Ninth Edition.

[d] Assume the net area of the plate's widest hole.

[e] $k_c = 4.05/(h/t)^{0.46}$ if $h > 70$. Otherwise, $k_c = 1.0$

Source: AISC Steel Manual, ASD, Ninth Ed., Table B5.1

For members with compact or noncompact sections with unbraced length in the compression region greater than L_c, the allowable bending stress in tension is determined from Equation (4.8). For members that are loaded in the plane which includes an axis of symmetry in their web, the allowable bending stress in compression is determined as the larger value from Equations (4.9) or (4.10) and (4.11)

$$F_b = \left[\frac{2}{3} - \frac{F_y(l/r_T)^2}{1530 \times 10^3 C_b}\right] F_y \tag{4.9}$$

when

$$\sqrt{\frac{102 \times 10^3 C_b}{F_y}} < \frac{l}{r_T} < \sqrt{\frac{510 \times 10^3 C_b}{F_y}}$$

$$F_b = \frac{170 \times 10^3 C_b}{(l/r_T)^2} \tag{4.10}$$

where

$$\frac{l}{r_T} > \sqrt{\frac{510 \times 10^3 C_b}{F_y}}$$

For any value of l/r_T

$$F_b = \frac{12 \times 10^3 C_b}{\frac{l_d}{A_f}} < 0.60F_y \tag{4.11}$$

Equation (4.11) applies only to sections with a compression flange that is solid and approximately rectangular in cross section and that has an area not less than the tension flange. The symbols used in Equations (4.9), (4.10), and (4.11) are listed below.

l = distance between cross sections braced against twist or lateral displacement of the compression flange (in.). For cantilevers braced only against twist at the support, l may be taken conservatively as the actual length.

r_T = radius of gyration of a section comprising the compression flange plus one-third of the compression web area, taken about an axis in the plane of the web (in.).

A_f = area of the compression flange (in.^{2}).

C_b = $1.75 + 1.05(M_1/M_2) + 0.3(M_1/M_2)^2$, but no more than 2.3. It is conservative to consider C_b as unity. For values of C_b less than 2.3, see Table 4.2

TABLE 4.2 Values of C_b for Use in Equations (4.9), (4.10), and (4.11)

M_1/M_2	C_b	M_1/M_2	C_b	M_1/M_2	C_b
− 1.00	1.00	−0.45	1.34	0.10	1.86
−0.95	1.02	−0.40	1.38	0.15	1.91
−0.90	1.05	−0.35	1.42	0.20	1.97
−0.85	1.07	−0.30	1.46	0.25	2.03
−0.80	1.10	−0.25	1.51	0.30	2.09
−0.75	1.13	−0.20	1.55	0.35	2.15
−0.70	1.16	−0.15	1.60	0.40	2.22
−0.065	1.19	−0.10	1.65	0.45	2.28
−0.60	1.23	−0.05	1.70	> 0.47	2.30
−0.55	1.26	0	1.75		
−0.50	1.30	0.05	1.80		

Source: AISC Manual, Ninth Edition, Table 6.

(Table 6, AISC Manual, Ninth Edition). M_1 is the smaller and M_2 the larger bending moment at the ends of the unbraced length, taken about the strong axis of the member, and where M_1/M_2, the ratio of end moments, is positive when M_1 and M_2 have the same sign (reverse curvature bending) and negative when they are of opposite signs (single curvature bending). When the bending moment at any point within an unbraced length is larger than that at both ends of this length, the values of C_b is taken as unity.

Members with compact sections, doubly symmetrical I- and H-shapes, solid round and square bars, and solid rectangular bars bent about their weaker axes have an allowable bending stress of

$$F_b = 0.75F_y \tag{4.12}$$

for $F_y < 65$ k/in.^{2}. For noncompact shapes bent about their minor axis, the allowable bending stress is

$$F_b = 0.60F_y \tag{4.13}$$

for $F_y < 65$ k/in.^{2}. For doubly symmetrical I- and H-shapes having noncompact flanges and $F_y > 65$ k/in.^{2}, the allowable bending stress about the weaker axes is

$$F_b = F_y\left(1.075 - 0.005\frac{b_f}{2t_f}\sqrt{F_y}\right) \tag{4.14}$$

Box members, rectangular tubes, and circular tubes that are compact have an allowable bending stress of

$$F_b = 0.66F_y \tag{4.15}$$

To qualify as a compact section, the box-shaped member must meet the requirements of compactness set before. In addition, its depth must not exceed six times its width, its flange be no greater than two times its web thickness, and its laterally unsupported length L_b be less than or equal to

$$L_c = \left(1950 + 1200\frac{M_1}{M_2}\right)\frac{b}{F_y} \tag{4.16}$$

except that it need not be less than 1200 (b/F_y). The conditions for M_1 and M_2 are the same as mentioned before. For box-type and tubular flexural members that meet the noncompact requirements previously outlined here, the allowable stress is

$$F_b = 0.60F_y \tag{4.17}$$

Allowable Shear Stress

The allowable shear stress for flexural members is

$$F_v = 0.40F_y \tag{4.18}$$

for h/t_w equal or less than $380/\sqrt{F_y}$. For $h/t_w > 380/\sqrt{F_y}$, the allowable shear stress in the web is

$$F_v = \frac{F_y}{2.89}(C_v) \tag{4.19}$$

F_v should not exceed $0.40F_y$.

$$C_v = \left[\frac{45{,}000k_v}{F_y(h/t_w)^2}\right]$$

when C_v is less than 0.8.

$$C_v = \frac{190}{h/t_w}\sqrt{\frac{k_v}{F_y}}$$

when C_v is more than 0.8.

$$k_v = 4.00 + \frac{5.34}{(a/h)^2}$$

when a/h is less than 1.0.

$$k_v = 5.34 + \frac{4.00}{(a/h)^2}$$

when a/h is more than 1.0.

t_w = thickness of web (in.)
a = clear distance between transverse stiffeners (in.)
h = clear distance between flanges at the section under investigation (in.)

Intermediate stiffeners are required whenever the ratio h/t_w exceeds 260 and the maximum web shear stress f_v is greater than the limit set by Equation (4.19). When required, the spacing of the intermediate stiffeners shall be such that the web shear does not exceed the value for F_v given by Equation (4.19) and

$$\frac{a}{h} < \left[\frac{260}{(h/t_w)}\right]^2 < 3.0 \tag{4.20}$$

4.4 DEFLECTIONS AND VIBRATIONS OF BEAMS IN BENDING

In addition to strength considerations, serviceability of the structural members in building plays a significant role. As beams take on long spans, deflection restrictions may control the design. Excessive deflections will cause damage to plastered ceilings and other nonstructural elements attached to the building frame. On flat roofs, a relatively large deflection can be catastrophic due to ponding. The deflected roof structure forms a bowl-shaped volume that will retain rain water unless properly drained. Thus, deflection increases due to the accumulation of rain water. This process continues, unless checked, leading to an unstable condition when collapse occurs. Various building codes limit the maximum live load deflections to 1/360 and 1/240 of the span for plastered and nonplastered ceilings, respectively.

For a symmetrical and uniformly loaded simply supported beam, the maximum deflection is

$$\Delta = \frac{5Wl^3}{384EI} \tag{4.21}$$

where

Δ = deflection (in.)
W = total uniform load, including weight of beam (k)
l = span of beam (in.)

For $E = 29{,}000$ k/in.^{2} and a specific value of F_b, this equation reduces to the expressions shown in Table 4.3. In this table, L = span (in feet) and d = depth of beam (in inches).

TABLE 4.3 Maximum Deflection of Beams

	F_b (k/in.^{2})	Deflection (in.)
$F_y = 36$ k/in.^{2}	23.8	$0.02458L^2/d$
	21.6	$0.02234L^2/d$
$F_y = 50$ k/in.^{2}	33	$0.03414L^2/d$
	30	$0.03103L^2/d$

Source: AISC Steel Manual, ASD, Ninth Ed., p. 2–32.

Example 4.1

Given the framing plane of Figure 4.4, design a typical beam using the ASD method, assuming the beam to be simply supported:

Design Data for 4-in. concrete slab

$$\text{Live load} = 50 \text{ lb/ft}^2$$

$$\text{Partitions} = 20 \text{ lb/ft}^2$$

Solution

Consider a 1-ft strip as shown in Figure 4.5a to calculate the uniform loads on a typical beam. Having determined the magnitude of the uniform

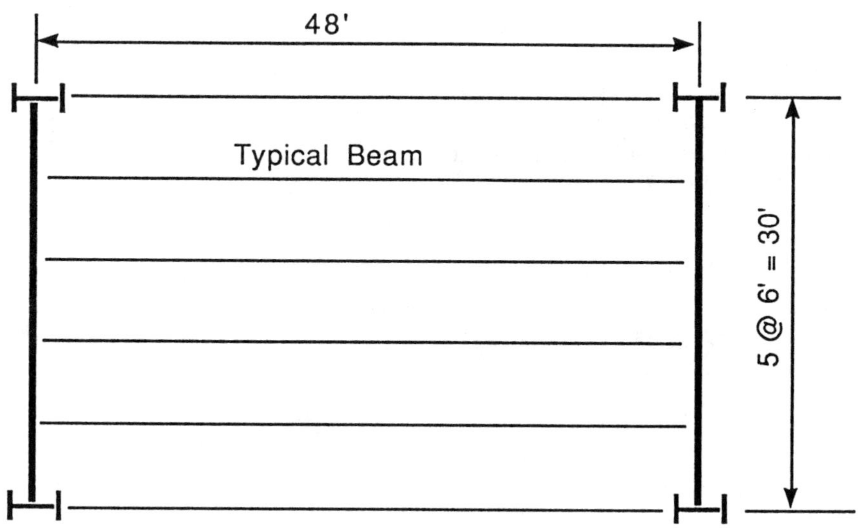

Figure 4.4 Typical floor bay.

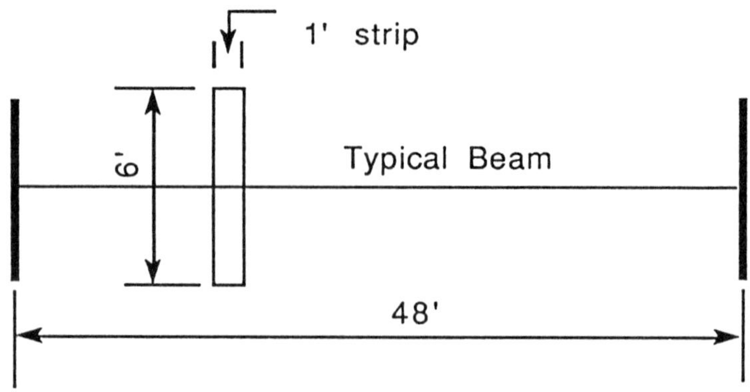

(a)

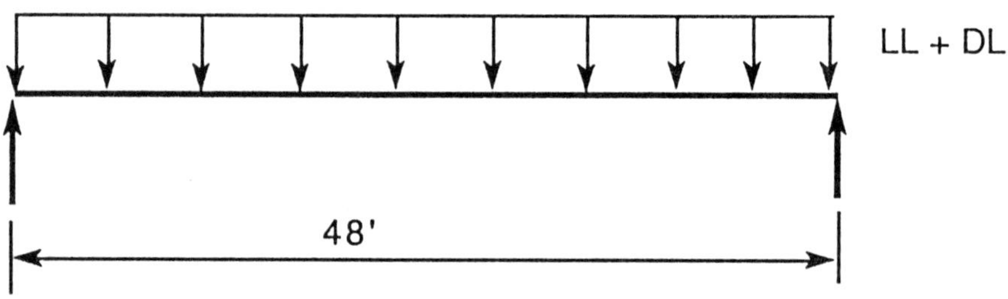

(b)

Figure 4.5 Load calculations on a typical beam.

load and placed it on the beam as shown in Figure 4.5b, we can readily obtain the maximum bending moment:

Dead load (DL)

$$\begin{aligned}
\text{Weight of the slab} &= \frac{4}{12} \times 1\text{ ft} \times 6\text{ ft} \times \frac{150}{1000} \\
&= 0.30\text{ k/ft} \\
\text{Assumed weight of beam} &= 0.05\text{ k/ft} \\
\text{Total } DL &= 0.35\text{ k/ft}
\end{aligned}$$

Live load (LL):

$$\begin{aligned}
\text{Live load on the strip} &= 1\text{ ft} \times 6\text{ ft} \times \frac{50}{1000} \\
&= 0.30\text{ k/ft}
\end{aligned}$$

Partition loads (PL)

$$\begin{aligned}\text{Partition loads on the strip} &= 1\text{ ft} \times 6\text{ ft} \times \frac{20}{1000}\\ &= 0.12\text{ k/ft}\\ \text{Total uniform load on the beam} &= 0.77\text{ k/ft}\end{aligned}$$

For a simply supported beam with a uniformly distributed load, the maximum bending moment occurs at the center of the span

$$\begin{aligned}M_{\max} &= \frac{wl^2}{8}\\ &= \frac{0.77 \times 48^2}{8}\\ &= 221.8\text{ k ft}\end{aligned}$$

In Equation (4.1), the stress is given for any point in the section. For the purpose of designing, it is required that the maximum stress which occurs at the point furthest from the neutral axis be determined. Let that distance be denoted by c. Then Equation (4.1) is modified as

$$f_b = \frac{Mc}{I}$$

If we let I/c be denoted by S_x, the section modulus, the above equation can be expressed as

$$f_b = \frac{M}{S_x}$$

Designing for the maximum allowable stress F_b, we find that the required section modulus for a beam to resist the given moment is

$$S_x = \frac{M}{F_b}$$

By assuming a compact section, the maximum allowable bending stress is $F_b = 0.66F_y$ from Equation (4.3). For structural steel with a yield point of 36 k/in.^{2}, $F_b = 24$ k/in.^{2}.

The required section for this case is

$$S_x = \frac{221.8 \times 12}{24}$$

$$= 110.9 \text{ in.}^3$$

Examine the allowable stress design selection table and find the nearest tabulated value of $S_x = 114$ in.^{3}, which corresponds to W24 × 55. Table 4.4 is taken from the allowable stress design selection table that appears in the AISC Manual, Ninth Edition, Section 2.

Since the above selection is based on the assumption that the section is compact, a check for compactness is in order. Table 4.5 extracted from the AISC Manual, Part 1, "Dimensions and Properties of W Shapes" is used to provide the check for compactness. The properties of W24 × 55 listed in Table 4.5 are

$$b_f = 7.005 \text{ in.}$$

$$\frac{b_f}{2t_f} = 6.9$$

$$\frac{d}{t_w} = 59.7$$

$$\frac{d}{A_f} = 6.66$$

and from Table 4.5

$$L_c = 7.0 \text{ ft}$$

The limiting width-thickness ratios for compression elements are given in Table 4.1 (AISC Manual, Table B5.1). A summary of those limits that are pertinent to this case is found in Table 4.6.

The selection of W24 × 55 confirms the compactness assumption made earlier. The compression flange must be supported laterally every 6 ft. In other words, place seven intermediate stiffeners at 6-ft intervals. The maximum shear occurs at the support

$$V_{\max} = \frac{wl}{2}$$

$$= 0.77 \times 48/2$$

$$= 8.5 \text{ k}$$

TABLE 4.4 Allowable Stress Design Selection for Shapes Used as Beams

F_y = 50 k/in.^{2}							F_y = 36 k/in.^{2}		
L_c (ft)	L_u (ft)	M_R (k-ft)	S_x (in.^{3})	Shape	Depth d (in.)	F_y'	L_c (ft)	L_u (ft)	M_R (k-ft)
8.1	**8.6**	**484**	**176**	**W24 × 76**	**$23\frac{7}{8}$**		**9.5**	**11.8**	**348**
9.3	20.2	481	175	W16 × 100	17		11.0	28.1	347
13.1	29.2	476	173	W14 × 100	$14\frac{3}{8}$	58.6	15.4	40.6	343
7.5	10.9	470	171	W21 × 83	$21\frac{3}{8}$		8.8	15.1	339
9.9	15.5	457	166	W18 × 86	$18\frac{3}{8}$		11.7	21.5	329
13.0	26.7	432	157	W14 × 99	$14\frac{1}{8}$	48.5	15.4	37.0	311
9.3	18.0	426	155	W16 × 89	$16\frac{3}{4}$		10.9	25.0	307
7.4	**8.5**	**424**	**154**	**W24 × 68**	**$23\frac{3}{4}$**		**9.5**	**10.2**	**305**
7.4	9.6	415	151	W21 × 73	$21\frac{1}{4}$		8.8	13.4	299
9.9	13.7	402	146	W18 × 76	$18\frac{1}{4}$	64.2	11.6	19.2	289
13.0	24.5	385	143	W14 × 90	14	40.4	15.3	34.0	283
7.4	**8.9**	**385**	**140**	**W21 × 68**	**$21\frac{1}{8}$**		**8.7**	**12.4**	**277**
9.2	15.8	369	134	W16 × 77	$16\frac{1}{2}$		10.9	21.9	265
5.8	**6.4**	**360**	**131**	**W24 × 62**	**$23\frac{3}{4}$**		**7.4**	**8.1**	**259**
7.4	**8.1**	**349**	**127**	**W21 × 62**	**21**		**8.7**	**11.2**	**251**
6.8	11.1	349	127	W18 × 71	$18\frac{1}{2}$		8.1	15.5	251
9.1	20.2	338	123	W14 × 82	$14\frac{1}{4}$		10.7	28.1	244
10.9	26.0	325	118	W12 × 87	$12\frac{1}{2}$		12.8	36.2	234
6.8	10.4	322	117	W18 × 65	$18\frac{3}{4}$		8.0	14.4	232
9.2	13.9	322	117	W16 × 67	$16\frac{3}{8}$		10.8	19.3	232
5.0	**6.3**	**314**	**114**	**W24 × 55**	**$23\frac{5}{8}$**		**7.0**	**7.5**	**226**
9.0	18.6	308	112	W14 × 74	$14\frac{1}{8}$		10.6	25.9	222
5.9	6.7	305	111	W21 × 57	21		6.9	9.4	220
6.8	9.6	297	108	W18 × 60	$18\frac{1}{4}$		8.0	13.3	214
10.8	24.0	294	107	W12 × 79	$12\frac{3}{8}$	62.6	12.8	33.3	212
9.0	17.2	283	103	W14 × 68	14		10.6	23.9	204

Source: Adapted from data in *Manual of Steel Construction*, 9th Ed., American Institute of Steel Construction.

TABLE 4.5 Section Dimensions and Properties

Designation	Area A (in.^{2})	Depth (in.)	Web t_w (in.)	Flange t_f (in.)	Flange b_f (in.)	I_x (in.^{4})	S_x (in.^{3})	S_y (in.^{3})
W24 × 76	22.4	23.92	0.440	0.680	8.990	2100	176	18.4
W16 × 100	29.4	16.97	0.585	0.985	10.425	1490	175	35.7
W14 × 109	32.0	14.32	0.525	0.860	14.605	1240	173	61.2
W21 × 83	24.3	21.43	0.515	0.835	8.355	1830	171	19.5
W18 × 86	25.3	18.39	0.480	0.770	11.090	1530	166	31.6
W14 × 99	29.1	14.16	0.485	0.780	14.565	1110	157	55.2
W16 × 89	26.2	16.75	0.525	0.875	10.365	1300	155	31.4
W24 × 68	20.1	23.73	0.415	0.585	8.965	1830	154	15.7
W21 × 73	21.5	21.24	0.455	0.740	8.295	1600	151	17.0
W18 × 76	22.3	18.21	0.425	0.680	11.035	1330	146	27.6
W14 × 90	26.5	14.02	0.440	0.710	14.520	999	143	49.9
W21 × 68	20.0	21.13	0.430	0.685	8.270	1480	140	15.7
W16 × 77	22.6	16.52	0.455	0.760	10.295	1110	134	26.9
W24 × 62	18.2	23.74	0.430	7.040	0.590	1550	131	9.80
W21 × 62	18.3	20.99	0.400	0.615	8.240	1330	127	13.9
W18 × 71	20.8	18.47	0.495	0.810	7.635	1170	127	15.8
W14 × 82	24.1	14.31	0.510	0.855	10.130	882	123	29.3
W12 × 87	25.6	12.53	0.515	0.810	12.125	740	118	39.7
W18 × 65	19.1	18.35	0.450	0.750	7.590	1070	117	14.4
W16 × 67	19.7	16.33	0.395	0.665	10.235	954	117	23.2
W24 × 55	16.2	23.57	0.395	0.505	7.005	1350	114	8.30
W14 × 74	21.8	14.17	0.450	0.785	10.070	796	112	26.6
W21 × 57	16.7	21.06	0.405	0.650	6.555	1170	111	9.35
W18 × 60	17.6	18.24	0.415	0.695	7.555	984	108	13.3
W12 × 79	23.2	12.38	0.470	0.735	12.080	662	107	35.8
W14 × 68	20.0	14.04	0.415	0.720	10.035	723	103	24.2

Source: Adapted from data in *Manual of Steel Construction*, ASD 9th Ed., 1989, by American Institute of Steel Construction.

TABLE 4.6 Summary of Limiting Width-Thickness Ratios for W24 × 55

Limits	$F_y = 36$
$\dfrac{b_f}{2t_f} = \dfrac{65}{\sqrt{F_y}}$	10.8 > 6.9
$\dfrac{d}{t_w} = \dfrac{640}{\sqrt{F_y}}$	106 > 59.7
$L_c = \dfrac{76 b_f}{\sqrt{F_y}}$	7.39 ft, use $L_b = 6.0$ ft
$L_c = \dfrac{20{,}000}{\dfrac{d}{A_f} F_y}$	6.95 ft, use $L_b = 6.0$ ft

From Equation (4.2), the shear stress is

$$f_v = \frac{18.5 \times Q_x}{I t_w}$$

Q_x, I, and t_w are obtained from the tables for W-shape dimensions and properties in Part 1 of the AISC Manual

$$Q_x = 67.1 \text{ in.}^3$$

$$I = 1350 \text{ in.}^4$$

$$t_w = 0.395 \text{ in.}$$

$$f_v = \frac{18.5 \times 67.1}{1350 \times 0.395}$$

$$= 2.33 \text{ k/in.}^2 < 0.40 F_y = 14.4 \text{ k/in.}^2$$

One last check remains and that is for the stability of the beam with regard to deflection. Building code requirements recommend that the live load deflection not exceed 1/360 of the span of the beam

$$\Delta_{\max} = \frac{48 \times 12}{360}$$

$$= 1.60 \text{ in.}$$

The maximum deflection due to live and partion loads at midspan is

$$\Delta = \frac{5Wl^3}{384EI} \times 1728$$

where

W = k/ft
l = length (ft)
E = Young's modulus of elasticity (k/in.^{2})
I = moment of inertia (in.^{4})

$$\Delta = \frac{5 \times (0.30 + 0.12) \times 48 \times (48)^3}{384 \times 29{,}000 \times 1350} 1728$$

$$\Delta = 1.281 \text{ in.} < 1.60$$

The same result may be obtained by using Table 4.3 after adjusting for the bending stress. Note that the partition load is included with the live load in the determination of the maximum deflection. This may not be necessary when the partitions constructed are of a permanent nature.

Thus, the selection of W24 × 55 has met the three requirements of bending stress, shear stress, and deflection serviceability.

4.5 DESIGN FOR FLEXURE: LRFD METHOD

LRFD is a method of proportioning structural elements in such a manner that no applicable limit state is exceeded when the structure is subjected to all possible factored loads during its life time. Strength limit is concerned with safety and applies to maximum load-carrying capacity.

The flexural design strength as determined by the limit state of lateral-torsional buckling is

$$M_u = \phi_b M_n \tag{4.22}$$

where

M_n = nominal strength
ϕ_b = resistance factor for flexure, 0.90

Compact Section Members with $L_b \leq L_r$

The nominal beam strength in bending for laterally unsupported compact section members bent about the major axis is

$$M_n = C_b\left[M_p - (M_p - M_r)\frac{L_b - L_p}{L_r - L_p}\right] \leq M_p \tag{4.23}$$

where

C_b = $1.75 + 1.05(M_1/M_2) + 0.3(M_1/M_2)^2 \leq 2.3$, where M_1 is the smaller and M_2 the larger end moment in the unbraced segment of the beam; M_1/M_2 is positive when the moments cause reverse curvature and negative when bent in single curvature

C_b = 1.0 for unbraced cantilevers and members where the moment within a significant portion of the unbraced segment is greater than or equal to the larger-segment end moments

L_b = distance between points braced against lateral displacement of the compression flange, or between points braced to prevent twist of the cross section

M_p = plastic bending moment (k-in.)

L_r = limiting laterally unbraced length for inelastic lateral-torsional buckling (in.)

M_r = limiting buckling moment when $\lambda = \lambda_r$ and $C_b = 1.0$ (k-in.)

λ = the width-thickness ratio

λ_r = the limiting width-thickness ratio for buckling

The limiting laterally unbraced length L_p for developing the full plastic bending a moment with $C_b = 1.0$, is

$$L_p = \frac{300r_y}{\sqrt{F_{yf}}} \tag{4.24}$$

for I-shaped members including hybrid sections and channels bent about their major axis.

For solid bars and box beams, the limiting laterally unbraced length is

$$L_p = \frac{3750r_y\sqrt{JA}}{M_p} \tag{4.25}$$

where

F_{yf} = specified minimum yield strength of the flange (k/in.^{2})

r_y = radius of gyration about the y axis (in.)

A = cross-sectional area (in.^{2})

J = torsional constant (in.^{4})

The values of L_p and M_r are obtained from the following expressions:

(a) For I-shaped members, doubly and singly symmetric with the compression flange larger than or equal to the tension flange, and channels loaded in the plane of the web

$$L_r = \frac{r_y X_1}{(F_{yw} - F_r)} \sqrt{1 + \sqrt{1 + X_2 (F_{yw} - F_r)^2}} \tag{4.26}$$

$$M_r = (F_{yw} - F_r) S_x \tag{4.27}$$

$$X_1 = \frac{\pi}{S_x} \sqrt{\frac{EGJA}{2}} \tag{4.28}$$

$$X_2 = 4 \frac{C_w}{I_Y} \left(\frac{S_x}{GJ} \right)^2 \tag{4.29}$$

where

S_x = section modulus about the major axis (in.^{3})
E = modulus of elasticity of steel (29,000 k/in.^{2})
G = shear modulus of elasticity of steel (11,000 k/in.^{2})
F_{yw} = yield stress of web (k/in.^{2})
I_y = moment of inertia about the y axis (in.^{4})
C_w = warping constant (in.^{6})
F_r = compressive residual stress in flange; 10 k/in.^{2} for rolled shapes, 16.5 k/in.^{2} for welded shapes)

(b) For singly symmetric, I-shaped sections with compression flange larger than the tension flange, use S_{xc} in place of S_x in Equations (4.27) through (4.29). S_x is the section modulus with respect to the compression flange.

(c) For a symmetric box section bent about the major axis and loaded in the plane of symmetry, M_r and L_r are determined from Equations (4.27) and (4.30), respectively.

(d) For solid rectangular bars bent about the major axis,

$$L_r = \frac{57{,}000 r_y \sqrt{JA}}{M_r} \tag{4.30}$$

$$M_r = F_y S_x \tag{4.31}$$

Compact Section Members with $L_b > L_r$

When the compression flange of compact sections bent about the major axis is not supported laterally, the nominal bending strength is determined from

$$M_n = M_{cr} \leq C_b M_r \tag{4.32}$$

where M_{cr} is the critical elastic moment and is determined from the following:

(a) For I-shaped members, doubly symmetric and singly symmetric with the compression flange larger than the tension flange (including hybrid members) and channels loaded in the plane of the web

$$M_{cr} = C_b \frac{\pi}{L_b} \sqrt{EI_y GI + \left(\frac{\pi E}{L_b}\right)^2 I_y C_w} \tag{4.33}$$

(b) For solid rectangular bars and symmetric box sections,

$$M_{cr} = \frac{57{,}000 C_b \sqrt{JA}}{\dfrac{L_b}{r_y}} \tag{4.34}$$

Design for Shear

The design strength for shear is based on action in the web. The area of the web A_w is taken as the product of the overall depth d and web thickness t_w. The shear strength is

$$V_u = \phi_v V_n \tag{4.35}$$

where

ϕ_v = nominal shear resistance factor
V_n = nominal shear

The nominal shear strength is determined from the following:

For $h/t_w \le 187\sqrt{k/F_{yw}}$

$$V_n = 0.60 F_{yw} A_w \tag{4.36}$$

For $187\sqrt{k/F_{yw}} < h/t_w \le 234\sqrt{k/F_{yw}}$

$$V_n = 0.60 F_{yw} A_w \frac{187\sqrt{\dfrac{k}{F_{yw}}}}{\dfrac{h}{t_w}} \tag{4.37}$$

For $h/t_w > 234\sqrt{k/F_{yw}}$

$$V_n = A_w \frac{26{,}400k}{\left(\frac{h}{t_w}\right)^2} \tag{4.38}$$

where

$$k = \text{the web buckling coefficient}$$

$$= 5 + \frac{5}{\left(\frac{a}{h}\right)^2} \tag{4.39}$$

When a/h exceeds 3.0, k is taken as 5.0. This applies to all rolled sections. Thus, the limit for h/t_w for a rolled section is 70.0 and Equation (4.36) applies when checking the nominal for a rolled section.

4.6 USE OF THE LOAD FACTOR DESIGN SELECTION TABLE Z_x FOR SHAPES USED AS BEAMS

Table Z_x in Part 3 of the LRFD Manual applies to the selection of braced beams with unbraced lengths not exceeding L_r. For beams with unbraced lengths exceeding L_p, it is more convenient to use the unbraced beam charts provided in the manual.

Once the ultimate moment M_u has been determined, the required plastic section modulus Z_x is obtained from

$$Z_x = \frac{12M_u}{\phi_b F_y} \tag{4.40}$$

Enter the column head Z_x in the table and find a value equal to or greater than the plastic section modulus required. The beam opposite the required Z_x and all beams above have sufficient capacity in bending based only on these parameters. The first beam appearing in boldface represents the most economical section. After the beam has been selected, the following check is required. If the lateral bracing of the compressive flange exceeds L_p, but is less than L_r, the design moment may be calculated by the following equation:

$$\phi_b M_n = C_b[\phi_b M_p - BF(L_b - L_p)] \le \phi_b M_p \tag{4.41}$$

where

$$BF = \frac{\phi_b(M_p - M_r)}{L_r - L_p} \tag{4.42}$$

If the bracing length L_b is significantly greater than L_p, i.e., $L_b > L_r$, it is recommended that the unbraced beam charts be used.

Example 4.2

Given the information in Example 4.1, use the LRFD method to select a typical beam.

Solution

Consider the partition load as a live load. The ultimate load is

$$
\begin{aligned}
w_u &= 1.2D + 1.6L \\
&= 1.2 \times 0.35 + 1.6 \times 0.42 \\
&= 1.09 \text{ k/ft}
\end{aligned}
$$

The ultimate bending moment is

$$
\begin{aligned}
M_u &= 1.09 \times \frac{48^2}{8} \\
&= 314.5 \text{ k/ft}
\end{aligned}
$$

From Equation (4.40), the required section modulus based on the LRFD method is

$$
\begin{aligned}
Z_x &= \frac{12M_u}{\phi_b F_y} \\
&= \frac{12 \times 314.5}{0.90 \times 36} \\
&= 116.5 \text{ in.}^3
\end{aligned}
$$

From the load factor design selection table, select W24 × 55 with the following properties:

$$
\begin{aligned}
BF &= 12.7 \text{ k} \\
L_r &= 16.6 \text{ ft} \\
L_p &= 5.6 \text{ ft} \\
\phi_b M_r &= 222 \text{ k-ft} \\
\phi_b M_p &= 362 \text{ k-ft} \\
h/t_w &= 54.6 \\
A_w &= 8.91 \text{ in.}^2
\end{aligned}
$$

Use seven intermediate stiffeners spaced equally, i.e., 5 ft center to center of the stiffeners. Since $L_b = 5$ ft $< L_p = 5.6$, there is no need to use the check as required by Equation (4.41)

$$\phi_b M_n = \phi_b M_p$$

$$\phi_b M_n > 314.5 \text{ k-ft} = M_u$$

If we check for shear, Equation (4.36) is used since $h/t_w = 54.6 < 70$

$$V_u \leq = \phi_v V_n = 0.6 F_{yw} A_w$$

$$\begin{aligned} V_u &= \frac{w_u L}{2} \\ &= \frac{1.09 \times 48}{2} \\ &= 26.2 \text{ k} \\ 0.6 F_{yw} A_w &= 0.6 \times 36 \times 8.91 \\ &= 192 \text{ k} > V_u = 26.2 \text{ k} \end{aligned}$$

The selection of $W24 \times 55$ proves satisfactory.

Alternate Solution

M_p Method

Knowing that the required factored moment $M_u = 314.5$ k-ft, find $\phi_b M_p$ in the load factor design selection table and choose a section with a value of 314.5 or larger. W24 × 55 provides a value of 362 k-ft and $L_b < L_p$.

Example 4.3

Determine the moment capacity of W27 × 84 for $F_y = 36$ k/in.^{2} with the compression flange braced at intervals of 10 ft ($C_b = 1.0$).

Solution

Examine the load factor design table and observe that for W27 × 84, $F_y = 36$ k/in.^{2}

$$\begin{aligned} \phi_b M_p &= 659 \text{ k-ft} \\ L_p &= 8.6 \text{ ft} \\ L_r &= 24.9 \text{ ft} \\ BF &= 15.0 \text{ k} \end{aligned}$$

From Equation (4.41),

$$\begin{aligned}\phi_b M_n &= C_b[\phi_b M_p - BF(L_b - L_p)] \leq \phi_b M_p \\ &= 1.0 \times [659 - 15.0 \times (10 - 8.6)] \\ &= 638 \text{ k-ft}\end{aligned}$$

Example 4.4

Select a beam of $F_y = 50$ k/in.^{2} steel subjected to service loads of $M_D = 156$ k-ft and $M_L = 257$ k-ft. Its compression flange is braced at intervals of 8.0 ft ($C_b = 1.0$).

Solution

Z_x Method

Assume that the shape is compact and $L_b \leq L_p$. The factored moment is

$$\begin{aligned}M_u &= 1.2M_D + 1.6M_l \\ &= 1.2 \times 156 + 1.6 \times 257 \\ &= 598.4 \text{ k-ft}\end{aligned}$$

The required section modulus is

$$\begin{aligned}Z_x &= \frac{12M_u}{\phi_b F_y} \\ &= \frac{12 \times 598.4}{0.90 \times 50.0} \\ &= 159.6 \text{ in.}^3\end{aligned}$$

Examine the load factor design selection table and note that for W24 × 68 with $F_y = 50$ k/in.^{2}, the maximum resisting moment $\phi_b M_n$ listed in the $\phi_b M_p$ column is 664 k-ft > $M_u = 598.4$ k-ft. Further note that

$$\begin{aligned}\phi_b M_n &= 664 \text{ k-ft} \\ L_p &= 6.6 \text{ ft} \\ L_r &= 17.4 \text{ ft} \\ BF &= 18.7 \text{ k}\end{aligned}$$

Since $L_p < L_b < L_r$,

$$\phi_b M_n = C_b[\phi_b M_p - BF(L_b - L_p)]$$
$$= 1.0[664 - 18.7(8.0 - 6.6)]$$
$$= 637.8 \text{ k-ft} \geq 598.4$$

Use W24 × 68.

Alternate Solution

M_p Method

With a given factored moment of 598.4 k-ft, review the selection table and note that in the column of $\phi_b M_n(\phi_b M_n \leq \phi_b M_p)$ values for W24 × 68, $F_y = 50$ k/in.^{2}, the value of $\phi_b M_p$ is 664 k-ft, which is greater than the given factored moment. This section has $L_p = 6.6$ ft, $L_r = 17.4$ ft, and $BF = 18.7$ k. Since $L_b = 8.0$ ft,

$$\phi_b M_n = 1.0[664.0 - 18.7(8.0 - 6.6)]$$
$$= 637.8 \text{ k-ft}$$

The selection of W24 × 68 is adequate.

4.7 SERVICEABILITY DESIGN CONSIDERATIONS AND THE LRFD METHOD

The concept of serviceability is developed in design to ensure that disruptions of a structure's functional use or damage to it during normal everyday use are extremely unlikely. Although malfunctions in the serviceability of a structure may not lead to its collapse, they can seriously impair its usefulness or in some cases incrue significant repair costs. The use of high-strength materials in design provides relatively flexible structures.

It is stated in the ASD method that serviceability checks are required for live loads, wind and earthquake forces, human activities such as walking and dancing, temperature fluctuations, and vibrations induced by machinery within the building. The LRFD specifications do not provide limiting deflections for individual members or structural assemblies. These limits depend on the function of the structure.

5

Torsion and Bending

5.1 INTRODUCTION

When the transverse forces acting on the beam do not pass through the shear center of its cross section, the beam will experience torsional stresses in addition to those caused by bending. Beams and girders that are placed on the periphery of the building support floor as well as wall loads. These members will experience stresses due to bending and, in addition, torsion. This chapter is intended to present a brief discussion of the theory of torsion and the use of tables and charts to determine the magnitude of stresses caused by torsion.

5.2 TORSIONAL STRESSES

In Figure 5.1a, the cross section of the member is subjected to pure torsional shear stresses, warping shear stresses, and warping normal stresses in addition to normal and shear bending stresses. Figure 5.1b is a vectorial representation of the beam shown in Figure 5.1a. When a structural member is subjected to loads applied outside the shear center of its cross section as shown above, the cross section tends to rotate about its longitudinal axis. See Figure 5.2.

The distribution of torsional stresses along the length of the beam shown in Figure 5.1 depends on the length of the beam, its torsional rigidity, and its end conditions. In Figure 5.3, several types of end con-

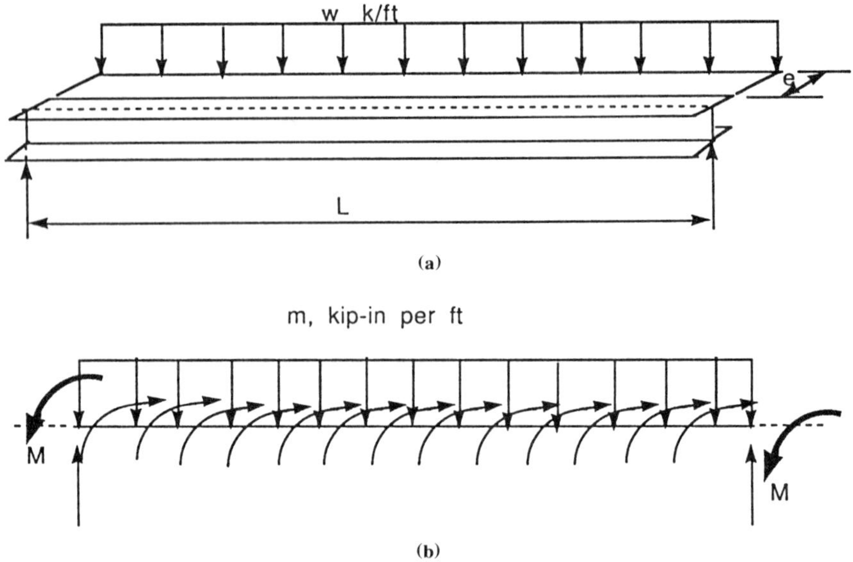

Figure 5.1 Uniform load applied outside of the plane of symmetry of a beam.

ditions are illustrated. Only ideal fixed, pinned, or free torsional end conditions will be considered in this text.

Pure Torsional Shear Stresses

Pure torsional shear stresses vary linearly across the thickness of an element of the cross section and act in a direction parallel to the edge of the element. See Figure 5.2c. At the two edges, the stresses are maximum and equal but of opposite direction. The stress at the edge of the element is determined by the formula

$$\tau_t = Gt\phi' \tag{5.1}$$

In a section that consists of several elements, pure torsional shear stresses will be largest in the thickest element of the cross section.

Warping Shear Stresses

Warping shear stresses are constant across the thickness of the element of the cross section, but vary in magnitude along the length of the element.

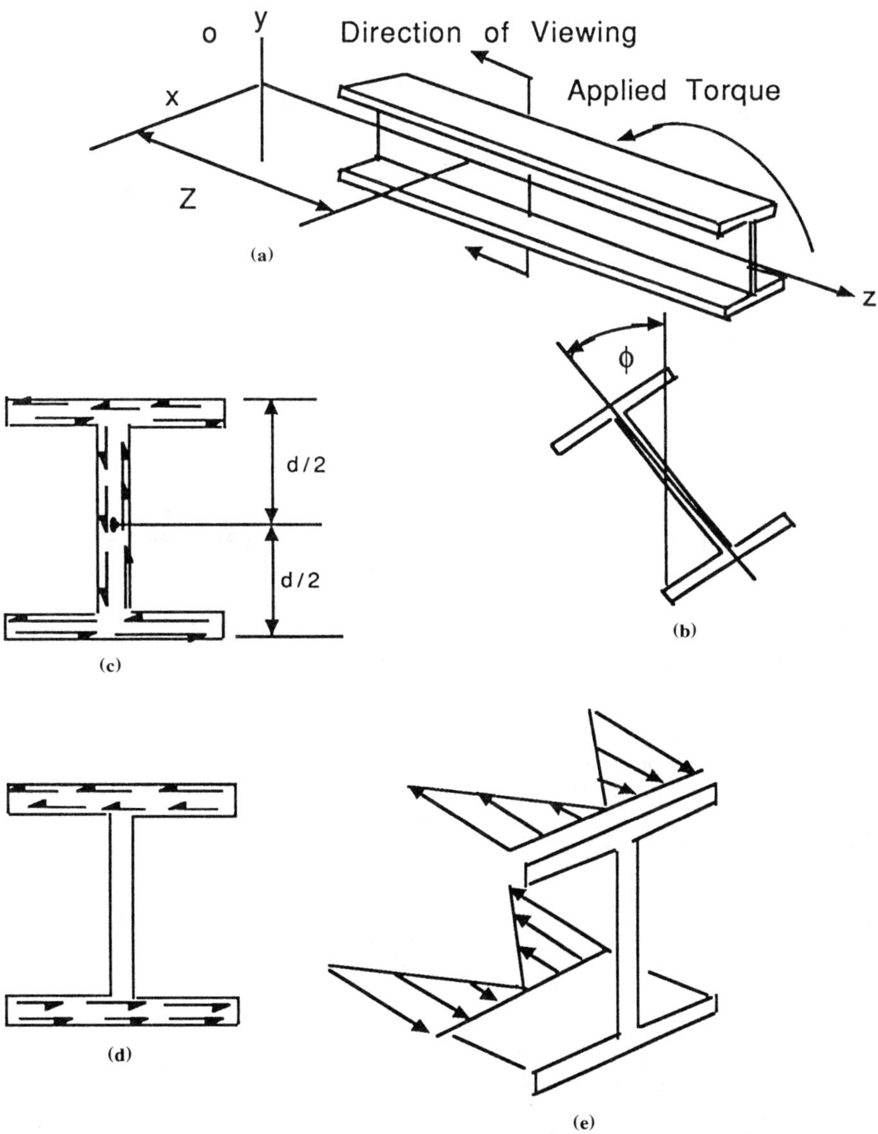

Figure 5.2 Torsional stresses and stress orientation. (a) General orientation of beam. (b) Positive angle of rotation. (c) Pure torsional shear. (d) Warping shear stresses. (e) Warping normal stresses.

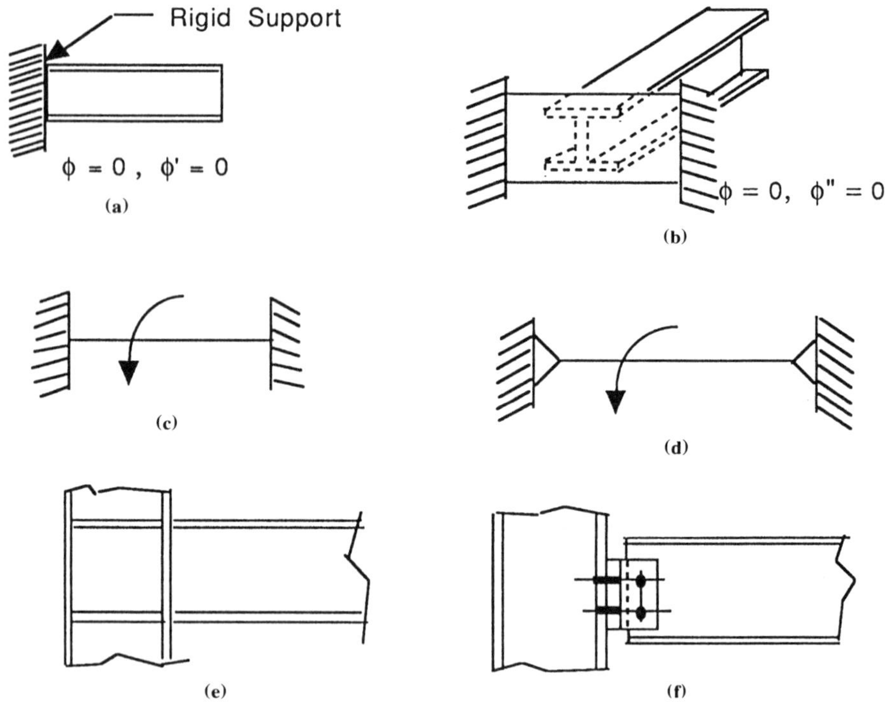

Figure 5.3 Bending stresses. (a) Ideal. (b) Ideal. (c) and (d) Schematic representations. (e) and (f) Structural connections.

See Figure 5.2d. The magnitude of the warping stresses is determined by the formula

$$\tau_{ws} = \frac{-ES_{ws}\phi'''}{t} \tag{5.2}$$

Warping Normal Stresses

Warping normal stresses act perpendicular to the surface of the cross section. See Figure 5.2e. They are constant across the thickness of an element of the cross section, but vary in magnitude along the length of the element. These stresses are given by the following formula:

$$\sigma_{ws} = EW_{ns}\phi'' \tag{5.3}$$

5.3 PLANE BENDING STRESSES

In addition to the stresses due to torsion, the section may be subjected to bending stresses σ_b and shear stresses τ_b due to plane bending already existing in the structural member. These stresses are obtained from Equations (4.1) and (4.2).

$$\phi_b = \frac{M_b y}{I} \tag{5.4}$$

$$\tau_b = \frac{VQ}{It} \tag{5.5}$$

These stresses are illustrated in Figure 5.4 for a wide flange and T-section members.

5.4 COMBINING TORSIONAL AND BENDING STRESSES

In order to determine the total stress condition of a structural member subjected to both torsion and bending, the stresses are added algebraically as long as they remain in the elastic range. It is imperative that the direction of the stresses be carefully noted. The positive direction of the torsional stresses is indicated in Figures 5.2 and 5.4. In the diagrams accompanying each figure, the stresses are shown acting on a section of the member located a distance Z from the left support and viewed in the direction indicated by the general orientation of Figure 5.2. In all the diagrams shown in Figure 5.4, the moment acts about the major axis of the cross section of the member and causes compression in the top flange and tension in the bottom one. Furthermore, the applied shear is assumed to act vertically downward along the minor axis of the cross section.

For wide-flange sections, σ_{ws} and σ_b are both at their maximum values at the edges of the flanges as shown in Figures 5.2 and 5.4. Likewise, there are two flange tips where these stresses are added regardless of the directions of the applied torsional and bending moments. Thus, for wide-flange shapes, σ_{ws} and σ_b should be added to determine the maximum longitudinal stress on the cross section. Also for wide-flange sections, the maximum values of τ_t, τ_{ws}, and τ_b in the flanges can be added at some point regardless of the direction of the applied torsional moment and vertical shear to give the maximum shear stress in the flange. The value of τ_b computed using Q_f from the torsional properties tables given in the Ninth Edition of the AISC Manual, Part 1, pp. 117 through 133, is the

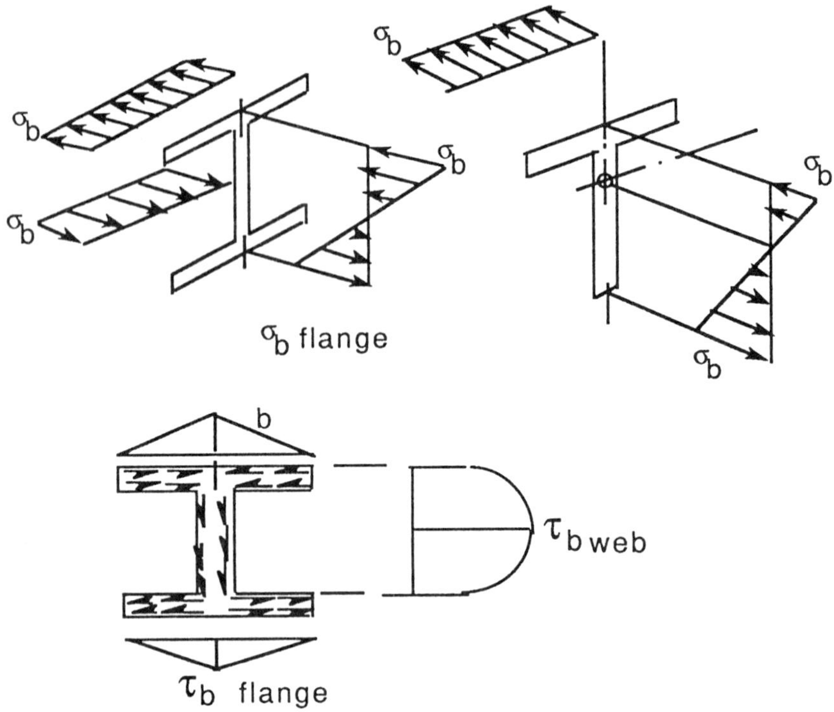

Figure 5.4 Torsional end conditions.

theoretical value at the edge of the web. It is within the accuracy of the method presented in this book to combine the resulting theoretical value with the torsional shearing stress calculated for the point at the intersection of the web and flange centerlines. In the web, the maximum value of the shear stress τ_b is added to the value of τ_t, regardless of the direction of loading, to give the maximum shear in the web.

5.5 TORSIONAL END CONDITIONS

The distribution of torsional stresses along the length of the member and the torsional rigidity depend on the end conditions. This book only considers ideal fixed, pinned, or free torsional end conditions. When the ideal conditions do not apply, it may be necessary to interpolate between conditions or, conservatively, to assume the worst possible conditions.

The fixed end condition is satisfied when rotation and warping of the cross section at the end of the member are prevented. A connection approximating this condition is shown in Figures 5.3a, 5.3c, and 5.3e. The pinned end condition occurs when the cross section at the end of the member is prevented from rotating, but is allowed to warp freely. The common simple structural connection shown in Figures 5.3b, 5.3d, and 5.3f is a reasonable example of this condition. For the free end condition, both rotation and warping of the cross section are unrestrained. The unsupported end of a cantilever beam illustrates this condition.

5.6 TORSIONAL LOADING AND END CONDITIONS

The series of charts given in Figures 5.5a through 5.5b provide the values for the angle of rotation and its derivatives [ϕ', ϕ'', and ϕ''' used in Equations (5.1), (5.2), and (5.3)] for any point along the length of the member. These terms are given in nondimensional form, i.e., each term is multiplied by a factor dependent on the torsional properties of the member and magnitude of the applied torsional moment. This product is called the *torsional function.*

The charts apply to 12 combinations of torsional loading and end restraints. A case chart listing is given in Figure 5.6. Each chart presents a series of curves for the torsional functions depending on different values of L/a, L being the length of the member and a the torsional property of its cross section.

For each case of loading, there are two charts. One provides values for ϕ and ϕ'', and the other for ϕ' and ϕ'''. Each chart shows the value of the torsional function (on the vertical scale) plotted against the location of the cross section (on the horizontal scale) from the left support.

How to Use the Charts

For the given cross section and span length, compute the value of L/a using L in inches and a as given in Table 5.1 for torsional properties, where $a = \sqrt{EC_w/GJ}$ in inches. With L/a, Z/L, and the case for torsional loading and end conditions known, enter the appropriate chart to obtain the torsional functions ϕ, ϕ', ϕ'', or ϕ'''. This result may then be used appropriately in Equations (5.1), (5.2), and (5.3).

Sign Convention

The case charts are based on applied torsional moments shown in the sketch accompanying each chart. In all cases, the torsional moment is

CASE 3

$\alpha = 0.5 \quad \phi', \phi'''$

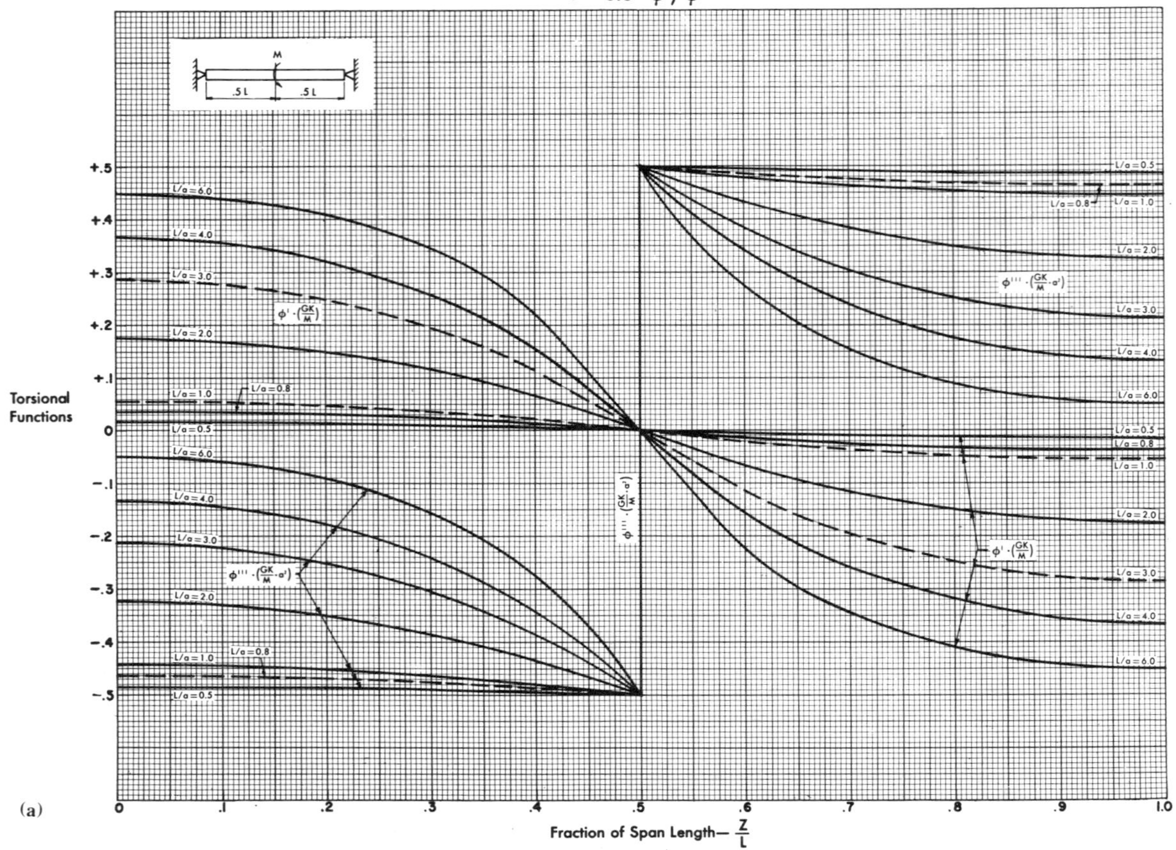

(a)

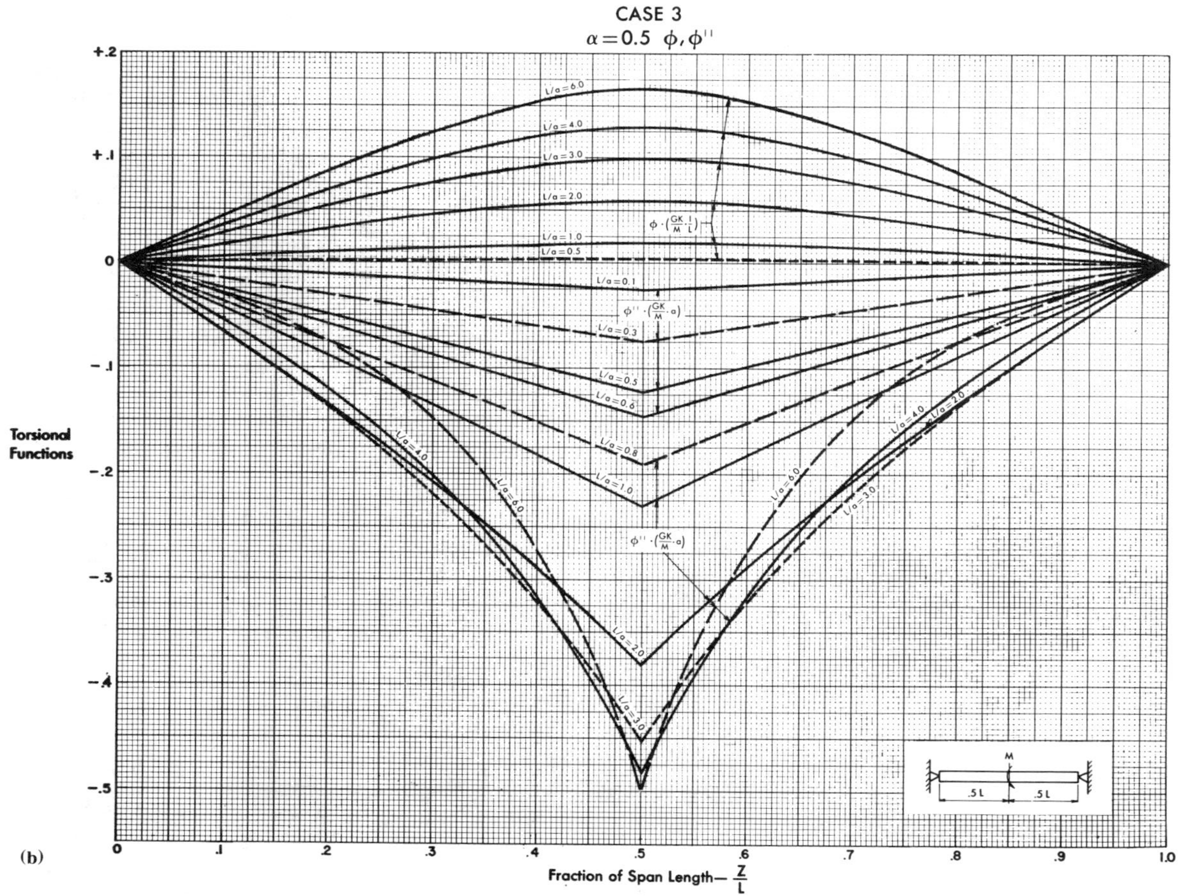

(b)

Figure 5.5 (a) Torsional functions. (b) Torsional functions. (Charts from Torsional Analysis of Steel Members by American Institute of Steel Construction.)

Case	Torsional Loading	Torsional End Restraints		Remarks	Page
		Left End	Right End		
1		Free $\phi = \phi'' = 0$	Free $\phi'' = 0$	Concentrated torques at ends of member with free ends	21
2	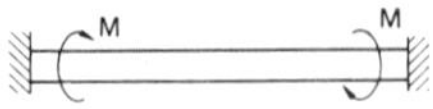	Fixed $\phi = \phi' = 0$	Fixed $\phi' = 0$	Concentrated torques at ends of member with fixed ends	22
3	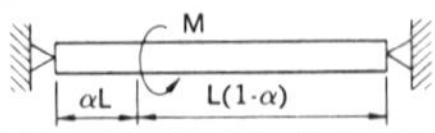	Pinned $\phi = \phi'' = 0$	Pinned $\phi = \phi'' = 0$	Concentrated torque at $\alpha = 0.1$, 0.3, or 0.5 on member with pinned ends	24
4	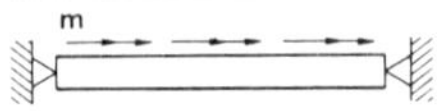	Pinned $\phi = \phi'' = 0$	Pinned $\phi = \phi'' = 0$	Uniformly distributed torque on member with pinned ends	30
5	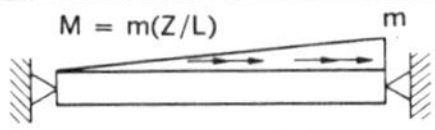	Pinned $\phi = \phi'' = 0$	Pinned $\phi = \phi'' = 0$	Linearly varying torque on member with pinned ends	32
6	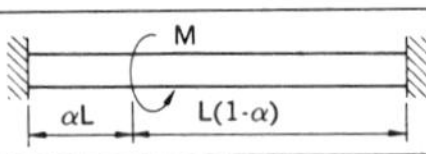	Fixed $\phi = \phi' = 0$	Fixed $\phi = \phi' = 0$	Concentrated torque at $\alpha = 0.1$, 0.3, or 0.5 on member with fixed ends	34
7	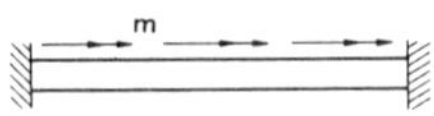	Fixed $\phi = \phi' = 0$	Fixed $\phi = \phi' = 0$	Uniformly distributed torque on member with fixed ends	40
8	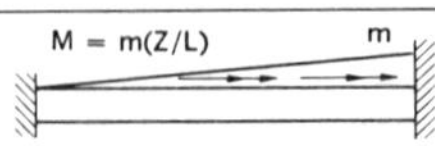	Fixed $\phi = \phi' = 0$	Fixed $\phi = \phi' = 0$	Linearly varying torque on member with fixed ends	42
9	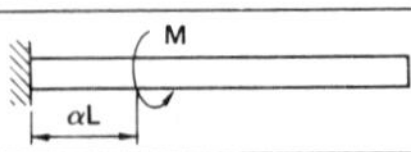	Fixed $\phi = \phi' = 0$	Free $\phi'' = 0$	Concentrated torque at $\alpha = 0.1$, 0.3, 0.5, 0.7, 0.9, or 1.0 on member with fixed and free ends	44
10	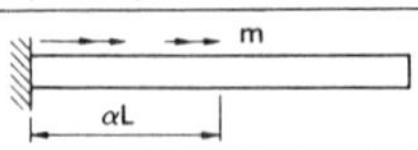	Fixed $\phi = \phi' = 0$	Free $\phi'' = 0$	Partial uniformly distributed torque along $\alpha = 0.1$, 0.3, 0.5, 0.7, 0.9, or 1.0 with fixed end and free end	56
11	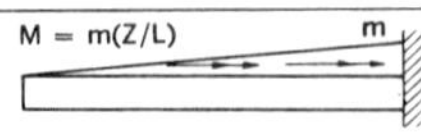	Free $\phi'' = 0$	Fixed $\phi = \phi' = 0$	Linearly varying torque on member with free and fixed ends	68
12	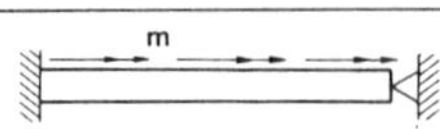	Fixed $\phi = \phi' = 0$	Pinned $\phi = \phi'' = 0$	Uniformly distributed torque on member with fixed and pinned ends	70

Figure 5.6 Case chart listing for beams in torsion. (Charts from Torsional Analysis of Steel Members by American Institute of Steel Construction.)

shown acting in a counterclockwise direction when viewed toward the left end of the member. This is considered to be a positive moment. If the applied torsional moment occurs in the opposite direction, it should be assigned a negative value for computational purposes. A positive stress or rotation computed with the formulas and torsional constants of this book acts in the direction shown in the cross-sectional view in Figure 5.2. A negative value indicates the opposite sense. In some of the case charts, the applied torsional moment is indicated by a vector notation using a line with two arrowheads as shown in Figure 5.6. This notation uses the

TABLE 5.1 Torsional Properties: W shapes

Section	Torsional Constant J (in.^{4})	Warping Constant C_w (in.^{6})	$\sqrt{\frac{EC_w}{GJ}}\ a$ (in.)	Normalize d Warping Constant W_{ns} (in.^{2})	Warping Statical Moment S_w (in.^{4})	Statical Q_f (in.^{3})	Moment Q_w (in.^{3})
W24 × 84	3.70	12,800	94.8	52.6	91.3	39.0	112.0
W24 × 62	1.71	4620	83.7	40.7	42.3	23.2	76.6
W24 × 55	1.18	3870	82.0	40.4	35.7	19.8	67.1
W21 × 62	1.83	5960	91.8	42.0	53.2	25.1	72.2
W18 × 46	1.22	1710	60.3	26.4	24.2	15.3	45.3
W12 × 50	1.78	1880	52.3	23.3	30.2	14.7	36.2

Source: Adapted from data in the *Manual of Steel Construction*, ASD 9th ed. 1989 American Society of Steel Construction.

"right-hand rule," whereby if the thumb of the right hand is extended and pointed in the direction of the vector, the remaining fingers of the right hand would curve in the direction of the moment.

5.7 APPLICATIONS

Example 5.1

A W12 × 50 section is used on a 20-ft span to support a 10-k load applied eccentrically at midspan as shown in Figure 5.7. Determine the stresses in the member and torsional rotation at midspan.

Solution

The eccentric load is resolved into a torsional moment and a load applied through the shear center of the section as shown in Figures 5.7c and 5.7d. The stresses due to plane bending and torsion are then determined independently.

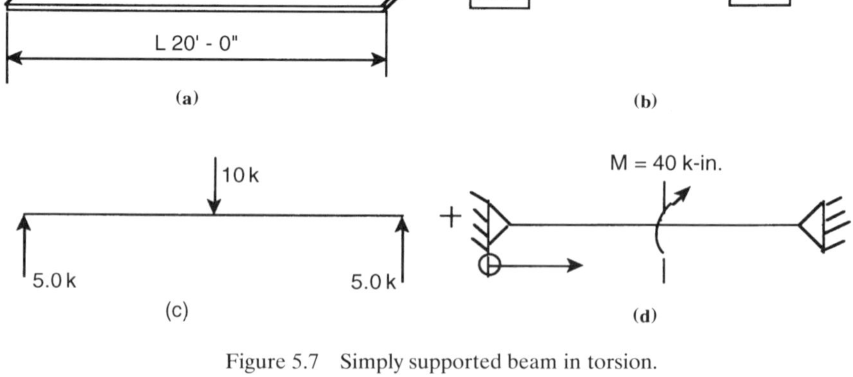

Figure 5.7 Simply supported beam in torsion.

The torsional properties of the section are obtained from Table 5.1. A similar table can be found in the AISC Manual.

$$J = 1.78 \text{ in.}^4$$
$$a = 52.3 \text{ in.}$$
$$C_w = 1880 \text{ in.}^6$$
$$W_{ns} = 23.3 \text{ in.}^2$$
$$S_{w_1} = 30.2 \text{ in.}^4$$
$$Q_f = 14.7 \text{ in.}^3$$
$$Q_w = 36.2 \text{ in.}^3$$
$$I_x = 394 \text{ in.}^4$$
$$S_x = 64.7 \text{ in.}^3$$
$$t_w = 0.370 \text{ in.}$$
$$d = 12.19 \text{ in.}$$
$$t_f = 0.640 \text{ in.}$$
$$b = 8.08 \text{ in.}$$

1. Plane Bending Stresses

(a) *Maximum longitudinal bending stresses.* These occur in the extreme fibers at midspan of the beam

$$M_b = \frac{PL}{4}$$
$$= \frac{10 \times 12 \times 12}{4}$$
$$= 600 \text{ k-in.}$$
$$\sigma_b = \frac{M_b}{S_x} = \frac{600}{64.7} = 9.20 \text{ k} \quad \text{(compression at top and tension at bottom)}$$

(b) *Maximum web shear stresses at the supports.*

$$\tau_b = \frac{VQ}{It}$$
$$= \frac{5 \times 36.2}{394 \times 0.370}$$
$$= 1.24 \text{ k/in.}^2$$

(c) *Maximum flange shear stresses.*

$$\tau_b = \frac{5.0 \times 14.7}{394 \times 0.640}$$

$$= 0.290 \text{ k/in.}^2$$

For this loading, the shear stresses are constant from the support to the load point.

2. *Torsional Stresses*

(a) *Torsional functions.* The web angle connection is assumed to provide a torsional pinned end condition. Case 3 with $\alpha = 0.50$ should therefore be used for this problem

$$\frac{l}{a} = \frac{20 \times 12}{52.3}$$

$$= 4.6$$

From the chart at the centerline ($Z/L = 0.5$),

$$\phi \times \frac{GJ}{M} \times \frac{1}{L} = +0.15$$

$$\phi = +0.15\frac{LM}{GJ}$$

$$\phi'' \times \frac{GJ}{M} \times a = -0.48$$

$$\phi'' = -0.48\frac{M}{GJa}$$

$$\phi' \times \frac{GJ}{M} = 0$$

$$\phi' = 0$$

$$\phi''' \times \frac{GJ}{M} \times a^2 = -0.50$$

$$\phi''' = -0.50\frac{M}{GJa^2}$$

At the support ($Z/L = 0$),

$$\phi \times \frac{GJ}{M} \times \frac{1}{L} = 0$$

$$\phi = 0$$

$$\phi'' \times \frac{GJ}{M} \times a = 0$$

$$\phi'' = 0$$

$$\phi' \times \frac{GJ}{M} = +0.40$$

$$\phi' = +0.40\frac{M}{GJ}$$

$$\phi''' \times \frac{GJ}{M} \times a^2 = -0.10$$

$$\phi''' = -0.10 \times \frac{M}{GJa^2}$$

$$\frac{M}{GJ} = \frac{-40}{11.2 \times 10^3 \times 1.78} = -2.01 \times 10^{-3}$$

In accordance with the sign convention of this book, the applied torsional moment is negative.

(b) *Pure torsional shear stresses.*

$$\tau_t = Gt\phi'$$

from Equation (5.1). At the centerline of the beam,

$$\phi' = 0$$

$$\tau_t = 0$$

At the support,

$$\text{Web} \quad \tau_t = 11.2 \times 10^3 \times 0.370 \times 0.40 \times (-2.01 \times 10^{-3})$$

$$= -3.3 \text{ k/in.}^2$$

$$\text{Flange} \quad \tau_t = 11.2 \times 10^3 \times 0.64 \times 0.40 \times (-2.01 \times 10^{-3})$$

$$= -5.8 \text{ k/in.}^2$$

Warping normal stresses. These are maximum at the tips of the flanges

$$\sigma_{ws} = EW_{ns}\phi''$$

At the centerline of the beam,

$$\sigma_{ws} = 29 \times 10^3 \times 23.3 \times (-0.48) \times \frac{-2.01 \times 10^{-3}}{52.3}$$

$$= 12.44 \text{ k/in.}^2$$

At the support,

$$\phi'' = 0$$

$$\sigma_{ws} = 0$$

Warping shear stresses. These are maximum at the center of the flanges

$$\tau_{ws} = \frac{ES_{ws}}{t}\phi'''$$

At the centerline of the beam,

$$\tau_{w_1} = 29 \times 10^3 \times \frac{30.2}{0.64} \times (-0.50) \times \frac{-2.01 \times 10^{-3}}{52.3^2}$$

$$= -0.50 \text{ k/in.}^2$$

At the support,

$$\tau_{w_1} = 29 \times 10^3 \times \frac{30.2}{0.64} \times (-0.10) \times \frac{-2.01 \times 10^{-3}}{52.3^2}$$

$$= -0.10 \text{ k/in.}^2$$

3. *Maximum Combined Stresses*

A summary of stresses is presented as shown below.

(a) The maximum normal stresses occur at midspan. Refer to Figure 5.8. At point 1,

$$\sigma_b = 12.44 + 9.27$$

$$= 21.71 \text{ k/in.}^2 \quad \text{(compression)}$$

At point 5,

$$\sigma_b = 12.44 + 9.27$$

$$= 21.71 \text{ k/in.}^2 \quad \text{(tension)}$$

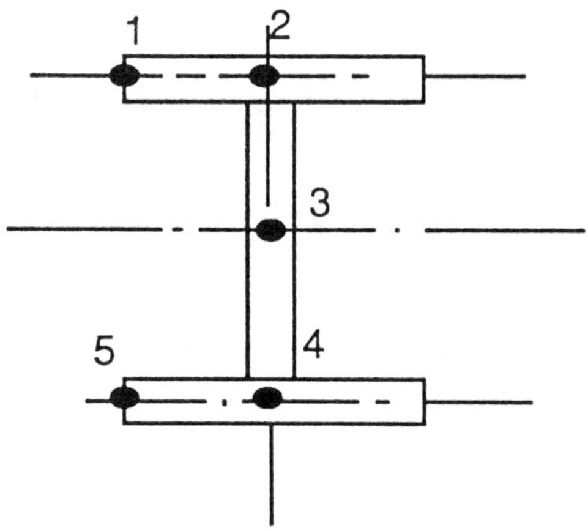

Figure 5.8 Stress location designation for beams in torsion.

(b) The maximum shear stresses occur at the support. At point 2,

$$\tau_{w_1} = -5.80 - 0.10 - 0.29$$
$$= -6.19\ \text{k/in.}^2 \quad \text{(acts toward the right)}$$

At point 4,

$$\tau_{w_4} = -6.19\ \text{k/in.}^2 \quad \text{(acts toward the left)}$$

At point 3,

$$\tau_{w_3} = -3.3 - 1.24$$
$$= -4.54\ \text{k/in.}^2 \quad \text{(acts downward)}$$

4. Torsional Rotation at Midspan

The angle of rotation at midspan is calculated as follows:

$$\phi = 0.15 \times 20 \times 12(-2.01 \times 10^{-3})$$
$$= -0.072\ \text{rad} \quad (4.126\ \text{deg})$$

For other cases the reader is referred to a publication by AISC, "Torsional Analysis of Steel Structures". There are twelve cases which include a variety of loading and support conditions as shown in Figure 5.6. Associated with these cases, there are fifty charts that provide the values of the torsional functions. This method of analysis is simple and can be accomplished with a minimum of calculations.

6

Design of Bracings for Wind and Earthquake Forces

6.1 INTRODUCTION

Lateral loads acting on a structure are mostly attributed to wind and earthquake forces. Ordinarily, in designing a building, members are first selected on the basis of gravity loads and then checked for a combination of loading patterns that include wind and/or earthquake forces. The magnitude of these forces is a matter of judgment and reliance on local and national code requirements. Rational analyses can be made using certain parametric assumptions and design criteria. The designer has the responsibility of using judiciously the guidelines of the code. Such action does not release him/her from the necessity of more elaborate analyses depending on specific building requirements and functions.

6.2 WIND FORCES

All exposed structures, including low-rise buildings, are affected by wind forces. Wind loads are a function of the site and the region in which the building is situated. The region determines the wind velocity as the fastest-mile wind speed in miles per hour calculated from the national meteorological records. Wind forces are resisted by a frame or assemblage of structural elements that work together to transfer wind load acting on the building as a whole to the foundation. Bracings, shear walls, roof and floor diaphrams are the structural components that conduct the transfer of the wind load to the foundation.

As wind blows against a building, the outer shell of the building experiences the effect of the wind force. Purlins in roof systems, girts, and wall studs receive the impact of the wind load against the building skin and transfer this load to the main frame system. Rigid connections, bracings, or a combination of both transfer these forces to girders and columns and eventually the foundation. Wind-resisting structural components act as a part of a structural system. However, they have to be designed as individually loaded members. The structural engineer would have to appropriate the load to these members and design them accordingly. A tributary area is the wind loading surface area of a definite structural element. When using braced frames in a building, the tributary area for the braced frame may come from several panels (see Figure 6.1).

The tributary area for the wind-resisting frame in the long direction is

$$A = \frac{75}{2} \times 147$$

$$= 5512.5 \text{ ft}^2$$

Consider an average wind pressure against the building of 25 lb/ft^2. The total wind load against the building is

$$P = 5512.5 \times \frac{25}{1000}$$

$$= 137.8 \text{ ks}$$

6.3 WIND VELOCITY PRESSURE

The design wind speed is converted to a velocity pressure from:

$$q_z = 0.00256 K_z (IV)^2 \tag{6.1}$$

where

V = basic mean design wind speed in mph (fastest mile) at a height above mean ground of 10 m (33 ft) for terrain exposure category C (see Table 6.1)

I = importance factor, a coefficient used to modify wind speed to provide a somewhat consistent level of risk based on usage (see Table 6.2)

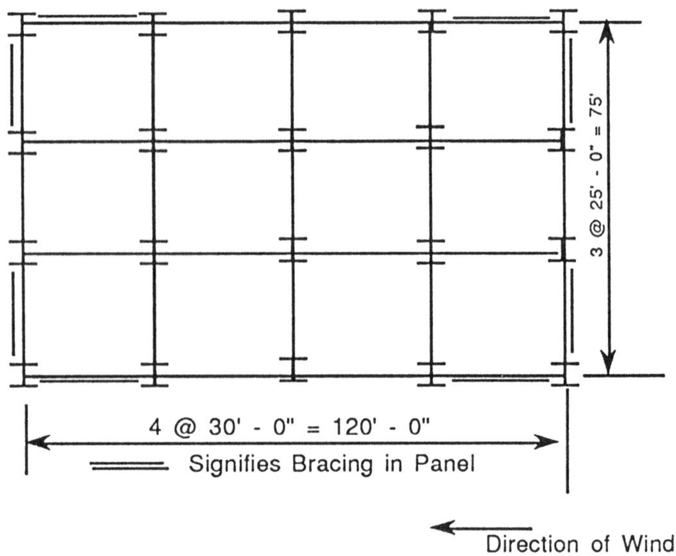

(a)

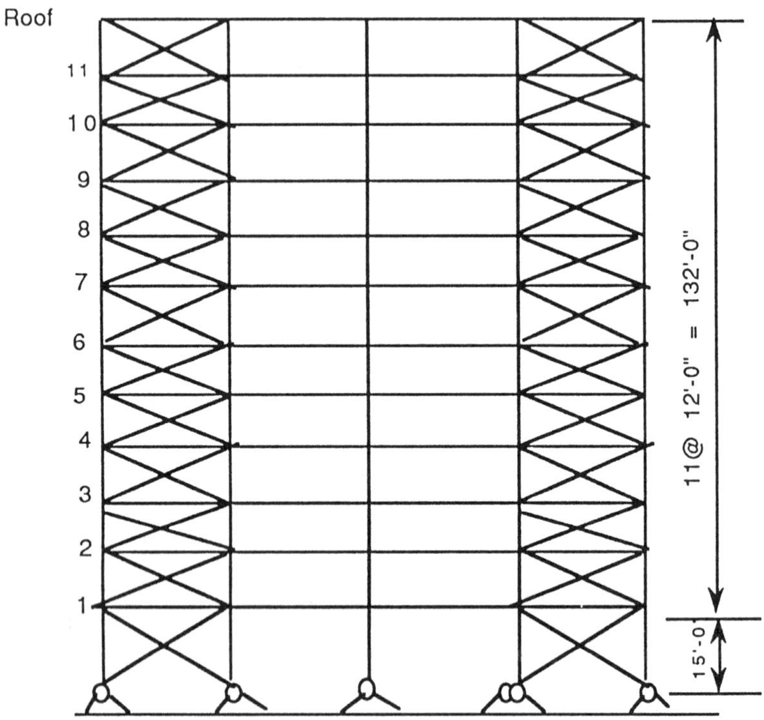

(b)

Figure 6.1 Wind against a multistory building. (a) Plan. (b) Elevation.

TABLE 6.1 Exposure Category Constants

Exposure Category	α	Z_g
A	3.0	1500
B	4.5	1200
C	7.0	900
D	10.0	700

Source: Table C6 from Minimum Design Loads for Buildings and Other Structures, 1990 by American Society of Civil Engineers.

K_z = velocity pressure exposure coefficient as a function of height (Table 6.3) or from Equations (6.2) and (6.3)

The constant 0.00256 reflects the mass density of air at standard conditions. The velocity pressure exposure coefficient K_z is given as a function of the height above the surface of the ground. It is obtained from

$$K_z = 2.58\left(\frac{z}{z_g}\right)^{2/\alpha} \quad \text{for} \quad z \geq 15 \text{ ft} \tag{6.2}$$

$$= 2.58\left(\frac{15}{z_g}\right)^{2/\alpha} \quad \text{for} \quad z \leq 15 \text{ ft} \tag{6.3}$$

where α and z_g are obtained from Table 6.1 and Figure 6.2.

TABLE 6.2 Importance Factors

	Importance Factor I	
Building Use Category	100 mi inland	Hurricane coastline
I	1.00	1.05
II	1.07	1.11
III	1.07	1.11
IV	0.95	1.00

Source: Table 5 from Minimum Design Loads for Buildings and Other Structures, 1990 by American Society of Civil Engineers.

TABLE 6.3 Velocity Pressure Exposure Coefficient K_z

	K_z			
Height z	Exposure A	Exposure B	Exposure C	Exposure D
0–15	0.12	0.37	0.80	1.20
20	0.15	0.42	0.87	1.27
25	0.17	0.46	0.93	1.32
30	0.19	0.50	0.98	1.37
40	0.23	0.57	1.06	1.46
50	0.27	0.63	1.13	1.52
60	0.30	0.68	1.19	1.58
70	0.33	0.73	1.24	1.63
80	0.37	0.77	1.29	1.67
90	0.40	0.82	1.34	1.71
100	0.42	0.86	1.38	1.75
120	0.48	0.93	1.45	1.81
140	0.53	0.99	1.52	1.87
160	0.58	1.05	1.58	1.92
180	0.63	1.11	1.63	1.97
200	0.67	1.16	1.68	2.01
250	0.78	1.28	1.79	2.10
300	0.88	1.39	1.88	2.18
350	0.98	1.49	1.97	2.25
400	1.07	1.58	2.05	2.31
450	1.16	1.67	2.12	2.36
500	1.24	1.75	2.18	2.41

Source: Table 6 from Minimum Design Loads for Buildings and Other Structures, 1990 by American Society of Civil Engineers.

The regional basic wind speed is given for exposure C at a height of 10 m (33 ft) above the mean elevation of the ground. The wind speed for other exposures is obtained from the following equation:

$$V_{33} = \sqrt{2.58}\left(\frac{33}{z_g}\right)^{1/\alpha} V_0 \tag{6.4}$$

where

V_0 = the wind speed for exposure C

For the purpose of assigning wind, snow, and seismic loads, certain categories are provided for buildings and other structures. These categories have an importance factor as given in Table 6.4. The importance

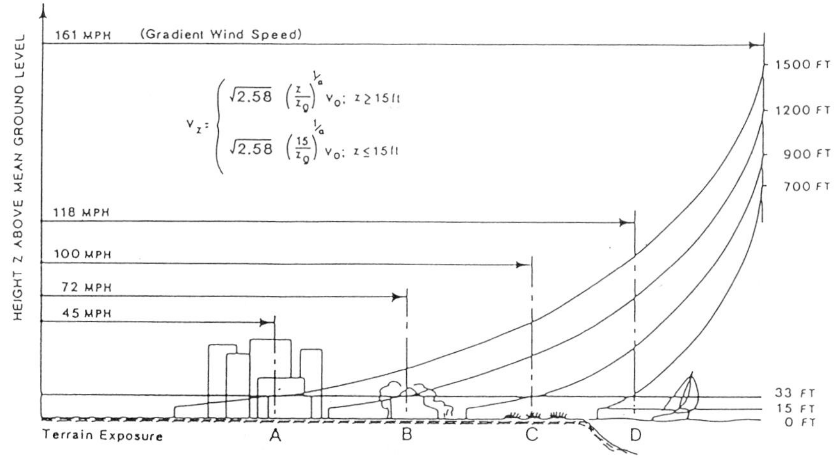

Figure 6.2 Wind velocity chart for the United States and Alaska. (Fig. A4.2 from Low Rise Building Systems Manual, 1986 by Metal Building Manufacturers Association, Inc.)

factors are given in Table 6.2. The numerical values of K_z are listed in Table 6.3 for up to 500 ft above mean ground level. For greater heights, use Equation (6.2).

6.4 SELECTION OF BASIC WIND SPEED (mph)

The selection of the basic wind speed depends on the site location of the building. Once the location of the building has been identified, the basic wind speed V_0 is obtained from Figure 6.3. To determine the design wind speed, it is not enough to know the general location of the building. The site must be identified relative to the exposure category as shown in Figure 6.2. Equation (6.4) is used to calculate V_{33}. From Figure 6.2, it is observed that the selection of an appropriate category will define the wind speed design value. Exposure categorization requires an understanding of the terrain immediately surrounding the building site.

6.5 EXTERNAL PRESSURES AND COMBINED EXTERNAL AND INTERNAL PRESSURES

The pressure on a building is given by

$$p_z = qGC_p \tag{6.5}$$

TABLE 6.4 Classification of Buildings and Other Structures in Terms of Importance When Used with Wind, Snow, and Seismic Loads

Type of Occupancy	Category
All buildings and structures except those listed below	I
Buildings and structures where the primary occupancy is one in which more than 300 people congregate in one area	II
Buildings and structures designated as essential facilities, including but not limited to Hospitals and other medical facilities having surgery or emergency treatment area Fire or rescue and police stations Primary communication facilities and disaster operation centers Power stations and other utilities required in an emergency Structures having critical national defense capabilities	III
Buildings and structures that represent a low hazard to human life or property in the event of failure, such as agricultural buildings, certain temporary facilities, and minor storage facilities	IV

Source: Table 1 from Minimum Design Loads for Buildings and Other Structures, 1990 by American Society of Civil Engineers.

where

q = velocity pressure [Equation (6.1)]
G = gust factor [Table 6.6 or Equations (6.6) through (6.10)]
C_p = pressure coefficient averaged over some time interval (Table 6.5)

The pressure coefficient C_p varies from one location to another in relation to the building. See Tables 6.5 and 6.6 and Figure 6.4.

IMPORTANT NOTES

1. For main wind-resisting systems, use building or structure height $h = z$.
2. Linear interpolation is acceptable for intermediate values of z.
3. For heights above ground of more than 500 ft, Equation (6.5) may be used.

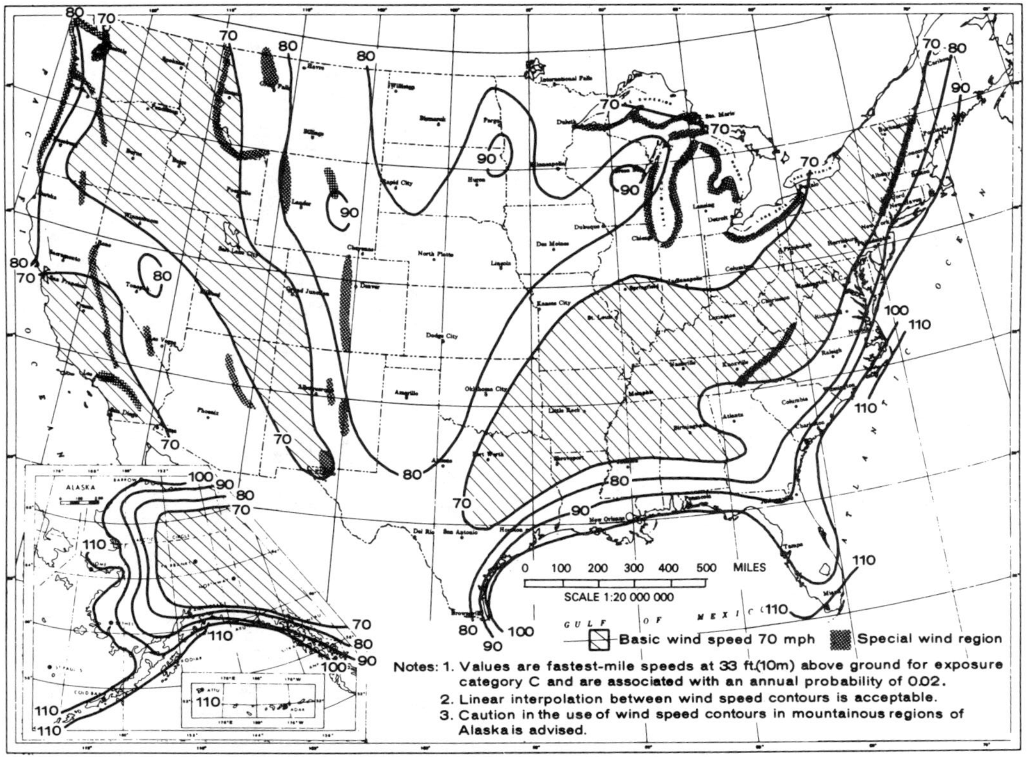

Figure 6.3 Wind speed variation with height for various site exposures. (Fig. 1 from Minimum Design Loads for Buildings and Other Structures, 1990 by American Society of Civil Engineers.)

TABLE 6.5 Wall Pressure Coefficients C_p

Surface	L/B	C_p	For use with
Windward wall	All values	0.8	q_z
	0–1	−0.5	
Leeward wall	2	−0.3	q_h
	≥ 4	−0.2	
Side walls	All values	−0.7	q_h

Source: Adapted from Minimum Design Loads for Buildings and Other Structures, 1990 by American Society of Civil Engineers.

4. The value of the gust response factor shall not be less than 1.0. The gust response factors listed in Table 6.7 are calculated from Equation (6.6)

$$G_z = 0.65 + 3.65T_z \tag{6.6}$$

where

$$T_z = \frac{2.35(D_0)^{1/2}}{(z/30)^{1/\alpha}} \tag{6.7}$$

The gust response factor for a building is calculated on the basis of the mean roof height of that building. The appropriate value can be obtained from Table 6.7. This value will be used for the entire wind-force-resisting system. For structural systems with a fundamental natural frequency of 1 Hz or less, the dynamic amplification becomes a factor that needs to be accounted for. For buildings and structures with a natural frequency $f < 1$ Hz, the gust response factor is

$$\overline{G} = 0.65 + \left[\frac{P}{\beta} + \frac{(3.32T_1)^2 S}{(1 + 0.002c)}\right]^{1/2} \tag{6.8}$$

For a latticed framework like television and radio towers, towers for power distribution, and any other similar structure, the gust response factor is determined from

$$\overline{G} = 0.65 + \left[\frac{1.25P}{\beta} + \frac{(3.32T_1)^2 S}{(1 + 0.002c)}\right]^{1/2} \tag{6.9}$$

TABLE 6.6 Roof Pressure Coefficients C_p for Use with q_h

Wind direction		*Windward*							Leeward
		Angle θ (deg)							
	h/L	0	10–15	20	30	40	50	> 60	
Normal to ridge	≤ 0.3	−0.7	0.2 −0.9	0.2	0.3	0.4	0.5	0.01θ	−0.7 for all values of h/L and θ
	0.5	−0.7	−0.9	−0.75	−0.2	0.3	0.5	0.10θ	
	1.0	−0.7	−0.9	−0.75	−0.2	0.3	0.5	0.10θ	
	> 1.5	−0.7	−0.9	−0.9	−0.9	−0.35	0.2	0.01θ	
Parallel to ridge	h/B or $h/L \leq 2.5$				−0.7				−0.7
	h/B or $h/L > 2.5$				−0.8				−0.8

Source: Figure 2 from Minimum Design Loads for Buildings and Other Structures, 1990 by American Society of Civil Engineers.

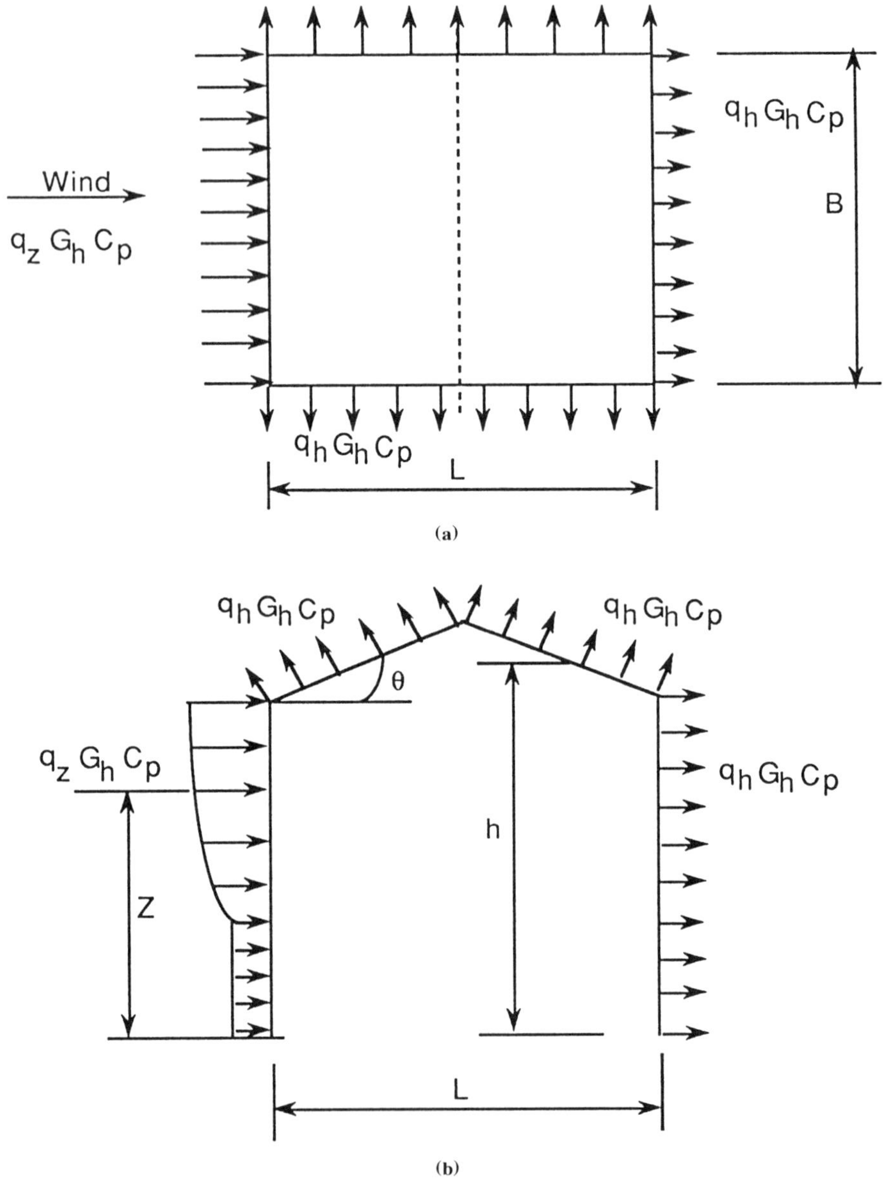

Figure 6.4 Wind drag coefficients for ordinary buildings. (a) Plan. (b) Elevation.

TABLE 6.7 Gust Response Factors G_h and G_z

Height above Ground Level z (ft)	G_h and G_z			
	Exposure A	Exposure B	Exposure C	Exposure D
0–15	2.36	1.65	1.32	1.15
20	2.20	1.59	1.29	1.14
25	2.09	1.54	1.27	1.13
30	2.01	1.51	1.26	1.12
40	1.88	1.46	1.23	1.11
50	1.79	1.42	1.21	1.10
60	1.73	1.39	1.20	1.09
70	1.67	1.36	1.19	1.08
80	1.63	1.34	1.18	1.08
90	1.59	1.32	1.17	1.07
100	1.56	1.31	1.16	1.07
120	1.50	1.28	1.15	1.06
140	1.46	1.26	1.14	1.05
160	1.43	1.24	1.13	1.05
180	1.40	1.23	1.12	1.04
200	1.37	1.21	1.11	1.04
250	1.32	1.19	1.10	1.03
300	1.28	1.16	1.09	1.02
350	1.25	1.15	1.08	1.02
400	1.22	1.13	1.07	1.01
450	1.20	1.12	1.06	1.01
500	1.18	1.11	1.06	1.00

Source: Table 8 from Minimum Design Loads for Buildings and Other Structures, 1990 by American Society of Civil Engineers.

where

h = building height (ft)
c = building width (ft)
f = building fundamental natural frequency (Hz)
β = structural damping factor (percentage of critical damping)
S = structure factor from Figure 6.5
s = surface friction factor obtained from Table 6.8
J = pressure profile factor as a function of ratio γ from Figure 6.5
γ = ratio obtained from Table 6.8
P = probability of exceeding design wind speed during a specified number of years n

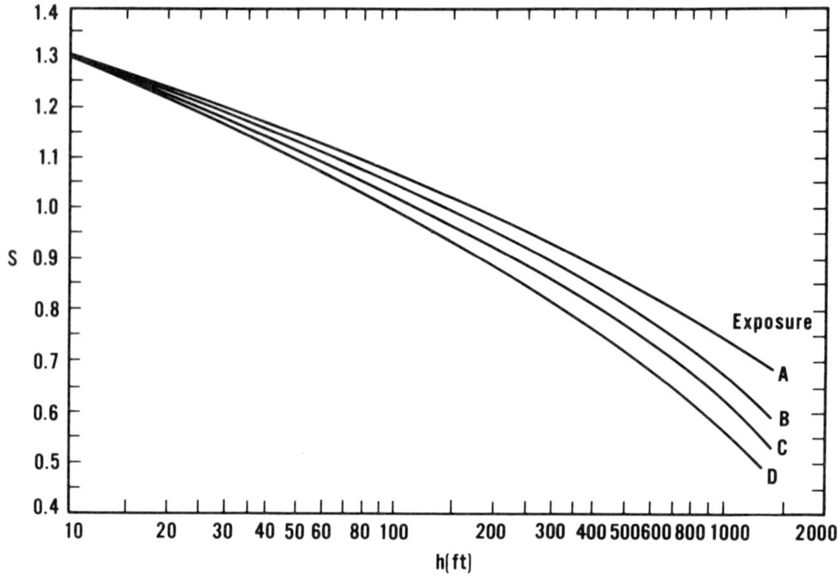

Figure 6.5 Structure size factor S. (Fig. C8 from Minimum Design Loads for Buildings and Other Structures, 1990 by American Society of Civil Engineers.)

Y = resonance factor as a function of γ and the ratio c/h from Figure 6.6a
D_0 = surface drag coefficient from Table 6.8
T_1 = exposure factor evaluated at two-thirds of the mean roof height

The values of P and $\bar{f}$ are obtained from Equations (6.10) and (6.11)

$$P = \bar{f}JV \tag{6.10}$$

$$\bar{f} = \frac{10.5fh}{sV} \tag{6.11}$$

TABLE 6.8 Parameters s, γ, and D_0

Exposure Category	s	γ	D_0
A	1.46	$8.20/h$	0.025
B	1.33	$3.28/h$	0.010
C	1.00	$0.23/h$	0.005
D	0.85	$0.02/h$	0.003

Source: Table C9 from Minimum Design Loads for Buildings and Other Structures, 1990 by American Society of Civil Engineers.

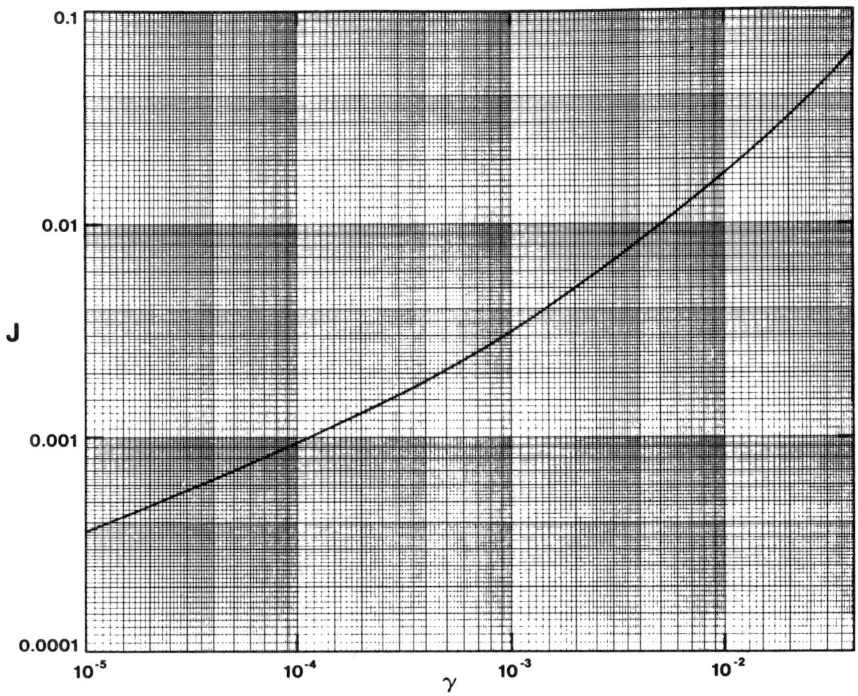

(a)

Figure 6.6a Pressure profile factor. (Fig. C6 from Minimum Design Loads for Buildings and Other Structures, 1990 by American Society of Civil Engineers.)

Example 6.1

The basic design wind speed for a certain site is 80 mph; the type of exposure is B.

$$\text{Building height } h = 450 \text{ ft}$$
$$\text{Building width } c = 80 \text{ ft}$$
$$\text{Building fundamental frequency } f = 0.56 \text{ Hz}$$
$$\text{Building damping coefficient } \beta = 0.025$$

Determine the dynamic-response factor.

Solution

$$\bar{f} = \frac{10.5fh}{sV}$$
$$= \frac{10.5 \times 0.56 \times 450}{1.33 \times 80}$$
$$= 24.868$$

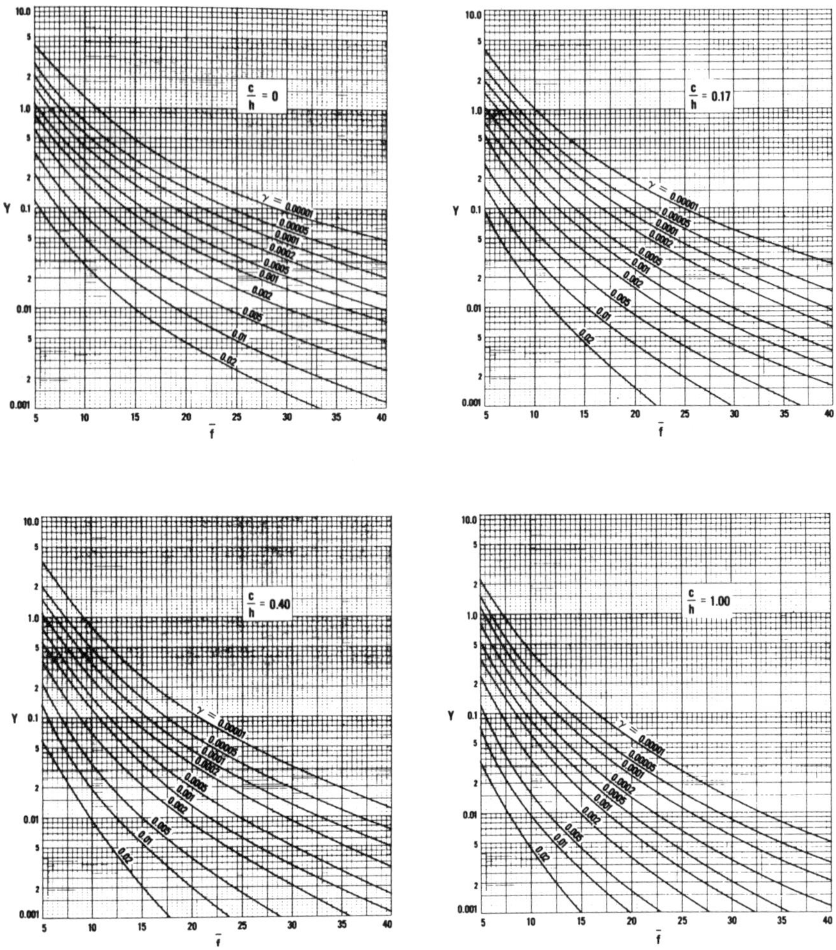

NOTE: The four sets of curves correspond to four different values of the ratio *c/h*.

(b)

Figure 6.6b Resonance factor. (Fig. C7 from Minimum Design Loads for Buildings and Other Structures, 1990 by American Society of Civil Engineers.)

For exposure *B*,

$$\gamma = \frac{3.28}{450} = 0.00729$$

$$\frac{c}{h} = \frac{80}{450}$$

$$= 0.178$$

Given the values of γ, $\bar{f}$, and c/h, use Figures 6.5 and 6.6 to obtain the parameters J, Y, and S

$$J = 0.0103$$
$$Y = 0.0030$$
$$S = 0.84$$

$$T_1 = \frac{2.35 \times (0.01)^{0.5}}{\left(\frac{300}{30}\right)^{1/4.5}}$$
$$= 0.141$$
$$P = 24.868 \times 0.0103 \times 0.0030$$
$$= 0.000768$$
$$= 0.65 + \left[\frac{0.000768}{0.025} + \frac{(3.32 \times 0.141)^2 \times 0.84}{1 + 0.002 \times 80}\right]^{1/2}$$
$$\overline{G} = 1.119$$

6.6 WIND PRESSURE PROFILE AGAINST BUILDINGS

The equivalent static wind pressure against closed buildings is determined from Equation (6.5), in which C_p represents the difference between positive pressure on the windward side and negative pressure on the leeward side. For values of these constants, refer to Table 6.5. Once the gust response factor has been determined, the equivalent static pressure against the building along its height can be calculated.

Example 6.2

In Example 6.1, the length of the building is 100 ft. The structure is located more than 100 mi from the coastline and is used as an office building. Calculate the pressure diagram along the height of the building.

Solution

From Table 6.1, the building category is *I*. Hence, the influence factor I is unity. From Equation (6.1),

$$q_z = 0.00256 K_z (IV)^2$$
$$= 0.00256 \times (1 \times 80)^2$$
$$= 16.4 K_z \text{ lb/ft}^2$$

and

$$p_z = 16.4K_z\overline{G} \times (C_p - C_n)$$

where C_p and C_n represent the positive and negative coefficients as given in Table 6.5. Then,

$$p_z = 16.4K_z \times 1.119 \times [0.8 - (-0.5)]$$
$$= 23.9K_z \text{ lb/ft}^2$$

With Table 6.3, the pressure diagram is completed as shown in Table 6.9. From this pressure distribution, the lateral forces at the level of each story can be determined.

6.7 ANALYSIS OF BRACED FRAMES FOR WIND FORCES

Buildings are subjected to wind forces frequently during their life time, but it is usually tall buildings (three or more stories high) that are designed

TABLE 6.9 Pressure Diagram

Height	Exposure B, K_z	p_z
0–15	0.37	8.8
20	0.42	10.0
25	0.46	11.0
30	0.50	12.0
40	0.57	13.6
50	0.63	15.1
60	0.68	16.2
70	0.73	17.4
80	0.77	18.4
90	0.82	19.6
100	0.86	20.6
120	0.93	22.2
140	0.99	23.7
160	1.05	25.1
180	1.11	26.5
200	1.16	27.7
250	1.28	30.6
300	1.39	33.2
350	1.49	35.6
400	1.58	37.8
450	1.67	39.9

with a special consideration given to wind loads. Wind forces are dynamic in nature. Some assumptions based on theoretical and experimental studies have been made available to simplify the evaluation of wind forces against buildings and other structures. One such assumption is that wind forces are transferred to the columns and from there to the foundation through a rigid floor structure and bracing frame system. The wind forces are considered concentrated loads applied against the braced frames at each story level. See Figure 6.7. Simple joint-by-joint analysis yields the distribution of the wind load in the frame system. Further, it is assumed that the bracing members will experience only tension. This assumption is made for simplifying the analysis. For all practical purposes, the results thus obtained are satisfactory for the design.

Example 6.3

Calculate the wind force against the building given in Example 6.2 for the same parameters in Example 6.1. The structure used is an office building.

Solution

The classification of the building is obtained from Table 6.4. It is classified as category I. The importance factor $I = 1.0$; it is determined from Table 6.2. For a B exposure, $\alpha = 4.5$ and $z_g = 1200$ ft; these values are obtained from Table 6.1. The wind velocity pressure as given by Equation (6.1) is

$$q_z = 0.00256K_z(IV)^2$$

Given the basic wind speed for the region (airport records), the equivalent basic wind speed for any other exposure is provided by Equation (6.4)

$$\begin{aligned} V_{33} &= \sqrt{2.58}\left(\frac{33}{z_g}\right)^{1/\alpha} V_0 \\ &= \sqrt{2.58}\left(\frac{33}{1200}\right)^{\frac{1}{4.5}} \times 80 \\ &= 57.8 \text{ mph} \end{aligned}$$

Then

$$\begin{aligned} q_z &= 0.00256(1 \times 57.8)^2 K_z \\ &= 8.56K_z \text{ lb/ft}^2 \end{aligned}$$

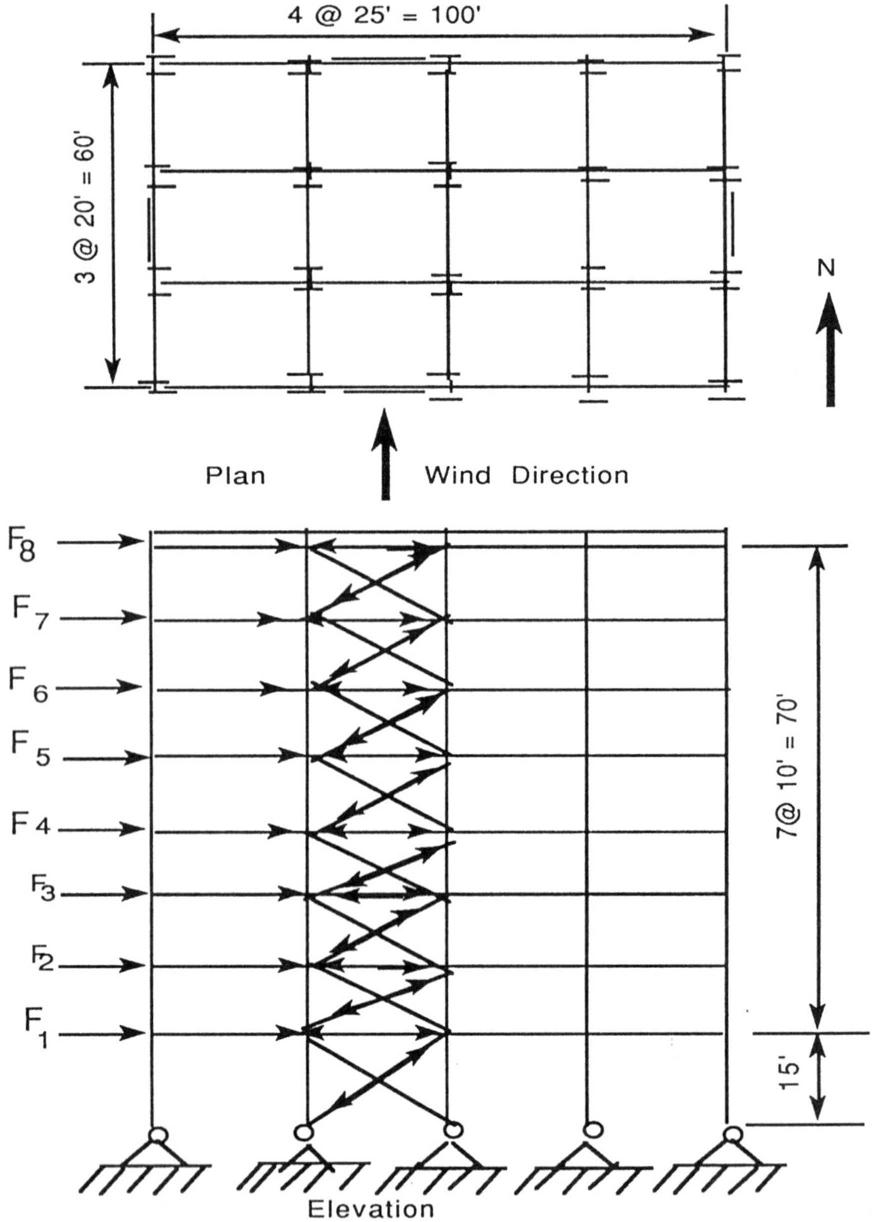

Figure 6.7 Transfer of wind forces into the building frame system of braced frames with standard bolt connections.

In Equation (6.5), the static wind pressure against the building is

$$p_z = q_z G C_p$$
$$= 8.56 K_z G C_p$$

The constant C_p is given in Table 6.5

$$C_p = [0.8 - (-0.5)]$$
$$= 1.33$$

Since the building has a basic fundamental frequency $f = 0.56$ Hz that is less than 1.0, the dynamic gust response factor G as determined in Example 6.1 is 1.119. The factor is used for all heights of the building. Thus, the wind static pressure against the building is written as

$$p_z = 8.56 \times 1.119 \times 1.13 K_z$$
$$= 12.45 K_z \ \text{lb/ft}^2$$

Consider that the wind pressure against the building conforms to a step pattern and these steps change at elevations of 30, 50, 100, and 200 ft. The elevation coefficient K_z is obtained from Table 6.3. A summary of the static wind pressure against the building is given below. The wind pressure per foot along the height of each exterior frame is obtained from the product of the wind pressure in pounds per square feet at that height and one-half the width of the building that is perpendicular to the direction of the wind:

z (ft)	K_z	p_z (lb/ft^2)	p_z (lb/ft)
30	0.50	6.2	186
50	0.63	7.8	234
100	0.86	10.7	321
200	1.16	14.4	

The wind pressure against the building is resisted by the two exterior frames since they are the only ones that are braced. Each frame receives one-half of the total wind load against the building. Thus, the pressure diagram for the wind load acting on each frame is the product of the wind pressure summarized above and one-half the width.

The calculation of the forces at the various story levels proceeds as follows:

$$F_1 = \frac{15 + 10}{2} \times \frac{186}{1000}$$
$$= 2.32 \text{ k}$$

$$F_2 = \frac{10 + 10}{2} \times \frac{186}{1000}$$
$$= 1.86 \text{ k}$$

$$F_3 = \frac{10 + 10}{2} \times \frac{234}{1000}$$
$$= 2.34 \text{ k}$$

$$F_4 = \frac{10 + 10}{2} \times \frac{234}{1000}$$
$$= 2.34 \text{ k}$$

$$F_5 = \frac{10 + 10}{2} \times \frac{321}{1000}$$
$$= 3.21 \text{ k}$$

$$F_6 = \frac{10 + 10}{2} \times \frac{321}{1000}$$
$$= 3.21 \text{ k}$$

$$F_7 = \frac{10 + 10}{2} \times \frac{321}{1000}$$
$$= 3.21 \text{ k}$$

$$F_8 = \frac{10}{2} \times \frac{321}{1000}$$
$$= 1.60 \text{ k}$$

The wind forces against each story are shown in Figure 6.8. The next step is to analyze the forces in the braced bay. The method of joints is used. Because braces can be very slender, it is assumed that they will carry tension forces only, so the frame is analyzed with one diagonal, which makes it a determinate structure. At every level, the horizontal component of the force in the bracing is equal to the sum of the wind forces above it. See Figure 6.9.

The horizontal components of the diagonals and the diagonal forces are shown in Table 6.10. The girders are subjected to axial forces due to wind.

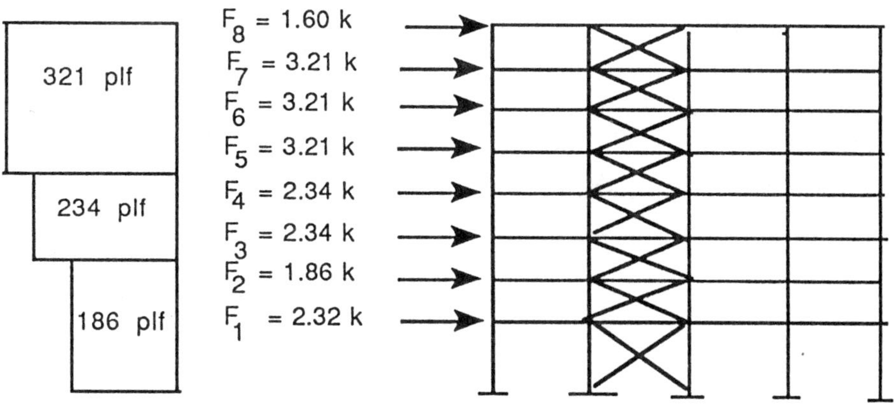

Figure 6.8 Wind shear against building floors.

The horizontal component of the diagonal force is

$$F_{jx} = \sum_{j=1}^{n} P_j \tag{6.12}$$

The force in the diagonal is

$$F_j = \frac{L_j}{L_{jx}} \sum_{j=1}^{n} F_{jx} \tag{6.13}$$

$$F_{jy} = F_{jx} \frac{L_{jy}}{L_{jx}} \tag{6.14}$$

The force in the column on the leeward side is

$$Fj_{cl} = \sum_{j=1}^{n} F_{jy} \tag{6.14}$$

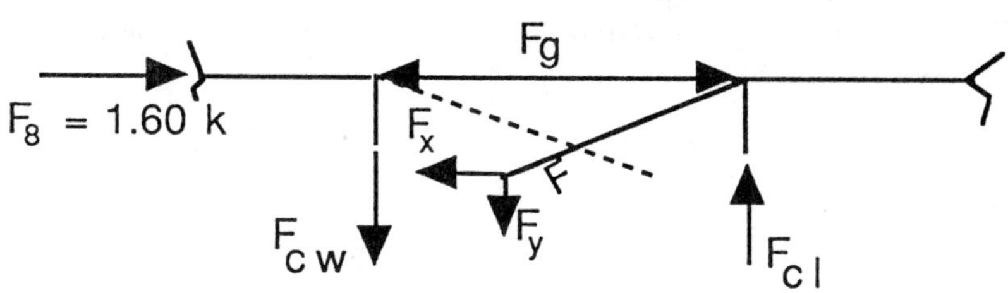

Figure 6.9 Resolution of wind shear in the braced frame.

TABLE 6.10 Diagonal Forces Due to Wind

Story	Wind Force (k)	F_x (k)	Diagonal Force F (k)
8th	1.60	1.60	
			1.72
7th	3.21	4.81	
			5.18
6th	3.21	8.02	
			8.64
5th	3.21	11.23	
			12.09
4th	2.34	13.57	
			14.61
3rd	2.34	15.91	
			17.14
2nd	1.86	17.77	
			18.45
1st	2.32	20.09	
			23.42

The force in the column on the windward side is

$$F_{jcw} = F_{j+1cl} \tag{6.15}$$

$$F_{ncw} = 0.0 \tag{6.16}$$

$$F_{jg} = F_{jx} \tag{6.17}$$

where the symbols in Equations (6.12) through (6.16) define the members as indicated in Figure 6.10.

Example 6.4

A 21-story building is located in the vicinity of O'Hare Airport in Chicago, Illinois. The structure is an office building. The plan and braced frame of the building are given in Figure 6.11. Calculate the wind pressure profile against the building.

Solution

The basic wind speed for the location is 75 mph (Figure 6.2). The building category is I (Table 6.4). The importance factor based on the building category is 1.05 (the coastline). The C_p factor is the sum of the

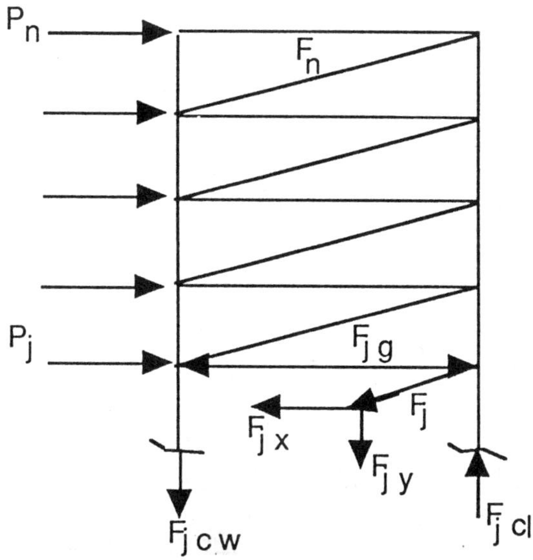

Figure 6.10 Braced frame nomenclature.

absolute values for the windward and leeward side coefficients (Table 6.5). The wind pressure at any height z is

$$p_z = 0.00256K_z(IV)^2C_pG_z$$

where K_z and G_z are obtained from Tables 6.3 and 6.6, respectively

$$\begin{aligned} p_z &= 0.00256K_z(1.05 \times 75)^2 1.3G_z \\ &= 20.639K_zG_z \text{ lb/ft}^2 \end{aligned}$$

By using Tables 6.3 and 6.6 in the equation for the wind pressure, the pressure profile can be obtained as shown below. To determine the pressure at each story level due to the parabolic pressure distribution, Simpson's rule is used as indicated

$$P_z = \frac{h}{6} \times (p_{z_{2m-2}} + 4p_{z_{2m-1}} + p_{z_{2m}}) \tag{6.18}$$

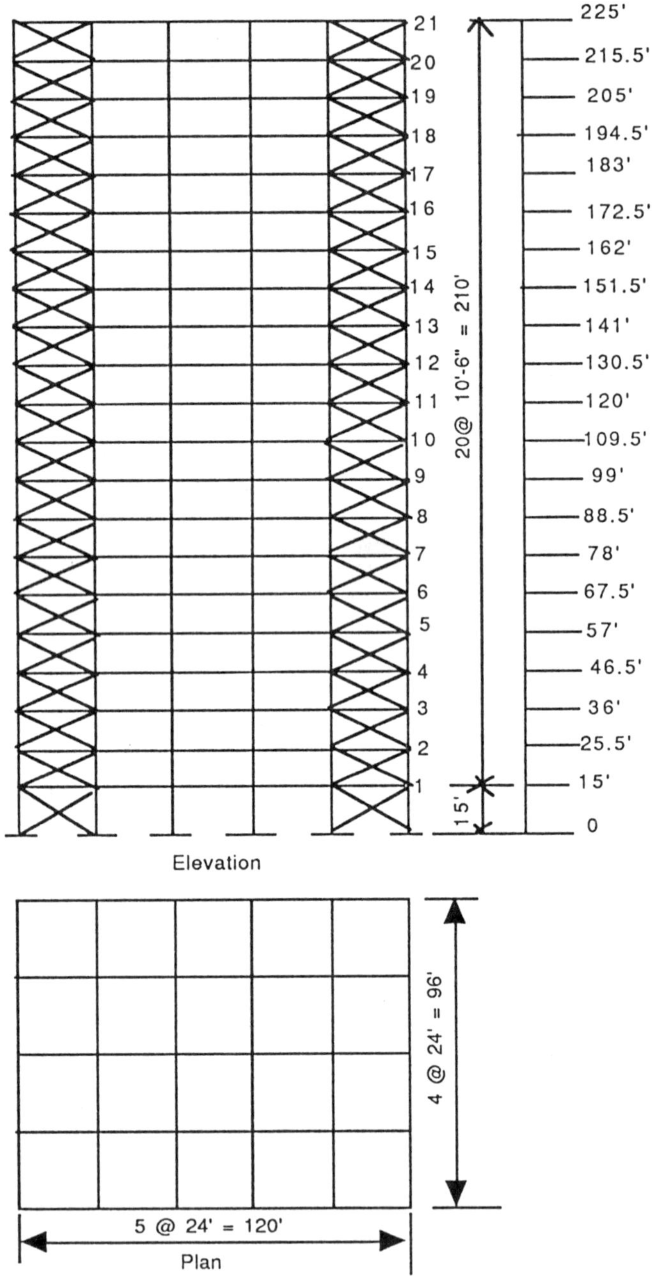

Figure 6.11 Twenty-one-story building frame system.

TABLE 6.11 Pressure Profile for Building in Example 6.4

z (ft)	K_z	G_z	p_z (lb/ft^2)
0–15	0.80	1.32	21.8
25.5	0.93	1.27	24.4
36	1.02	1.22	25.7
46.5	1.10	1.22	27.7
57	1.17	1.20	29.0
67.5	1.23	1.19	30.2
78	1.28	1.18	31.2
88.5	1.34	1.17	32.4
99	1.38	1.16	33.0
109.5	1.42	1.16	34.0
120	1.45	1.15	34.4
130.5	1.48	1.14	34.8
141	1.52	1.14	35.8
151.5	1.55	1.13	36.1
162	1.58	1.13	36.8
172.5	1.60	1.12	37.0
183	1.64	1.12	37.9
193.5	1.66	1.11	38.0
204	1.69	1.11	38.7
214.5	1.71	1.10	38.8
225	1.74	1.10	39.5

where

m = a number that indicates the number of a particular story; $m = 1, n$

n = number of stories in the building

P_z = intensity of wind pressure (lb/ft) along the width of the building and at the level of that particular story

p_z = the pressure (lb/ft^2), as given in Table 6.11, including the interpolated values at half-story height for each story6.11

h = the story height (ft)

The wind pressure profile is summarized in Table 6.11. The wind force against each story along the height of the building is shown in Table 6.12.

6.8 INTRODUCTION TO SEISMIC DESIGN

An earthquake causes ground motions at the base of the structure. Depending on the type of the structure, its response to ground motion will determine the magnitude of displacements and level of stresses in the frame of the structure. Thus, seismic design will require the determination

TABLE 6.12 Wind Forces

n	p_n (lb/ft^2)	W (lb/ft^2)	m	P_m (k)
0	21.8			
1	21.8	22.0	1	22.2
2	23.1			
3	24.4	24.3	2	24.5
4	25.0			
5	25.7	25.8	3	26.0
6	26.7			
7	27.7	27.6	4	27.8
8	28.4			
9	29.0	29.0	5	29.2
10	29.6			
11	30.2	30.2	6	30.4
12	30.7			
13	31.2	31.2	7	31.4
14	31.8			
15	32.4	32.4	8	32.7
16	32.7			
17	33.0	33.0	9	33.3
18	33.5			
19	34.0	34.0	10	34.3
20	34.2			
21	34.4	34.4	11	34.7
22	34.6			
23	34.8	34.8	12	35.1
24	35.1			
25	35.8	35.7	13	36.0
26	36.0			
27	36.1	36.1	14	36.4
28	36.4			
29	36.8	36.8	15	37.1
30	36.9			
31	37.0	37.0	16	37.3
32	37.4			
33	37.9	37.8	17	38.1
34	38.0			
35	38.0	38.1	18	38.4
36	38.4			
37	38.7	38.7	19	39.0
38	38.8			
39	38.8	38.9	20	39.2
40	39.2			
41	39.5	32.9	21	33.2

of seismic forces that act on the structure and the selection of member sizes to resist these forces in order to keep deflections within prescribed limits. The various seismic zones in the United States are shown in Figure 6.12.

The determination of seismic forces can be achieved by an equivalent static force procedure or a dynamic analysis approach. In this book, the former procedure will be used. However, the design engineer must use sound judgment on when the dynamic approach should be used in the prediction of earthquake forces against a building.

6.9 EQUIVALENT STATIC FORCE PROCEDURE

In this method, a step-by-step procedure is outlined in accordance with the ANSI building code. The following symbols and notations will be used throughout the procedure:

C	= numerical coefficient for the base share
C_p	= numerical coefficient used in determining the lateral force for parts or portions of buildings or other structures
D	= the dimension of the building in the direction of the earthquake
d_i	= deflection at level i relative to the base due to the applied lateral forces
F_i, F_n, F_x	= lateral force applied to level i, n, or x, respectively
F_p	= lateral force on a part of the structure and in the direction under consideration
F_t	= that portion of V, the total shear force at the base due to earthquake motion, considered concentrated at the top of the structure in addition to F_n
g	= acceleration due to gravity
h_i, h_n, h_x	= height above the base to level i, n, or x, respectively
I	= occupancy importance factor
K	= numerical coefficient related to stiffness of the structure
N	= the total number of stories above the base to level n
S	= numerical coefficient for site-structure resonance
T	= fundamental elastic period of vibration of structure in seconds per cycle in the direction under consideration
T_s	= characteristic site period
V	= total lateral force or shear at the base of the building due to earthquake motion
W	= the total dead load and applicable portions of other loads
Z	= seismicity coefficient

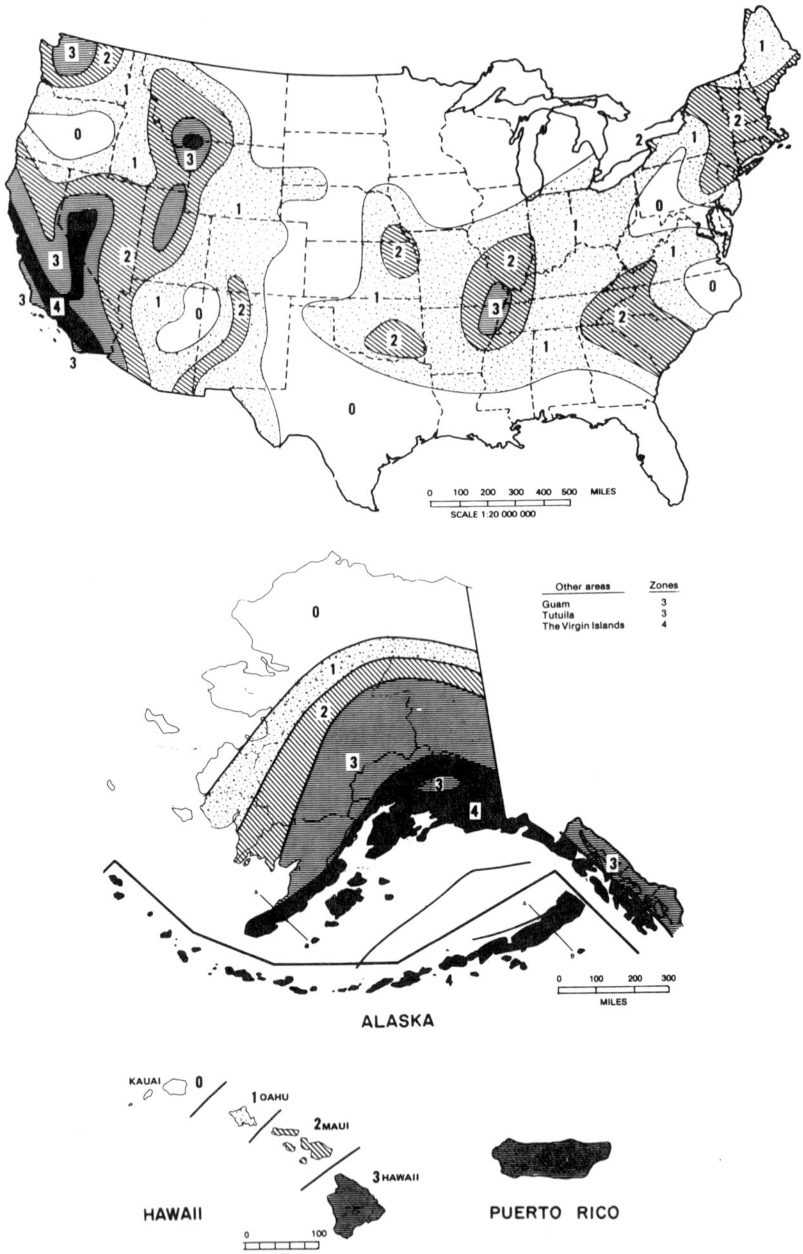

Figure 6.12 Seismic zone chart for the United States and its territories. (Courtesy of American Society of Civil Engineers, Minimum Design Loads for Buildings and Other Structures.)

TABLE 6.13 The Seismic Zone Coefficient Z

Seismic Zone from Figure 6.12	Z
4	1
3	3/4
2	3/8
1	3/16
0	1/8

Source: Table 21 from Minimum Design Loads for Buildings and Other Structures, 1990 by American Society of Civil Engineers.

The predicted earthquake forces against buildings are a function of the intensity of the earthquake, site characteristics, type of materials and method of construction used in the building, and function of the building. The total lateral seismic force against a building and in the direction of the seismic motion is expressed by the following formula:

$$V = ZIKCSW \tag{6.19}$$

The factors in Equation (6.19) have to be determined for each building. They can be obtained from Tables 6.13 through 6.16. The product CS need not exceed 0.14, or for soil profile type III in seismic zones 3 and 4, it need not exceed 0.11.

The value of C depends on the dynamic characteristics of the structural system and is obtained from the following equation:

$$C = \frac{1}{15\sqrt{T}} \tag{6.20}$$

The value of C must not exceed 0.12.

TABLE 6.14 Occupancy Importance Factor I

Category	I
I. All other buildings not in II and III	1.0
II. Assembly of 300 people or more	1.25
III. Essential facilities	1.5
IV. 14	NA

Source: Table 22 from Minimum Design Loads for Buildings and Other Structures, 1990 by American Society of Civil Engineers.

TABLE 6.15 Horizontal Stiffness Factor K for Buildings and Other Structures

Type of Lateral-Force-Resisting Building System	K
Bearing wall systems	
Walls provide for all or the major part of the vertical loads and resist the seismic force:	
Reinforced concrete shear walls or braced frames	1.33
Masonry walls	1.33
One-, two, or three-story light wood or metal frame-wall systems	1.00
Building frame systems	
A structural system with an essentially complete space frame providing seismic force resistance by shear walls or braced frames	1.00
Moment-resisting frame systems	
A structural system with an essentially complete space frame providing support for vertical loads. Seismic forces are resisted by moment-resisting frames:	
For ordinary steel frames	1.00
For special frames	0.67
For intermediate reinforced concrete frames	1.25
Dual systems	
A complete space frame to support vertical loads; special moment-resisting frames and shear walls or braced frames to resist seismic forces:	
Special moment-resisting frame	0.80
Intermediate moment-resisting frame	1.00
Elevated tanks	
Tanks plus full contents, where tanks are supported on their own tower and on a building	2.50
Structures other than buildings	2.00

Source: Table 23 from Minimum Design Loads for Buildings and Other Structures, 1990 by American Society of Civil Engineers.

T is the period of the structural frame system resisting the seismic horizontal forces and is a function of the structural material used in the frame and type of construction used in assembling the frame system. The

TABLE 6.16 Soil Profile Coefficients S

Soil Profile Type	S
Type I: Rock	1.0
Type II: Stiff clay greater than 200 ft in thickness	1.2
Type III: Soft to medium stiff clays and sand	1.5

Source: Table 24 from Minimum Design Loads for Buildings and Other Structures, 1990 by American Society of Civil Engineers.

determination of the period of a structural system is a complex problem. It can be computed using the Raleigh–Ritz method expressed in the following formula:

$$T = \sqrt{\left(\sum_{i=1}^{n} w_i \delta_i^2\right) \Big/ g\left[\sum_{i=1}^{n-1} F_i \delta_i + (F_t + F_n)\delta_n\right]} \tag{6.21}$$

in which the values of F_i, F_t, δ_i, and d_n are determined from the base shear V distributed according to Equations (6.22), (6.23), and (6.24).

The distribution of the lateral shear force V is expressed as

$$V = F_t + \sum_{i=1}^{n} F_i \tag{6.22}$$

The force F_t is a concentrated force applied at the top of the building to account for the parapet and any other appertenances that may exist on the roof. This force is expressed as a function of the period of the frame system and the total lateral force acting against the building. It is given by the following equation:

$$F_t = 0.07TV \tag{6.23}$$

The remainder of the lateral force is distributed linearly between the top and bottom of the building as follows:

$$F_n = \frac{(V - F_t)h_n}{\sum_{i=1}^{n} w_i h_i} \tag{6.24}$$

In the absence of a rational analysis for the determination of the period T as shown above, it can be evaluated based on the type of frame under consideration. For braced frames, an approximate equation can be used for moderately high buildings

$$T = \frac{0.05H}{\sqrt{D}} \tag{6.25}$$

where

H = the height of the building above the ground
D = the width of the building in the direction of the earthquake

For a 100% ductile-moment frame resisting the lateral forces, the period of the building is

$$T = 0.10N \tag{6.26}$$

where

N = the number of stories

It has been found that Equation (6.26) yields poor results for low-rise buildings. It is reasonably accurate for buildings in the 40-story range. Otherwise, one must use the Raleigh–Ritz equation. As an initial approximation, Teal's formula (Edward Teal, Teal's Method or Formula, AISC Engineering Journals, Second and Fourth Quarters, 1975) provides reasonable results

$$T = \frac{1}{4}\sqrt{\frac{\Delta}{C_1}} \tag{6.27}$$

where

$C_1 = ZICS$
Δ = the lateral deflection at the top of the building

One can begin by assuming a deflection at the top of the building that is equal to two-thirds of the maximum allowable drift stipulated by the local building code. Use that value in Equation (6.27). By using an average value for S between 1 and 1.5, which is the range presented in Table 6.16, and IZ, as known from the site and function of the building, a relationship between the period T and factor C is established from Equations (6.27) and (6.20). The result is expressed as follows:

$$T = \frac{1}{4}\sqrt{\frac{4.3}{\left(ZIS\frac{1}{15\sqrt{T}}\right)}} \tag{6.28}$$

Knowing the values of Z, I, and S will provide a value for T. Use this value in the procedure to obtain V and the subsequent forces F_t and F_i against the building. With these forces at hand, the deflection of the frame can be obtained. Use the deflection in the Raleigh–Ritz formula and compare the results for T with the one obtained from Equation (6.28). For any significant difference, the calculated T can be used to determine a new set of deflections to calculate a value for T. The process converges rapidly.

Example 6.5

Consider a 10-story hospital in Los Angeles, California. It is essentially a steel space frame with braced frames in the N–S direction and ductile-moment frames in the E–W direction. The roof and floors are constructed of reinforced concrete slab ($3\frac{1}{4}$ in. with a 3-in. metal deck) placed on open steel joists. Floor-to-floor height is 10 ft, 6 in. The plan and elevation of the building are shown in Figure 6.13. Determine the distribution of the seismic forces against the building in the N–S orientation. The site consists of sand and moderately stiff clay.

Solution

Loads

1. Roof

(a) Roofing and insulation	8.0 lb/ft^2
(b) Metal deck	3.0
(c) Concrete slab	45.0
(d) Ceiling	5.0
(e) Steel framing	8.0
Total roof dead load	69.0 lb/ft^2
(f) Live load (reducible)	20.0 lb/ft^2

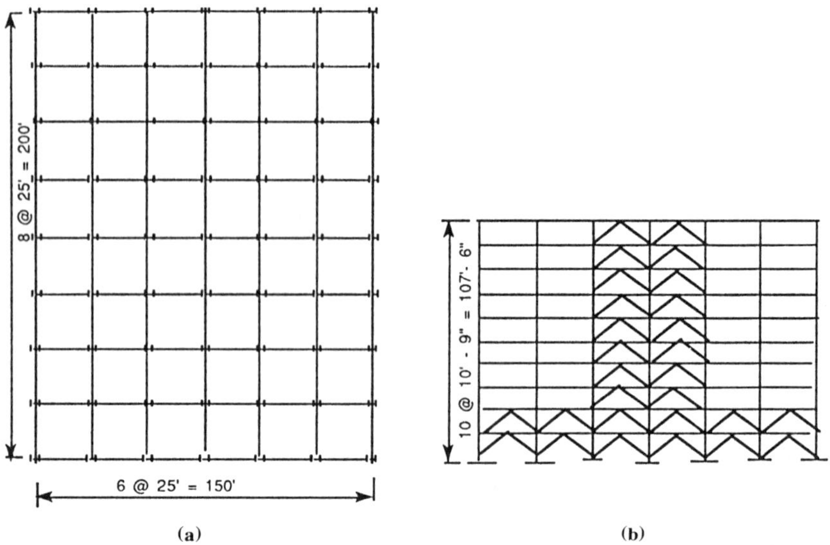

Figure 6.13 Ten-story building for seismic design. (a) Plan. (b) Elevation N–S.

2. Typical Floor Loading

(a)	Metal deck	3.0 lb/ft^2
(b)	Concrete slab	45.0
(c)	Ceiling	5.0
(d)	Partitions	20.0
(e)	Steel framing with fire retardant	13.0
	Total dead load	86.0 lb/ft^2
(f)	Live load (reducible) average	85.5 lb/ft^2

3. Curtain Walls

(a)	Average weight	15.0 lb/ft^2

Lateral Force Analysis

From Equation (6.19), we write the expression for the seismic force against the building

$$V = ZIKCSW$$

$$Z = 1.0$$

Los Angeles is in region 4; see Figure 6.12.

$$I = 1.5$$

The building is a hospital; see Table 6.14.

$$K = 1.00$$

The frame is steel and braced; see Table 6.15.

$$S = 1.5$$

The soil is category 3; see Table 6.16.

$$T = \frac{0.05H}{\sqrt{D}}$$

The frame is braced

$$T = \frac{0.05 \times 107.5}{\sqrt{150}}$$
$$= 0.44 \text{ sec}$$

Then from Equation (6.20),

$$C = \frac{1}{15\sqrt{0.44}}$$
$$= 0.10$$

The product of CS is 0.15. However, in zones 3 and 4, it need not exceed 0.14. Hence, the seismic force against the building is

$$V = 1 \times 1.5 \times 1 \times 0.14W$$
$$= 0.21W$$

Calculate the total dead load of the building. Note that the building dimensions to the building lines are 202.5 ft × 152.5 ft.

Typical Floor Load

Floor:	$202.5 \times 152.5 \times 0.086$	= 2656 k
Curtain Walls:	$2(202.5 + 152.5)0.015 \times 10.75$	= 114 k
	Total per floor	= 2770 k

TABLE 6.17 Distribution of Seismic Force Against One Typical Frame

Level	Height (ft)	Weight w_n (k)	$w_n h_n$ (k/1000)	$1/\Sigma w_n h_n$	$(w_n h_n/\Sigma w_n h_n)$ $\times$(V-ft)	F_i
R	107.5	2220	238.65	0.150	207.2	251.2
9	96.75	2770	280.00	0.176	243.0	243.0
8	86.00		238.22	0.150	207.2	207.2
7	75.25		208.44	0.131	180.9	180.9
6	64.5		178.66	0.112	154.7	154.7
5	53.75		148.88	0.094	129.8	129.8
4	43.00		119.11	0.075	103.6	103.6
3	32.25		3389.0	0.056	77.3	77.3
2	21.50		59.56	0.037	51.1	51.1
1	10.75		29.78	0.019	26.2	26.2
0			—	—	—	

Roof Load

Roof	$202.5 \times 152.5 \times 0.069$	$= 2131$ k
Wall and 3-ft parapet	$2(202.5 + 152.5)8.375 \times 0.015$	$=$ 89 k
	Total for roof	$= 2220$ k

Thus, the total weight of the building

$$9 \times 2770 + 2220 = 27{,}150 \text{ k}$$

The seismic force against the building is

$$\begin{aligned} V &= 0.21 \times 27{,}150 \\ &= 5701.5 \text{ k} \end{aligned}$$

Let the building be braced in the short direction at the 1st, 4th, 6th, and 9th line. Then, the seismic force resisted by one single frame is one-fourth the total horizontal force (5701.5/4 = 1425 k). The force F_t is

$$\begin{aligned} F_t &= 0.07 \times 0.44 \times 1425 \\ &= 44 \text{ k} \\ V - F_t &= 1425 - 44 \\ &= 1381 \text{ k} \end{aligned}$$

The distribution of this force is accomplished by using Equation (6.24). Arrange the results in tabular form. See Table 6.17. With these results the frame can be analyzed structurally.

7

Connections

7.1 INTRODUCTION

The behavior of connections is the least understood in the field of structural engineering. Most connections are designed on the basis of experience rather than theoretical analysis. Rational analysis of joints is intricate and highly indeterminate. Most failures in steel structures occur because of inadequate connections. It is common practice in design engineering to leave the detailing of connections to the fabricator, hence adding to the confusion of who is responsible in the event of failure. However, the structural engineer who designs the structure is responsible for the design of the connections. It is of the utmost importance then that the engineer be proficient in connection design.

Until fairly recently, welding and riveting dominated the field of connections. Modern steel structures are constructed with welded connections, bolted ones with high-strength or common bolts, or a combination of both. Figure 7.1 is a representation of a column splice that is designed to transfer axial as well as bending loads.

7.2 TYPES OF CONNECTIONS

There are several types of connections, the use of which depends on the magnitude and importance of the job. Figure 7.2 shows various types of

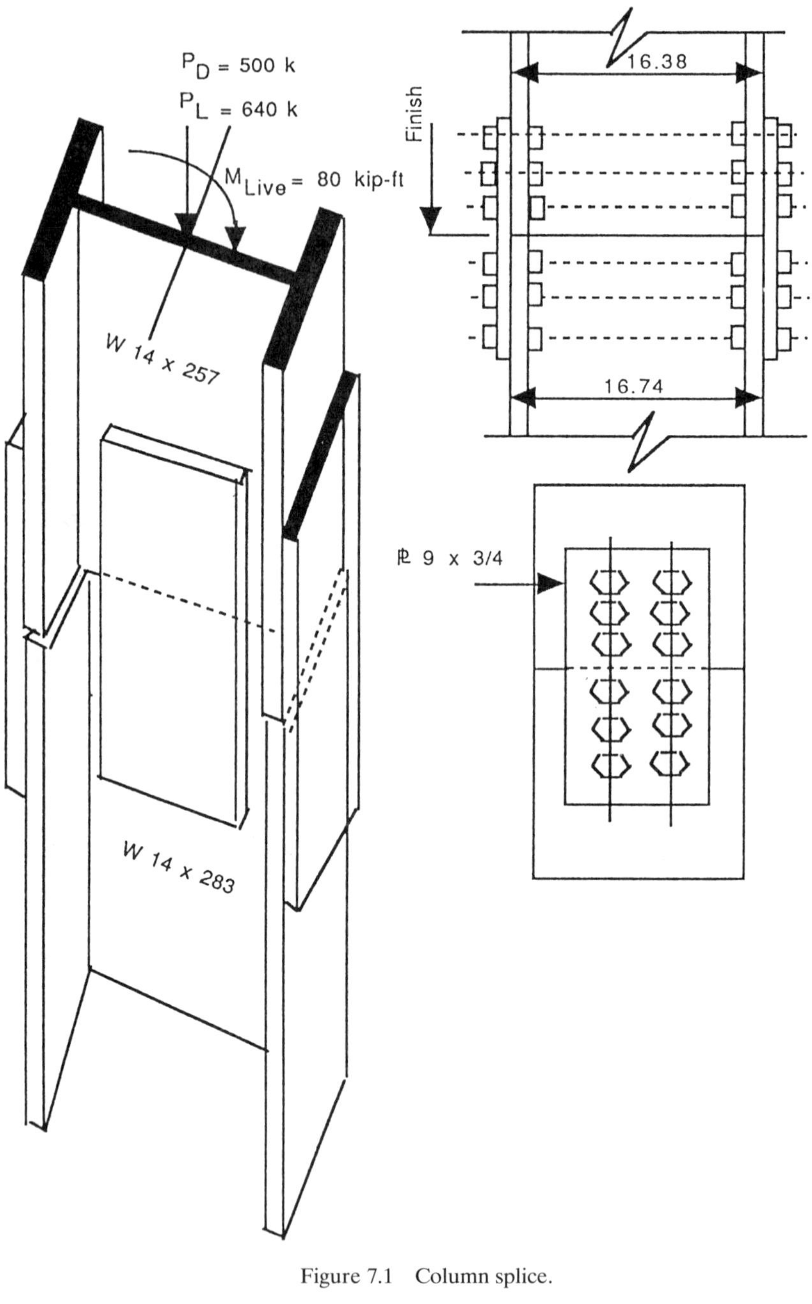

Figure 7.1 Column splice.

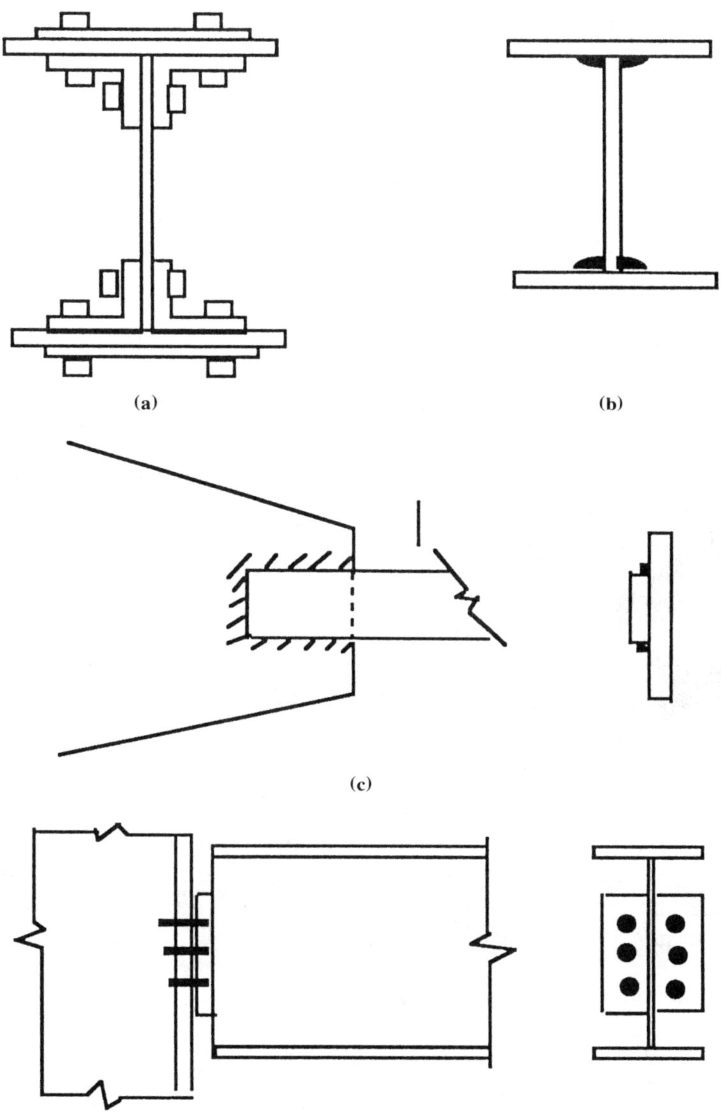

(a) (b) (c) (d)

Figure 7.2 Standard column connections.

connections that are used currently in the building industry or have been employed in the past.

For standard connections, the "AISC Specifications for Structural Steel Buildings," Section J4 (ASD) stipulates that the effective area to resist failure due to shear and tension when present is obtained by deducting the area of the standard holes along the minimum net failure surface. A standard hole is defined by the nominal fastener diameter plus 1.8 in. ($\frac{1}{16}$ for clearance and $\frac{1}{16}$ for cracks due to punching).

7.3 FRAMED BEAM CONNECTION: BOLTED

The standard practice in the design engineering of building is to show the reactions for simply supported beams on the contract drawing. Fabricators then can specify the standard connection in accordance with the AISC specifications to meet the load requirements. However, the engineer must be familiar with the basic assumptions behind this practice and be able to verify it by analysis. The AISC specifications require that several conditions be checked to identify the allowable capacity of the framed beam connection. These conditions are bolt shear, bolt bearing on the connecting material, beam web tear-out known as block shear, shear on the net area of the connecting angles or plates, and local bending stresses due to eccentricity and other factors.

Example 7.1

The diagram in Figure 7.3 represents an interior bay of a floor system. The live and dead loads are 80 and 55 lb/ft^2, respectively. The girder size that is required to carry these loads adequately is W33 × 118. Design the beam column standard connection. See Figure 7.4.

Solution

The reaction at the end of the girder is 48 k.

1. *Check for shear in the bolts.* From the AISC Manual, Table I-D, "Shear," for A325-N bolts with double shear, the capacity of one bolt is 18.6 k

$$n = \frac{48}{18.6}$$

$$= 3$$

$$P_a = 55.8 \text{ k} > 48$$

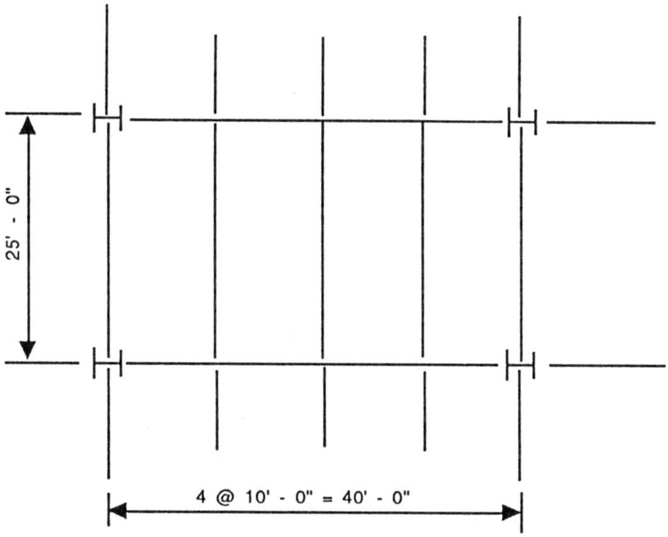

Figure 7.3 Typical interior bay framing system.

2. *Check the net area in the angles*. Use Table II-C and a $\frac{1}{4}$-in. angle and $\frac{3}{4}$-in.-diameter bolt

$$P_a = 52.7 \text{ k}$$

3. *Check the bearing capacity should slip occur*. By using Table I-F at $F_u = 58 \text{ k/in.}^2$ and with a $\frac{3}{4}$-in.-diameter bolt and $l_v = 1\frac{1}{2}$ in. or greater, the allowable load for one fastener with a 1-in. thick plate is 52.2 k

$$\begin{aligned} P_a &= 52.2 \times 3 \times 0.55 \\ &= 86.1 \text{ k} \end{aligned}$$

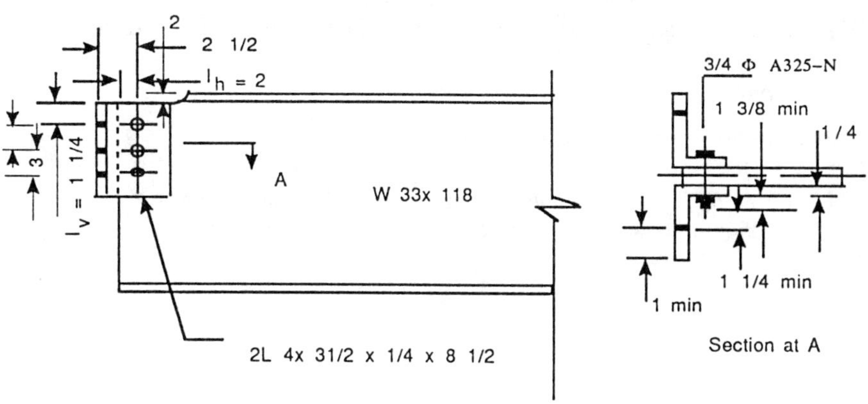

Figure 7.4 Typical beam-to-column connection.

Since the beam is coped, block shear must be checked. The resistance to block shear according to AISC ASD, Section J4, is

$$R_{BS} = (C_1 + C_2)F_u t$$

$$l_v = 1\tfrac{1}{2} \text{ in.} \qquad \text{and} \qquad l_h = 2 \text{ in.}$$

From Table I-G, the coefficients C_1 and C_2 are 1.38 and 0.99, respectively. Then the allowable block shear resistance is

$$\begin{aligned} R_{BS} &= (1.38 + 0.99) \times 58 \times 0.55 \\ &= 137.5 \text{ k} > 48 \end{aligned}$$

The connection as given is adequate. Other allowances must be established to conform to the AISC requirements. These are listed as follows:

(a) Insertion and tightening clearances can be checked in the table of assembling clearances (AISC Manual, Part 4).

(b) Need for reinforcement at the coped section in bending.

(c) Adequacy of this connection with respect to the supporting member.

(d) If a smaller gage on the outstanding leg of the connection angles is required, the bolts must be staggered with those in the web. An even stagger will permit the use of a 5-in. gage with $3\frac{1}{2} \times 3\frac{1}{2} \times \frac{1}{4} \times 0$ ft, 10 in. The stagger portion of the table of stagger for tightening (AISC Manual, Part 4) is used in verifying clearances.

Example 7.2

The diagram in Figure 7.5 represents an interior bay of a library stackroom's floor system. The live and dead loads are 150 and 65 lb/ft^2 respectively. The girder size required to carry these loads adequately is W40 × 324. Design the beam column standard connection. See Figure 7.6.

Solution

The reaction at the end of the girder is 171 k.

1. *Check for shear in the bolts*. From the AISC Manual, Table I-D, "Shear," for A325-N bolts with double shear, the capacity of one bolt is 18.6 k

$$\begin{aligned} n &= \frac{171}{18.6} \\ &= 10 \\ P_a &= 186 \text{ k} > 171 \end{aligned}$$

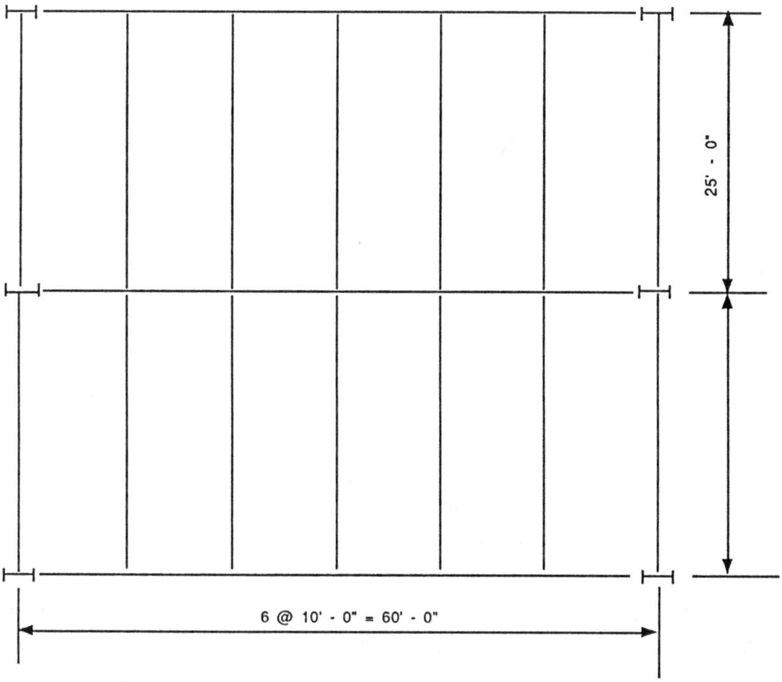

Figure 7.5 Plan view of a typical standard column-girder connection.

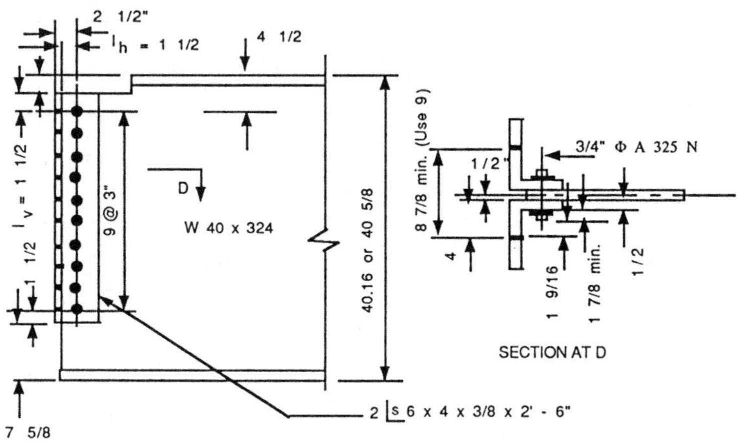

Figure 7.6 Typical beam column standard connection.

2. *Check the net area in the angles*. Use Table II-C and a $\frac{3}{8}$-in. angle and $\frac{3}{4}$-in.-diameter bolt

$$P_a = 279 \text{ k}$$

3. *Check the bearing capacity should slip occur*. By using Table I-F at $F_u = 58 \text{ k/in.}^2$ and with a $\frac{3}{4}$-in.-diameter bolt and $l_v = 1\frac{1}{2}$ in. or greater, the allowable load for one fastener with a 1 in. thick plate is 52.2 k

$$\begin{aligned} P_a &= 52.2 \times 10 \times 1.00 \\ &= 522 \text{ k} \end{aligned}$$

Since the beam is coped, block shear must be checked. The resistance to block shear according to AISC ASD, Section J4, is

$$R_{BS} = (C_1 + C_2)F_u t$$

$$l_v = 1\tfrac{1}{2} \text{ in.} \qquad \text{and} \qquad l_h = 2 \text{ in.}, \quad n = 10$$

From Table I-G, the coefficients C_1 and C_2 are 1.38 and 5.58, respectively. Then the allowable block shear resistance is

$$\begin{aligned} R_{BS} &= (1.38 + 5.58) \times 58 \times 1.00 \\ &= 403.6 \text{ k} > 171 \end{aligned}$$

The connection as given is adequate. Other allowances must be established to conform to the AISC requirements. These are listed as described above.

Example 7.3

Given the beam as shown in Figure 7.7, the selection of W36 × 393 satisfies the requirements of the beam design. See Figure 7.8. Design a

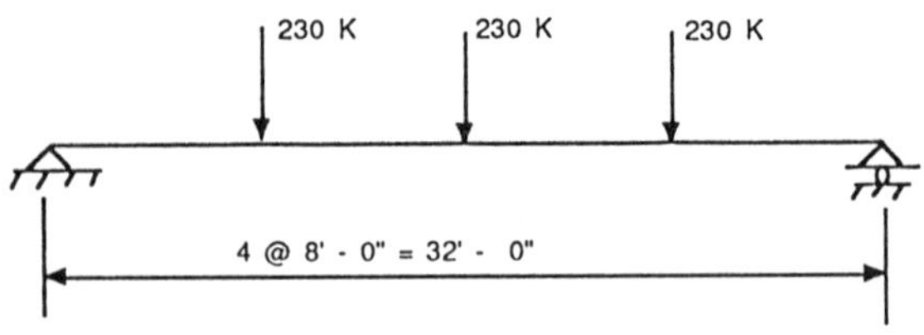

Figure 7.7 Large concentrated loads on a 32-ft girder.

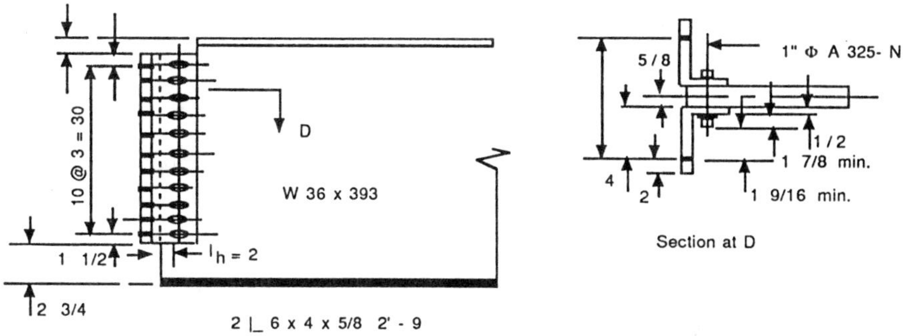

Figure 7.8 Standard column-girder connection.

standard connection in accordance with the AISC specifications. We are given the following data given:

Beam size: W36 $\times$ 393, $t_w = 1.220$ in. ($1\frac{1}{4}$)
A 572 Grade 50 ($F_y = 50$ k/in.^{2} and $F_u = 65$ k/in.^{2})
Beam reaction = 351 k
Bolts: 1-in.-diameter A325-N
Holes: $1\frac{1}{16} \times 1\frac{5}{16}$ in. (short slots), long axis perpendicular to transmitted force

Use $2\angle s$ $6 \times 4 \times \frac{5}{8} \times 2$ ft, 9 in.

Solution

1. *Check for shear in the bolts*. A 1-in.-diameter bolt A325-N in double shear has a capacity of 33 k (Table I-D). The number of required bolts is

$$n = \frac{351}{33.0}$$

$$= 11$$

$$P_a = 11 \times 33$$

$$= 363 \text{ k} > 351 \text{ k}$$

2. *Check the net area in the angles*. Use Table II-C and a $\frac{5}{8}$-in. angle and 1-in.-diameter bolt. The table is based on an allowable shear of $0.3E_u$ (17.4 k/in.^{2} for A36 angles) for the net area of the area of the two angles.

Net area of the two angles A_n

$$A_n = 2 \times [33 \times \tfrac{5}{8} - 11(1 + \tfrac{1}{16}) \times \tfrac{5}{8}]$$
$$= 26.641 \text{ in.}^2$$
$$P_a = 17.4 \times 26.641$$
$$= 463.5 \text{ k} > 351 \text{ k}$$

Check the capacity of the A36 connection angles using Table I-F with $F_u = 58$ k/in.^{2} and $l_v = 1\frac{1}{2}$-in. edge distance. The top bolt has a material bearing value of 43.5 k/in. and each succeeding bolt has a material bearing value of 69.6 k/in. The allowable bearing on the connection angles is then

$$P_a = \tfrac{5}{8} \times (43.5 + 10 \times 69.6) \times 2$$
$$= 924 \text{ k} > 351 \text{ k}$$

3. *Check the beam web tear-out.* From Table I-G with $l_v = 1\frac{1}{2}$ in. and $l_h =$ 2 in., $C_1 = 1.45$ and $C_2 = 4.81$

$$R_{BS} = (1.45 + 4.81) \times 581.22$$
$$= 443 \text{ k} > 351 \text{ k}$$

The connection as given is adequate. Other allowances must be established to conform to the AISC requirements. These are listed as described above.

7.4 FRAMED BEAM CONNECTION: WELDED E70XX ELECTRODES FOR COMBINATION WITH TABLE II AND TABLE III CONNECTIONS

Connections may have bolted and welded parts requiring the use of Table II with Table III from the AISC Manual, Part 4.* To accommodate typical gages, angle leg widths will generally be $4 \times 3\frac{1}{2}$ in., the 4 in. being the outstanding leg. There are two cases for this type of connection. Case I is shown in Figure 7.9 in which the weld type A is shown in the web of the girder, while the other legs of the pair of angles are bolted to the flanges of the column. In case II (see Figure 7.10), the opposite is true. The pair of angles are welded to the flanges of the column, and the other set of legs is bolted to the web of the girder.

---

*All tables mentioned here refer to those in AISC ASD Manual, Ninth Edition.

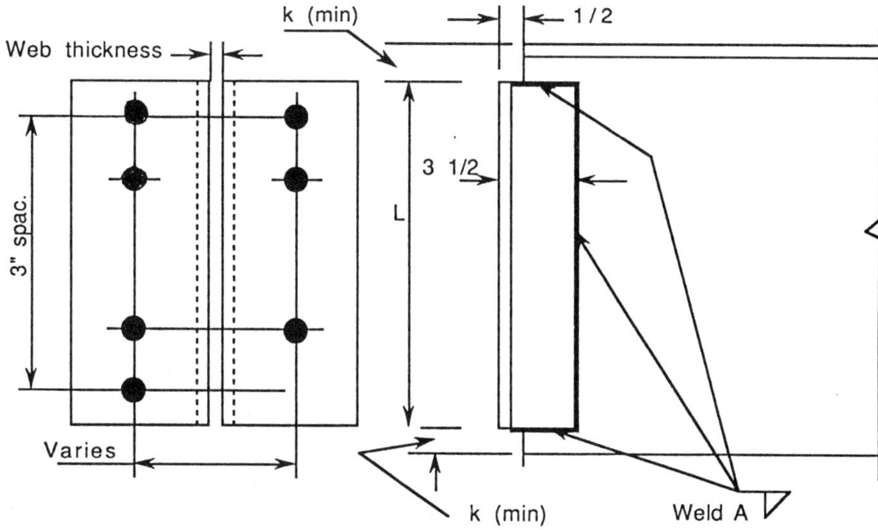

Figure 7.9 Typical welded connection: Case I.

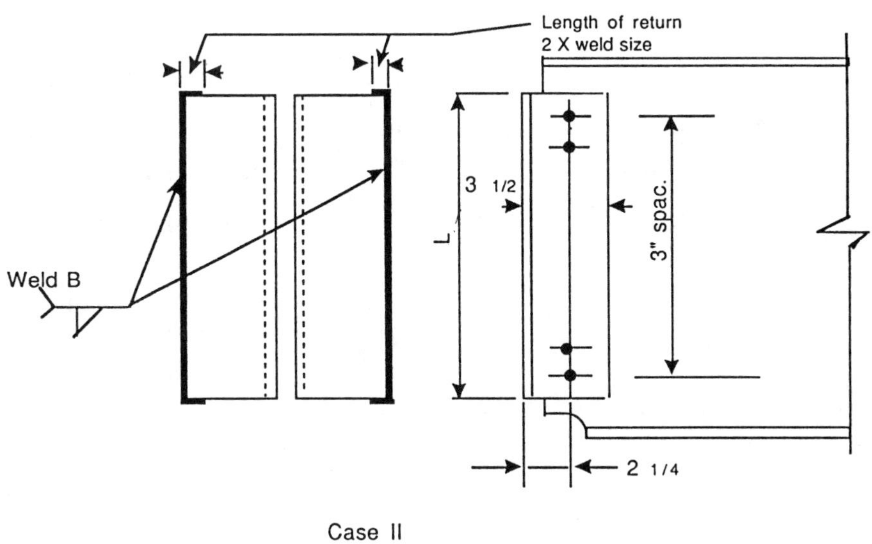

Figure 7.10 Typical welded connection: Case II.

The following AISC recommendations are used in the design of welded connections for cases I and II:

1. Width of web legs in case I may be reduced optionally from $3\frac{1}{2}$ to 3 in.
2. Width of outstanding legs in case II may be reduced optionally from 4 to 3 in. for values of $L = 5\frac{1}{2}$ in. through 1 ft, $5\frac{1}{2}$ in.
3. Angle thickness is equal to weld size plus $\frac{1}{16}$ in., or thickness of angle from the applicable Table II whichever is greater.
4. Angle length L must be as tabulated in Table III.
5. When holes for bolts are used, investigate the bearing capacity of the supporting member.
6. Welded connections type A and B can be chosen from Table IV. This table provides greater economy and allows more flexibility in the section of angle lengths and connection capacities.
7. Allowable capacity for weld A is based on the center solution presented in the AISC ASD Manual for the development of Tables XIX through XXVI, "Eccentric Loads on Weld Groups > For Weld B." Traditional vector analysis is used.

Example 7.4

Given the frame shown in Figure 7.11 and floor loads of 120 lb/ft^2 total dead and live loads, design a connection of the Case I type. Use A-36 steel,

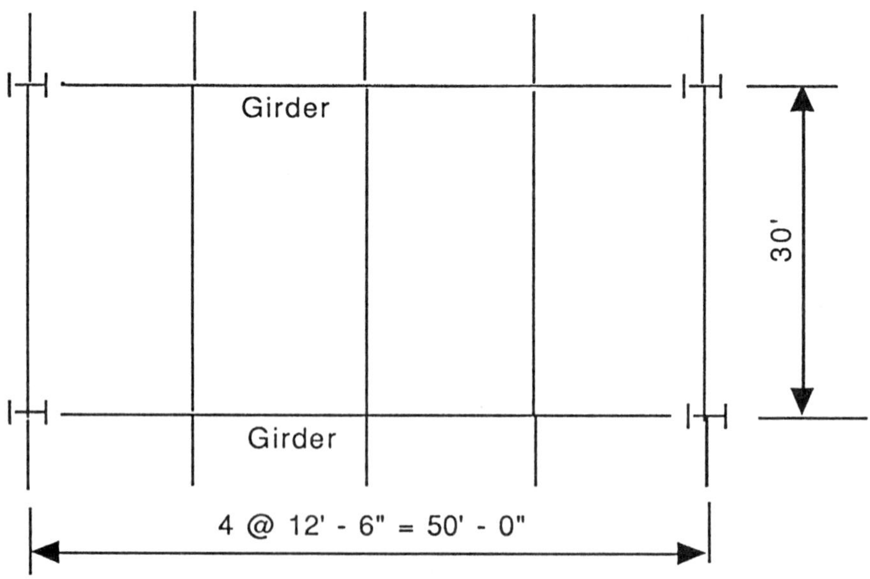

Figure 7.11 Typical floor framing system.

$\frac{7}{8}$-in.-diameter A325-N bolts and E70XX electrodes for combination with Table II and Table III connections.

Solution

We are given the following data:

Beam: W27 × 161 (not coped)

$$t_w = 0.660 \text{ in.}$$
$$F_y = 36 \text{ k/in.}^2$$
$$F_v = 14.5 \text{ k/in.}^2$$

Reaction: 90 k
Bolts: $\frac{7}{8}$-in.-diameter ASTM A325-N
Welds: E70XX

Examine the data under weld *A* and note that a value of 92.1 k satisfies the reaction. It requires $\frac{5}{16}$-in. welds and $8\frac{1}{2}$ in. long angles. The thickness of the thinner part to be welded must be at least larger than the weld (AISC ASD Manual, Specifications, Section J2-2b). Use an angle thickness of $\frac{3}{8}$ in. minimum. The 0.660-in. web thickness exceeds the minimum required 0.64 in., so the increase in capacity is $(0.660/0.640) \times 92.1 = 95$ k.

Note in Table II-A that the angle provides three rows and 75.8-k capacity. Hence, use 11- $\frac{1}{2}$-in. angle length with four rows and a capacity of 101 k.

DATA

Two angles: $\angle 4 \times 3\frac{1}{2} \times \frac{3}{8} \times 11\frac{1}{2}$ in.
$F_y = 36 \text{ k/in.}^2$
Eight $\frac{7}{8}$-in.-diameter ASTM A325-N bolts (threads included in shear plane)
$\frac{5}{16}$-in. fillet weld, E70XX

Example 7.5

We are given the following data:

Beam: W18 × 40 (not coped)

$$t_w = 0.315 \text{ in.}$$

Reaction: 22.5 k
Bolts: $\frac{7}{8}$-in.-diameter ASTM A325-N
Welds: E70XX

Solution

Note in Table II-A that two rows of $\frac{7}{8}$-in.-diameter ASTM A325-N bolts provide a capacity of 50.5 k and require an angle of $\angle 4 \times 3\frac{1}{3} \times \frac{3}{8} \times 7$ in. In Table III, an $8\frac{1}{2}$ in. long angle requires a $\frac{3}{16}$-in. weld and minimum web thickness of 0.38. The reduced capacity for the web is calculated as follows: $(0.315/0.38) \times 55.3 = 45.8$. The thickness of the angle must be larger than the size of the weld by $\frac{1}{61}$ in. Hence, use an angle thickness of $\frac{1}{4}$ in.

DATA

Two angles: $\angle 4 \times 3\frac{1}{2} \times \frac{1}{4} \times 8\frac{1}{2}$
$F_y = 36$ k/in.^{2}
Six $\frac{7}{8}$-in.-diameter ASTM A325-N bolts
$\frac{3}{16}$-in. fillet weld, E70XX

For more complicated connections, the AISC ASD Manual provides ample resource material for design purposes.

8

Anchor Bolts and Baseplates

8.1 INTRODUCTION

Compressive stresses for structural steel members are much higher than the corresponding values of masonry or reinforced concrete on which the steel members may be supported. In order to avoid crushing of the concrete or masonry from bearing stresses that are imposed by the steel members, an intermediate material is used to spread the compressive stresses in the steel to an acceptable level for the concrete of masonry. The analysis for the determination of the stress bearing on the support under a column or beam is a very complex problem. Instead, a simple assumption is used in which the stress is considered to spread uniformly on the support.

8.2 DESIGN OF COLUMN BASEPLATES

All superstructures must be anchored to their support system to prevent sliding or overturning. Columns might rest on pile caps, reinforced concrete mats, individual concrete footings or pads. Floor frames that are supported on bearing walls have to have some type of anchoring system to prevent uplift or horizontal movement. See Figure 8.1 for a beam-bearing wall support. In each of these cases, the transfer of stress from the superstructure to its support system requires rational design so as not to exceed the bearing capacity of these systems.

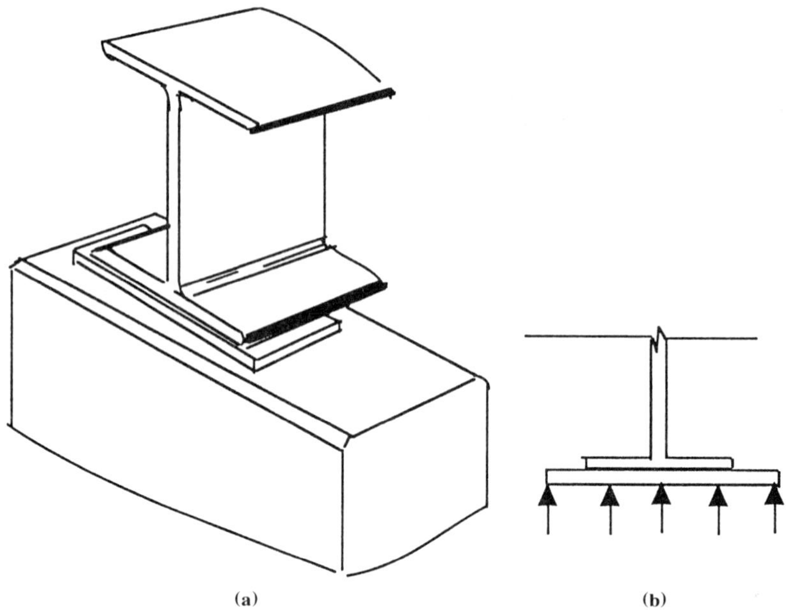

Figure 8.1 Beam baseplate.

For the transfer of loads from the column to its base, the AISC ASD Manual outlines a procedure that is rational and simple to follow. Figure 8.2 presents the dimensions used in the baseplate design. The notation used below is the same as in the AISC ASD Manual

P = total column load (k)
$A_1 = B \times N$ = gross area of the baseplate (in.^{2})
A_2 = full cross-sectional area of the supporting concrete member
F_b = allowable bending stress in the steel plate (k/in.^{2})
F_p = allowable bearing pressure on support (k/in.^{2})
f_p = actual bearing pressure (k/in.^{2})
f'_c = compressive strength of concrete (k/in.^{2})
t_p = thickness of the plate (in.)

For relatively large values of m and n, the baseplate is designed as a cantilever beam, fixed at the edges of a rectangle whose dimensions are outlined by the dashed lines in Figure 8.2. The column load is assumed to be distributed uniformly on the supporting pedestal. This is a simplification

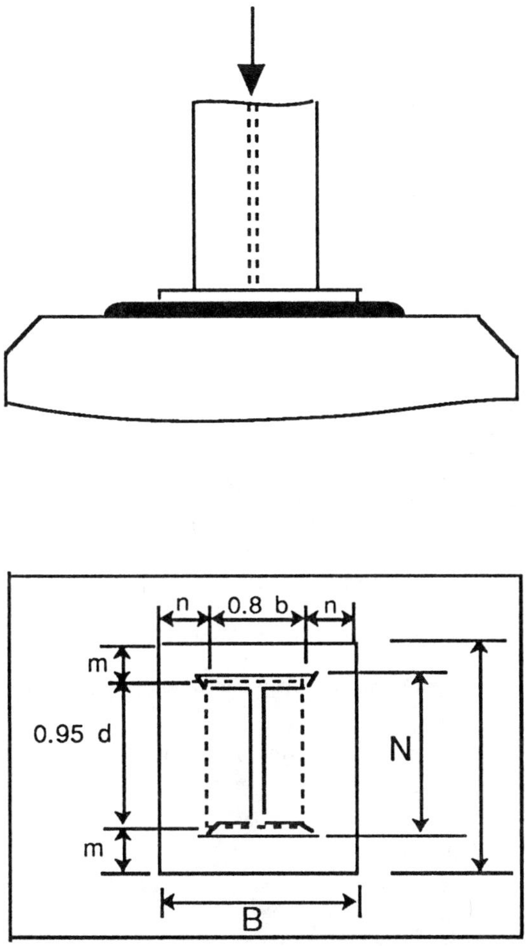

Figure 8.2 Column baseplate: AISC procedure.

that yields conservative results and has been proven to be reliable. See Figure 8.3.

The thickness of the plate is obtained from the following equations:

$$t_p = 2m\sqrt{\frac{f_p}{F_y}} \tag{8.1}$$

$$t_p = 2n\sqrt{\frac{f_p}{F_y}} \tag{8.2}$$

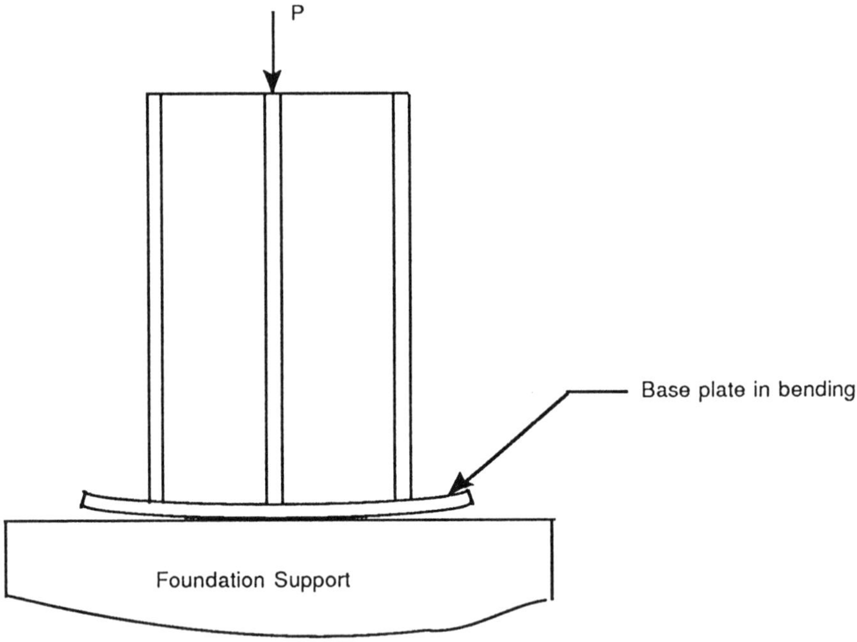

Figure 8.3 Baseplate foundation interaction.

Use the larger value of t_p as obtained from the above equations.

The baseplate is optimized when m and n are equal. This condition warrants that

$$N = \sqrt{A_1} + \Delta \tag{8.3}$$

and

$$\Delta = 0.5(0.95d - 0.80b_f) \tag{8.4}$$

$$B = \frac{A_1}{N} \tag{8.5}$$

For small values of m and n, the model presented in Figure 8.4 is used in the stress analysis of the baseplate. The column load is assumed to be transferred to the concrete over the shaded area, including the area of the cross section of the column

$$A = 2(d + b - 2L)L \tag{8.6}$$

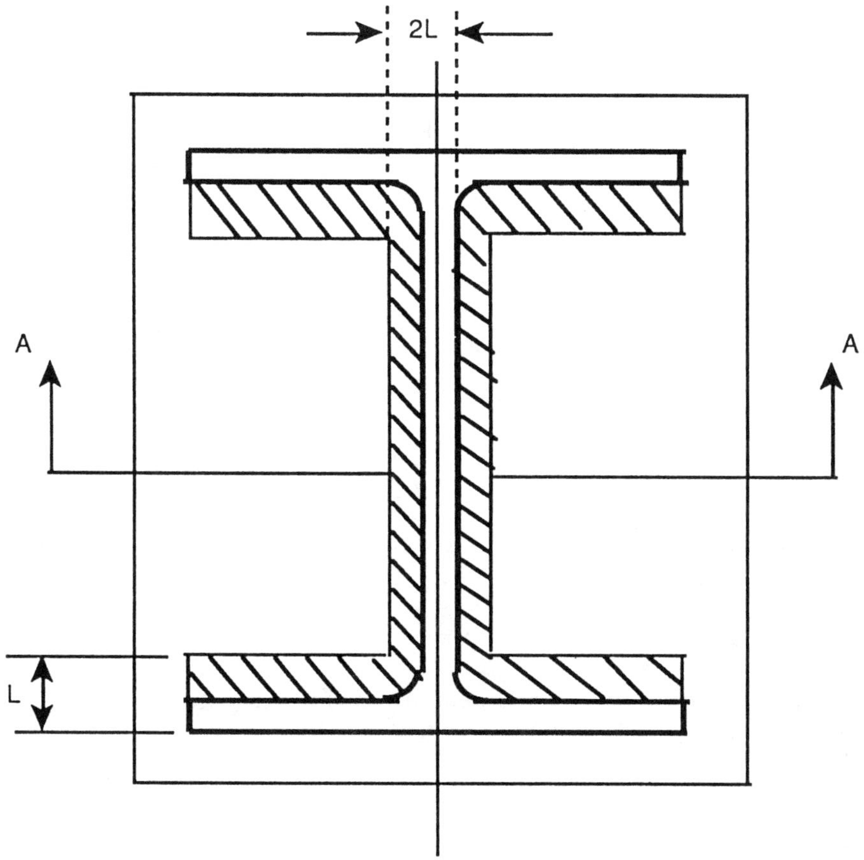

(a)

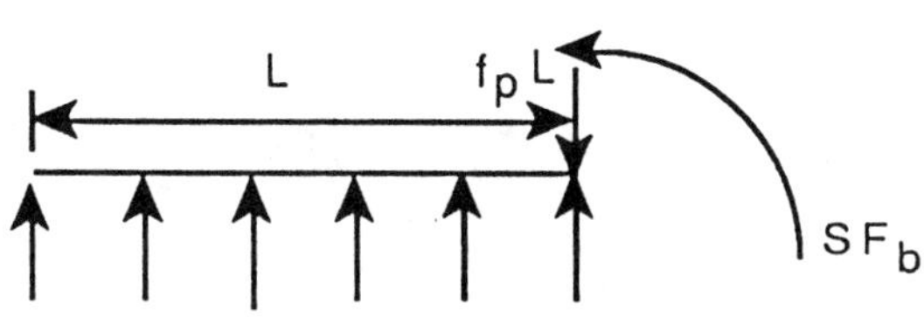

(b)

Figure 8.4 Stress action pattern on a baseplate.

where

d = depth of the column (in.)
b = flange width of column section (in.)

Equating the sum of the pressure over the shaded area A to the column load yields

$$F_p = \frac{P}{2(d + b - 2L)L} \tag{8.7}$$

Assuming a cantilever model as shown in Figure 8.4b and considering a strip of the plate 1 in. wide, we determine the bending moment to be

$$M_{\max} = f_p \frac{L^2}{2} \tag{8.8}$$

and

$$M_{\max} = SF_b \tag{8.9}$$

where

$$S = \frac{1}{6}t^2 \tag{8.10}$$

Letting $f_p = F_b$ and combining Equations (8.7) through (8.10) yield

$$t = L\sqrt{\frac{3f_p}{F_b}} \tag{8.11}$$

The allowable bearing strength of concrete is

$$F_p = 0.35f_{c'} \tag{8.12}$$

but

$$F_p \leqq 0.70f_{c'} \tag{8.13}$$

The limits for the areas A_1 and A_2 are given in the AISC ASD Manual as follows:

$$A_1 \geqq \frac{P}{0.70f_{c'}} \tag{8.14}$$

$$A_2 \geqq \frac{P}{0.175f_{c'}} \tag{8.15}$$

$$A_1 \geqq \frac{1}{A_2}\left(\frac{P}{0.35f_{c'}}\right)^2 \tag{8.16}$$

The procedure for designing column baseplates is as follows.

Step 1. Calculate

(a) $A_1 = \dfrac{P}{0.70 f_{c'}}$

(b) $A_1 = \dfrac{1}{A_2}\left(\dfrac{P}{0.35 f_{c'}}\right)^2$

where

$$A_2 = \frac{P}{0.175 f_{c'}}$$

Use the larger of these two.

Step 2. Determine N and B

$$N = \sqrt{A_1} + \Delta$$

where

$$\Delta = 0.5(0.95d - 0.80b)$$

and

$$B = \frac{A_1}{N}$$

Step 3. Determine the actual bearing pressure on concrete

$$f_p = \frac{P}{B \times N}$$

Step 4. Determine m, n, and n'

$$m = 0.5(N - 0.95d)$$

$$n = 0.5(B - 0.80b)$$

$$n' = \frac{b - t_w}{2}\sqrt{\frac{1}{1 + 3.2\alpha^3}}$$

where

$$\alpha = \frac{b - t_w}{2(d - 2t_f)}$$

The subscripts w and f refer to the web and flange, respectively.

Step 5. Use the larger of the values m, n, or n' to solve for t by using the appropriate formula

$$t_{p} = m\sqrt{\frac{f_p}{0.25F_y}}$$

$$= n\sqrt{\frac{f_p}{0.25F_y}}$$

$$= n'\sqrt{\frac{f_p}{0.25F_y}}$$

Example 8.1

We are given a column W12 × 120, of A-36 steel, carrying a concentric load of 590 k. The concrete strength is 3 k/in.^{2}. Design the baseplate and specify the dimensions for the pedestal under the plate.

Solution

List the section properties

$$d = 13.12 \text{ in.}$$
$$b = 12.32 \text{ in.}$$
$$t_w = 0.710 \text{ in.}$$
$$t_f = 1.105 \text{ in.}$$

Begin the procedure.

Step 1. Calculate

(a)

$$A_1 = \frac{590}{0.70 \times 3}$$
$$= 281 \text{ in.}^2$$

(b)

$$A_2 = \frac{590}{0.175 \times 3}$$
$$= 1157 \text{ in.}^2$$

$$A_1 = \frac{1}{1157}\left(\frac{590}{0.35 \times 3}\right)^2$$
$$= 273 \text{ in.}^2$$

Use $A_1 = 281$ in.^{2}.

Step 2. Determine N and B

$$\Delta = 0.5(0.95 \times 13.12 - 0.80 \times 12.32)$$
$$= 1.30 \text{ in.}$$
$$N = \sqrt{281} + 1.30$$
$$= 18.06 \quad (\text{use } N = 18 \text{ in.})$$
$$B = \frac{281}{18}$$
$$= 15.6 \quad (\text{use } B = 16 \text{ in.})$$

Step 3. Calculate f_p

$$f_p = \frac{590}{16 \times 18}$$
$$= 2.05 \text{ k/in.}^2 < 0.70 f_{c'} = 2.1 \text{ k/in.}^2$$

Step 4. Calculate m, n, and n'

$$\alpha = \frac{12.32 - 0.710}{2(13.12 - 2 \times 1.105)}$$
$$= 0.53$$
$$n' = \frac{12.32 - 0.710}{2}\sqrt{\frac{1}{1 + 3.2 \times 0.53^3}}$$
$$= 4.78 \text{ in.}$$
$$n = 0.5(16 - 0.80 \times 12.32)$$
$$= 3.07 \text{ in.}$$
$$m = 0.5(18 - 0.95 \times 13.12)$$
$$= 2.77 \text{ in.}$$

Use $n' = 4.78$ in.

Step 5. Calculate t_p

$$t_p = 4.78\sqrt{\frac{2.05}{0.25 \times 36}}$$
$$= 2.2813 \text{ in.} \quad (\text{use } t_p = 2\tfrac{5}{16})$$

Use the following baseplate: 1 ft, 4 in. × 1 ft, 6 in. × $2\frac{5}{16}$.

The area of the pier is calculated in Step 1 and found to be 1157 in.^{2}. Use the dimensions of 30 in. × 35 in. for an area of 1150 in.^{2}.

9

Built-Up Beams: Plate Girders

9.1 INTRODUCTION

A plate girder is distinguished from a rolled beam by the web slenderness ratio h/t_w. When spans are long and loads are relatively large, it is more economical to use plate girders rather than rolled sections. With the introduction of railroads in the middle of the nineteenth century, riveted plate girders were commonly used in bridge designs spanning 50 to 150 ft. Later with the introduction of welding, riveted plate girders became less and less frequently used. Plate girders may have a constant section throughout their span or a variable section. Several plate girder sections are shown in Figure 9.1

9.2 DESIGN OF PLATE GIRDERS BY ASD METHOD

In addition to meeting the bending stress requirements, in the design of plate girders, flange and web instability requirements must be met to ensure that the girder reaches its optimum strength. Slenderness limitations are applied whenever no stiffeners are used in the girder or when stiffeners are placed at a distance greater than $1\frac{1}{2}$ times h, the distance between the flanges

$$\frac{h}{t_w} \leq \frac{14{,}000}{\sqrt{F_{yf}(F_{yf} + 16.5)}} \tag{9.1}$$

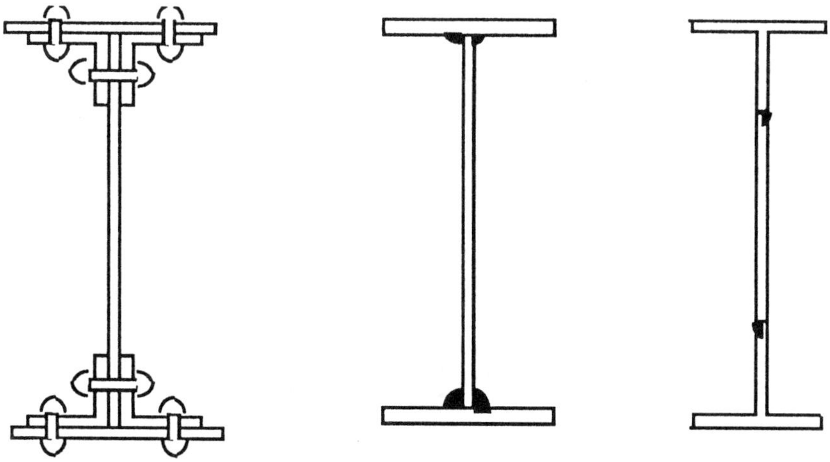

(a) (b) (c)

Figure 9.1 Typical plate girder construction.

for $a/h \geq 1\frac{1}{2}$

$$\frac{h}{t_w} \leq \frac{2000}{\sqrt{F_{yf}}} \tag{9.2}$$

for $a/h < 1\frac{1}{2}$,

where

h = distance between the flanges
t_w = thickness of the web
a = distance between the stiffeners
F_{yf} = steel yield stress in the flange

Allowable Bending Stress Requirements

The maximum bending stress in the compression flange $F_{b'}$ is given by

$$F_{b'} \leq F_b R_{PG} R_e \tag{9.3}$$

for $(h/t_w) \geq (760/\sqrt{F_b})$,

where

F_b = applicable bending stress given in Chapter 4 (AISC Manual, Ninth Edition, Chapter F) (k/in.^{2})

$$R_{PG} = 1 - 0.0005\frac{A_w}{A_f}\left(\frac{h}{t} - \frac{760}{\sqrt{F_b}}\right) \leq 1.0 \tag{9.4}$$

$$R_e = \frac{\left[12 + \frac{A_w}{A_f}(3\alpha - \alpha^3)\right]}{\left(12 + 2 \times \frac{A_w}{A_f}\right)} \leq 1.0 \tag{9.5}$$

For nonhybrid girders, $R_e = 1.0$ and

A_w = area of web at the section under consideration (in.^{2})

A_f = area of compression flange (in.^{2})

$$\alpha = 0.6\frac{F_{yw}}{F_b} \leq 1.0$$

Allowable Shear Stress with Tension Field Action

The maximum web shear stress under any loading applied to the plate girder may not exceed the value given in Equations (4.18) or (4.19). For nonhybrid plate girders stiffened in accordance with the provisions of Section G4, if the coefficient $C_v \leq 1.0$, the allowable shear including tension field is given by

$$F_v = \frac{F_y}{2.89}\left[C_v + \frac{1 - C_v}{1.15\sqrt{1 + (a/h)^2}}\right] \leq 0.40F_y \tag{9.6}$$

where

$$C_v = \frac{45{,}000k_v}{F_y\left(\frac{h}{t_w}\right)^2} \text{ when } C_v \text{ is less than 0.8}$$

$$= \frac{190}{\frac{h}{t_w}}\sqrt{\frac{k_v}{F_y}} \text{ when } C_v \text{ is greater than 0.8}$$

$$k_v = 4.00 + \frac{5.34}{\left(\frac{a}{h}\right)^2} \text{ when } a/h \text{ is less than 1.0}$$

$$k_v = 5.34 + \frac{4.00}{\left(\frac{a}{h}\right)^2} \text{ when } a/h \text{ is greater than 1.0}$$

t_w = thickness of web (in.)
a = clear distance between transverse stiffeners (in.)
h = clear distance between flanges at the section under investigation (in.)

Transverse Stiffeners

When the ratio h/t_w exceeds 260 and the maximum web shear stress f_v is greater than that allowable (as given by the AISC Manual, Section F4-2), stiffeners are used to provide web and flange stability. The spacing of the stiffeners is given by

$$\frac{a}{h} \leq \left[\frac{260}{(h/t_w)}\right]^2 \quad \text{and} \quad 3.0 \tag{9.7}$$

When stiffeners are provided in pairs, the total gross area of the stiffener is

$$A_{st} = \frac{1 - C_v}{2}\left[\frac{a}{h} - \frac{(a/h)^2}{\sqrt{1 + (a/h)^2}}\right] YDht_w \tag{9.8}$$

where

Y = ratio of yield stress of web steel to that of stiffener steel
Y = 1.0 for the case when the same steel is used in the web and stiffeners
D = 1.0 for stiffeners used in pairs
= 1.8 for single angle stiffeners
= 2.4 for single plate stiffeners

The maximum allowable shear stress in the stiffener is

$$f_{vs} = h\sqrt{\left(\frac{F_y}{340}\right)^3} \tag{9.9}$$

Bolts or rivets used to connect stiffeners to the web are spaced not more than 12 in. on center. For weld connections, the intermittent fillet welds are used at a maximum clear distance between the welds not more than 16 times the web thickness nor more than 10 in.

Stiffeners may run from the compression flange to a point not closer than 4 in. nor further than 6 in. from the near toe of the fillet weld at the tension flange. When single stiffeners are used, they shall be attached to the compression flange to prevent torsion in the web. Lateral bracing must be attached to the compression flange to transmit 1% of the total flange stress, unless the flange is made only of angles.

The moment of inertia of a single or double intermediate stiffener is limited as follows:

$$I_{st} \geq \left(\frac{h}{50}\right)^4 \tag{9.10}$$

Combined Shear and Tension Stress

The bending stress in the web of a plate girder is limited to a maximum of $0.60F_y$ or

$$f_b = \left(0.825 - 0.375\frac{f_v}{F_v}\right)F_y \tag{9.11}$$

where

f_v = computed average web shear stress (total shear divided by web area) (k/in.^{2})

F_v = allowable web shear stress according to Equation (9.6)

When the plate girder is built of steel with a yield point greater than 65 k/in.^{2}, the allowable shear stress may not be more than the value given in Equation (4.18) or (4.19), whichever is smaller.

9.3 APPROXIMATE METHOD FOR SELECTION OF TRIAL SECTION

To minimize the time for the design of plate girders, an approximate method of stress calculation is presented in the following paragraphs. It is observed that in plate girders the moment of inertia of the section is provided mostly by the flanges. The contribution of the web theoretically is given by $\frac{1}{12}t_w h^3$, and t_w and h are the thickness of the web and clear distance between the flanges, respectively. Noting that the area of the web is $t_w h$, we can write the expression for the moment of inertia as $\frac{1}{12}A_w h^2$. If a fraction of the web area is lumped with the flange area, the contribution of that is $A_w/n(h/2)^2$. Then the moment of inertia of the section is the sum of the contribution of the flanges and web

$$I = 2 \times A_f\left(\frac{h}{2}\right)^2 + \frac{1}{12}A_w h^2 \tag{9.12}$$

Note that the second term on the right in Equation (9.11) can be written as

$$\frac{1}{12}A_w h^2 = 2 \times \frac{A_w}{6}\left(\frac{h}{2}\right)^2 \tag{9.13}$$

which yields $n = 6$.

Substituting Equation (9.13) into (9.12) and simplifying result in

$$\begin{aligned} I &= 2 \times A_f\left(\frac{h}{2}\right)^2 + 2 \times \frac{A_w}{6}\left(\frac{h}{2}\right)^2 \\ &= \left(A_f + \frac{A_w}{6}\right) \times 2\left(\frac{h}{2}\right)^2 \end{aligned} \tag{9.14}$$

The section modulus S_x is given by $I/(d/2)$, where $d \simeq h$ for deep girders. Then,

$$S_x = \left(A_f + \frac{A_w}{6}\right)h \tag{9.15}$$

The bending stress is limited to $0.60F_y$ in order to accommodate the combined shear and tension stress in the web, as provided in Equation (9.10) (AISC Manual, Section G5-1).

Example 9.1

Design a welded plate girder shown in Figure 9.2 to support a uniform load of 2.8 k/ft and two concentrated loads of 80 k located 20 ft from each end. The compression flange will be supported laterally only at points of the concentrated loads.

Solution

1. Trial Section Selection

For no reduction in the compression flange stress,

$$\frac{h}{t_w} \le \frac{760}{\sqrt{F_b}}$$

as noted in the AISC Manual, Section G2. Using A36, we obtain

$$\frac{h}{t_w} = 162$$

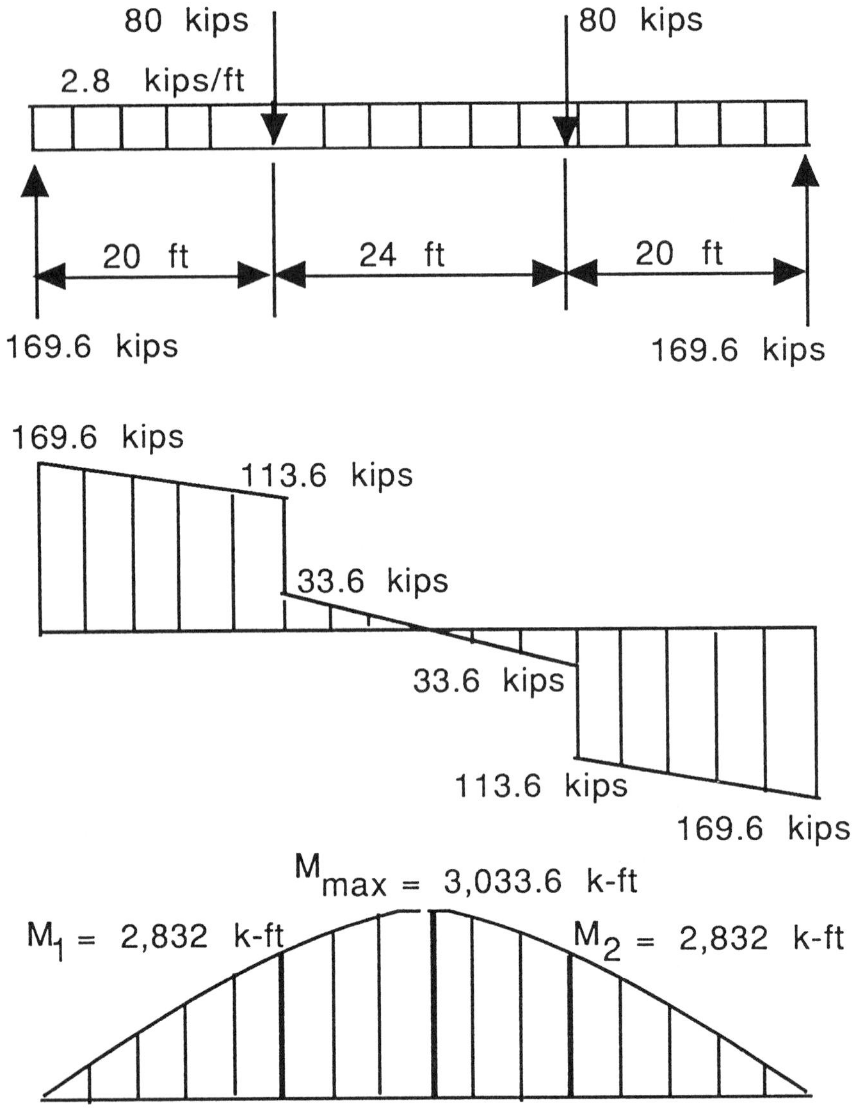

Figure 9.2 Plate girder loading, shear, and bending moment diagrams.

The ratio of the outstanding projection of a plate from the web of the girder to its thickness is given by

$$\frac{b}{2t_f} = \frac{95}{\sqrt{F_y}} \quad \text{(AISC Manual, Section B5-2)}$$

For A36 steel, $b/t_f = 31.6$. Using the approximation of $t_f = 2t_w$ yields $h/b \geq 2.6$.

From Equation (9.14), we obtain

$$S_x = \left(A_f + \frac{A_w}{6}\right)h = \left(A_f + \frac{1}{6}\frac{h^2}{162}\right)h$$

Letting $t_f = 2t_w$ and $b = h/3$, we can express the above relationship for the section modulus as

$$S_x = \left(2 \times \frac{h}{3} \times \frac{h}{162} + \frac{1}{6}\frac{h^2}{162}\right)h$$

Solving for h yields

$$h = 5.793\sqrt[3]{S_x}$$

Given the maximum moment of 3033.6 k-ft and $F_b = 0.60F_y$ (A-36 steel, $F_b = 22\ \text{k/in.}^2$), the required modulus is

$$S_x = \frac{3033.6 \times 12}{22} = 1654.7\ \text{in.}^3$$

$$h = 5.793\sqrt[3]{S_x} = 68.5\text{, say, } 70\ \text{in.}$$

Then,

$$t_w = \frac{70}{162} = 0.432\ \text{in.} \quad \text{(use 0.4375 in.)}$$

$$t_f = 2t_w = 0.875\ \text{in.}$$

$$b = \frac{h}{3} \quad \text{or} \quad b = 23.3\ \text{in.} \quad (\text{use } b = 24\ \text{in.})$$

The section is shown in Figure 9.3.

Another fast way of determining the required height of the web, thus meeting AISC requirements, is a graphical approach. Having determined

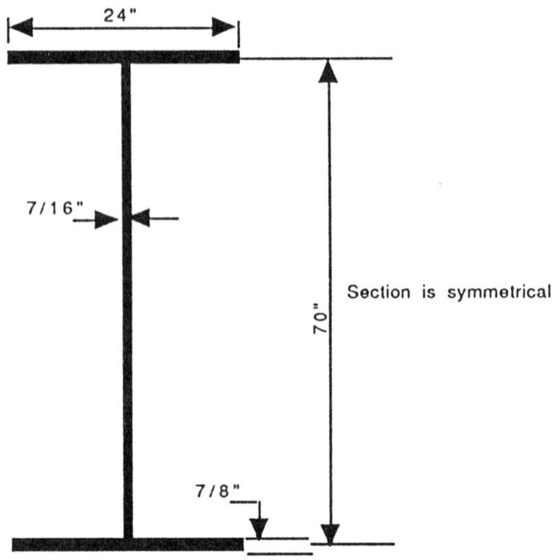

Figure 9.3 Plate girder section.

the required section modulus based on the given loads and span of the girder and choosing a ratio of h/t_w as an average value between the lower and upper limits specified by the AISC specifications, examine the graph shown in Figure 9.12. Go vertically from the S_x axis to intersect the curve for the appropriate h/t_w ratio. Then project a straight line horizontally to intersect the h axis. This determines the value of h in inches.

2. *Section Properties*

The calculation of the section properties is shown in Table 9.1.

(a) Calculate the moment of inertia.

(b) The furnished section modulus is

$$\frac{63{,}955}{35} = 1827.3 \text{ in.}^3$$

TABLE 9.1 Section Properties for Example 9.1

Section	A (in.^{2})	y (in.)	$\sum_{i=1}^{n} A_i y_i^2$	I_0 (in.^{4})	I_{gr} (in.^{4})
Web $\frac{7}{16} \times 70$	30.62			12,505	12,505
1 flange $\frac{7}{8} \times 24$	21.00	35.00	25,725	1	25,725
1 flange $\frac{7}{8} \times 24$	21.00	35.00	25,725	1	25,725
Moment of inertia					63,955

3. Check Flange Stresses

Check bending at midspan

$$\text{Weight of section} = \frac{72.62}{144} \times \frac{590}{1000}$$

$$= 0.298 \text{ k/ft}$$

$$\text{Moment due to weight of beam} = 0.298 \times \frac{64^2}{8}$$

$$= 152.6 \text{ k/ft}$$

$$f_b = \frac{(3033.6 + 152.6) \times 12}{1827.3}$$

$$= 20.9 \text{ k/in.}^2 < 22$$

4. Check Spacing Between Stiffeners

$$I_{0y} = 0.875 \times \frac{24^3}{12}$$

$$= 1008 \text{ in.}^3$$

where

I_{0y} = moment of inertia of the flange about the y axis

$r_T = \sqrt{\frac{1008}{21}}$

= 6.928 in.

= radius of gyration of the flange

$M_{max} > M_1$ and M_2; then $C_b = 1$.

$$\frac{l}{r_T} = \frac{24 \times 12}{6.928}$$

$$= 42.6 < 53\sqrt{C_b} = 53$$

No intermediary stiffeners are needed between the panel points. Allowable stresses are based on lateral buckling criteria

$$F_b = 0.60F_y$$

$$= 21.6 \text{ k/in.}^2$$

Reduced allowable bending stress in the compression flange is

$$R_{PG} = 1 - 0.0005\left(\frac{A_w}{A_f}\right)\left(\frac{h}{t} - \frac{760}{\sqrt{F_b}}\right)$$

$$= 1 - 0.0005 \times \frac{21.0}{30.6}\left(\frac{70}{0.4375} - \frac{760}{\sqrt{21.6}}\right) = 1.0$$

Since the section is nonhybrid, $R_e = 1.0$

$$F_{b'} = F_b = 21.6 \text{ k/in.}^2 > 20.9 \text{ k/in.}^2$$

5. Check Bending Stress in 20-ft Panel

Maximum bending stress

$$f_b = \frac{2832 \times 12}{1827.3}$$

$$= 18.6 \text{ k/in.}^2$$

$$C_b = 1.75 + 1.05\frac{M_1}{M_2} + 0.3\left\{\frac{M_1}{M_2}\right\}^2$$

where

$$M_1 = 0$$

Then,

$$\frac{M_1}{M_2} = 0$$

and

$$C_b = 1.75$$

$$\frac{l}{r_T} = \frac{20 \times 12}{6.928}$$

$$= 34.6 < 53\sqrt{C_b} = 70.1$$

Allowable stress in the 20-ft panel is based on lateral buckling criteria

$$F_b = 0.60F_y$$

$$= 21.6 \text{ k/in.}^2 > 18.6 \text{ k/in.}^2$$

No stress reduction in the compression flange is needed.

6. Stiffener Requirements

Bearing Stiffeners

(a) Bearing stiffeners are required at the girder ends.

(b) Check the bearing under concentrated loads. Assume point bearing and $\frac{1}{4}$-in. web-to-flange welds. Local web yielding is as follows: When the web load to be resisted is a concentrated load producing tension or compression, applied at a distance from the member end that is greater than the depth of the member,

$$\frac{R}{t_w(N + 5k)} \leq 0.66F_y$$

where

R = concentrated load or reaction (k)
t_w = thickness of web (in.)
N = length of bearing (not less than k for end reactions)
k = distance from outer face of flange to web toe of fillet (in.)
$k = \frac{7}{8} + \frac{1}{4} = 1\frac{1}{8}$ in. (AISC Manual, Equation K1-2)

$$\frac{80}{0.4375(0 + 5 \times 1.125)} = 32.5 \text{ k/in.}^2$$
$$> 0.66 \times 36$$

Provide bearing stiffeners under concentrated loads.

Intermediate Stiffeners

(a) Check the shear stress in an unstiffened end panel

$$\frac{h}{t_w} = 160$$

$$\frac{a}{h} = \frac{20 \times 12}{70}$$
$$= 3.429$$

From Equation (4.19), the maximum allowable shear is

$$F_v = \frac{F_v}{2.89} \times C_v$$

$$C_v = \frac{45{,}000k_v}{F_y(h/t_w)^2} \quad \text{(when } C_v \text{ is less than 0.8)}$$

$$K_v = 5.34 + \frac{4}{3.429^2} = 5.68$$

$$C_v = \frac{45{,}000 \times 5.68}{36 \times 160^2} = 0.277$$

$$F_v = \frac{36 \times 0.277}{2.89} = 3.45 \text{ k/in.}^2$$

$$f_v = \frac{169.6}{0.4375 \times 70} = 5.52 \text{ k/in.}^2 > 3.45 \text{ k/in.}^2$$

Provide intermediate stiffeners.

(b) For the end panel stiffener (tension field action not permitted), by using Equation (4.19 or Table 1-36 from the AISC Manual, a value of a/h is found to be equal to unity. Hence, $a = 70$ in. or 5 ft, 10 in.

(c) Check for additional stiffeners. The shear at the first intermediate stiffener is

$$V = 169.6 - \left(2.8 \times \frac{70}{12}\right)$$

$$= 153.3 \text{ k}$$

$$f_v = \frac{153.3}{30.62}$$

$$= 5.01 \text{ k/in.}^2$$

Using the same procedure as in part (b) yields the value for a

$$\frac{a}{h} = 1.2$$

$$a = 70 \times 1.2$$

$$= 84 \text{ in.} \quad \text{or} \quad 7 \text{ ft}, 0 \text{ in.}$$

The maximum a/h from Equation (9.7) and the AISC Manual, Section F-5, is

$$\frac{a}{h} = \left(\frac{260}{h/t}\right)^2$$

$$= 2.64 > 1.2$$

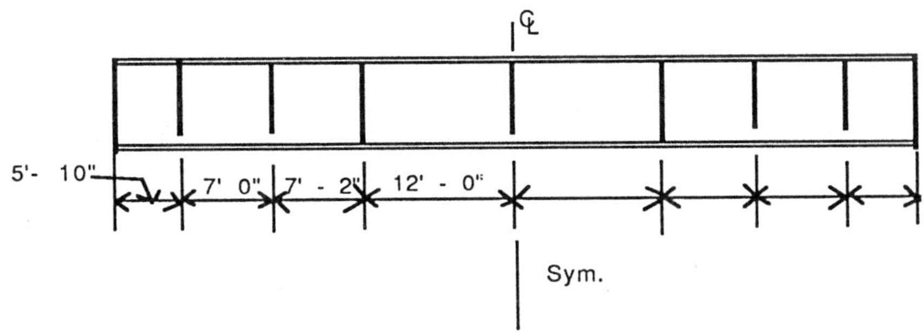

Figure 9.4 Stiffener spacing for plate girder.

Check the distance between the second stiffener and the concentrated load for the maximum allowable a/h

$$\frac{a}{h} = \frac{86}{70}$$
$$= 1.23 < 2.64$$
$$V = 113.6 \text{ k}$$
$$f_v = \frac{113.6}{30.62}$$
$$= 3.71 \text{ k/in.}^2 < 4.9 \text{ k/in.}^2 \quad \text{(AISC Manual, Table 1-36)}$$
$$F_v = 4.9 \text{ k/in.}^2$$

Figure 9.4 shows the beam with stiffeners.

Combined Shear and Tensile Stress

Check the interaction at the concentrated load in the tension field panel. Plate girder webs that depend on tension field action, as provided in Equation G3-1 in the AISC Manual, shall be proportioned such that the bending tensile stress, due to moment in the plane of the girder web, shall not exceed $0.60F_y$ nor

$$F_b = \left(0.825 - 0.375\frac{f_y}{F_y}\right)F_y$$
$$= \left(0.825 - 0.375 \times \frac{3.71}{4.9}\right) \times 36$$
$$= 19.48 \text{ k/in.}^2$$
$$f_b = 18.6 \text{ k/in.}^2 < 19.48 \text{ k/in.}^2$$

Stiffener Size

For an Intermediate Stiffener

(a) *Area required (single plate stiffener).* The size of the stiffener is determined from Equation (9.8) (AISC Manual, Section G4-1)

$$A_{st} = \frac{1 - C_v}{2}\left[\frac{a}{h} - \frac{\left(\frac{a}{h}\right)^2}{\sqrt{1 + \left(\frac{a}{h}\right)^2}}\right] YDht_w$$

Since the shear force is maximum at the support, the size of the stiffener is controlled by the shear at the support. The ratio of the distance between the support and first intermediate stiffener and the height of the web $a/h = 1$

$$k_v = 5.34 + \frac{4}{\left(\frac{a}{h}\right)^2} = 9.34$$

$$C_v = \frac{45{,}000}{36(160)^2} \times 9.34 = 0.456$$

For the same material in the web and the stiffener, $Y = 1$. For a single plate stiffener, $D = 2.4$

$$A_{st} = \frac{1 - 0.456}{2}\left(1 - \frac{1}{\sqrt{2}}\right) \times 1 \times 2.4 \times 30.62$$

$$= 5.85 \text{ in.}^2$$

Try one bar $\frac{7}{8} \times 8$

$$A_{st} = 7.0 \text{ in.}^2 > 5.85 \text{ in.}^2$$

(b) *Check the width-thickness ratio.*

$$\frac{8}{0.875} = 9.1 < 15.8$$

(c) *Check the moment of inertia.*

$$I_{\text{required}} = (70/50)^4 = 3.84 \text{ in.}^4$$

$$I_{\text{proved}} = \tfrac{1}{3}(0.875 \times 8^3) = 149.3 \text{ in.}^4 > 3.84 \text{ in.}^4$$

(**d**) *The minimum length required.* From the AISC Manual, Section G-4, the length of the stiffener is $70 - \frac{1}{4} - (6 \times 0.4375) = 67.125$ in. ($67\frac{1}{8}$). Use for intermediate stiffeners one plate $\frac{7}{8} \times 8$ in. $\times$ 5 ft and $7\frac{1}{8}$ in. fillet-weld to the compression flange and web.

Design Bearing Stiffeners (AISC Manual, Section K1.8)

For the end of the girder stiffeners, design for the end reaction. Try two plates, $\frac{7}{8} \times 8$ in. bars.

(**a**) *Check the width-thickness ratio (AISC Manual, Section B5.1).*

$$\frac{8}{0.875} = 9.1 < 15.8$$

(**b**) *Check the compressive stress (AISC Manual, Section K1.8).*

$$I = 0.875 \times \frac{16.4375^3}{12} = 323.8 \text{ in.}^4$$

$$A_{\text{eff}} = 2 \times 8 \times 0.875 + 12 \times 0.4375^2$$

$$= 16.3 \text{ in.}^2$$

$$r = \sqrt{\frac{323.8}{16.3}} = 4.46 \text{ in.}$$

$$\frac{Kl}{r} = \frac{0.75 \times 70}{4.46}$$

$$= 11.8$$

Allowable stress

$$F_a = 21.06 \text{ k/in.}^2$$

$$f_a = \frac{169.6}{16.3} = 10.40 \text{ k/in.}^2 < 21.06 \text{ k/in.}^2$$

Use for the bearing stiffeners at the end of the girder and under the concentrated loads, two plates $\frac{7}{8} \times 8$ in. $\times$ 5 ft. and $9\frac{3}{4}$ fillet-weld. See Figure 9.5.

Example 9.2

Repeat the same problem as in Example 9.1, but use a ratio of

$$\frac{h}{t_w} = 322 \quad \text{(AISC Manual, Section G1-1)}$$

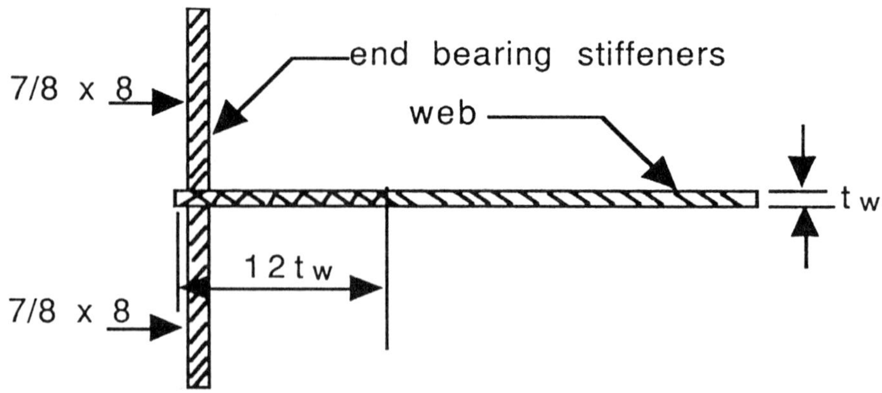

Figure 9.5 End bearing stiffener for plate girder.

Solution

1. Trial Section Selection

Allow a reduction in the compression flange

$$\frac{h}{t_w} = \frac{14{,}000}{\sqrt{F_y(F_y + 16.5)}} = 322 \quad \text{(AISC Manual, Section G1-1)}$$

Recall that the required $S_x = 1654.7$ in.^{3}. Given a certain girder section, its section modulus is

$$S_x = \left(A_f + \frac{1}{6}A_w\right)h$$

Using the ratios of $h/t_w = 322$ and $b = h/6$ yields

$$h = 8.62\sqrt[3]{S_x} = 8.62\sqrt[3]{1654.7} = 101.7 \text{ in.}$$

Use $h = 96$ in.

$$t_w = \frac{96}{322} = 0.2981 \text{ in.} \quad (\text{use } t_w = \tfrac{5}{16} = 0.3125 \text{ in.})$$

$$\frac{h}{t_w} = \frac{96}{0.3125} = 307$$

$$t_f = 2t_w = \tfrac{5}{8} \text{ in.}$$

2. Section Properties

For the calculations of section properties, refer to Figure 9.6 and Table 9.2.

(a) *Section modulus from the above table.*

$$S_x = \frac{80{,}640}{48}$$
$$= 1680 \text{ in.}^4$$

(b) *Approximate method.*

$$S_x = \left(A_f + \frac{1}{6}A_w\right)h = (12.5 + 5) \times 96 = 1680 \text{ in.}^4$$

Note that both methods yield identical results.

3. Check Stresses

Check Flange Stress

(a) *Check the bending stress in a 24-ft panel.* The maximum bending stress at midspan is

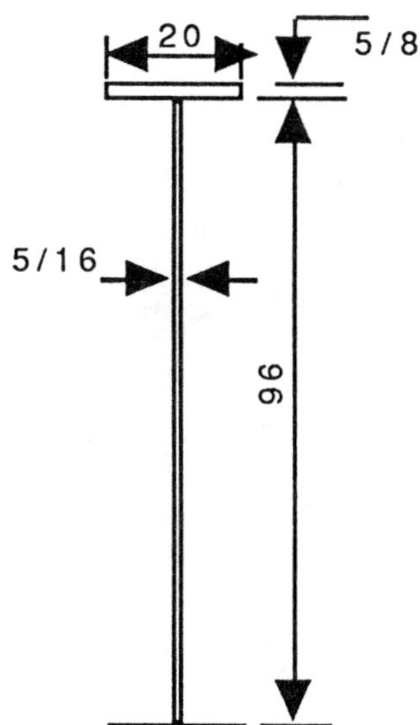

Figure 9.6 Plate girder section dimensions.

TABLE 9.2 Section Properties for Example 9.2

Section	A (in.^{2})	y (in.)	$\sum_{i=1}^{n} A_i y_i^2$	I_0 (in.^{4})	I_{gr} (in.^{4})
Web $\frac{5}{16} \times 96$	30.00			23,040	23,040
1 flange $\frac{5}{8} \times 20$	12.50	48.00	28,800	1	28,800
1 flange $\frac{5}{8} \times 20$	12.50	48.00	28,800	1	28,800
Moment of inertia					80,640

$$f_b = \frac{3033.6 \times 12}{1680} = 21.7 \text{ k/in.}^2$$

The moment of inertia of flange plus $\frac{1}{6}$ of the web about the Y-Y axis is obtained as follows:

$$I_{0y} = \frac{5}{8} \times \frac{20^3}{12} = 416.7 \text{ in.}^4$$

$$\left(A_f + \frac{A_w}{6}\right) = (12.5 + 5) = 17.5 \text{ in.}^2$$

$$r_T = \sqrt{\frac{416.7}{17.5}} = 4.88 \text{ in.}$$

$M_{\max} > M_1$ and M_2 then, $C = 1.0$.

$$\frac{l}{r_T} = \frac{24 \times 12}{4.88} = 59.0$$

$$\sqrt{\frac{102 \times 10^3 C_b}{F_y}} = 53$$

$$\sqrt{\frac{510 \times 10^3 C_b}{F_y}} = 119$$

$$\sqrt{\frac{102 \times 10^3 C_b}{F_y}} \le \frac{l}{r_T} \le \sqrt{\frac{510 \times 10^3 C_b}{F_y}}$$

$$F_b = \left[\frac{2}{3} - \frac{F_y (l/r_T)^2}{1530 \times 10^3 C_b}\right] F_y = 0.66 F_y$$

Allowable stress based on lateral buckling criteria is

$$F_b = 0.60F_y = 21.6\ \text{k/in.}^2$$

The reduced allowable stress bending in the compression flange is

$$R_{PG} = 1 - 0.0005\frac{A_w}{A_f}\left(\frac{h}{t_w} - \frac{760}{\sqrt{F_b}}\right)$$

$$= 1 - 0.0005\frac{30}{12.5}\left(\frac{96}{0.3125} - \frac{760}{\sqrt{21.6}}\right) = 0.828$$

Since the same material is used in the web and flanges, $R_e = 1.0$.
The reduced allowable stress in the compression flange is

$$F_b' = F_b R_{PG} R_e = 21.6 \times 0.828 \times 1.0 = 17.88\ \text{k/in.}^2 < 21.7\ \text{k/in.}^2$$

The size of the section must be increased. This can be done by changing the thickness of the web to $\frac{7}{16}$. See Figure 9.7. The new section modulus is

$$(1.25 + \tfrac{1}{6} \times 0.4375 \times 96) \times 96 = 1872\ \text{in.}^3$$

$$f_b = \frac{3033.6 \times 12}{1872} = 19.45\ \text{k/in.}^2$$

$$\frac{h}{t_w} = \frac{96}{0.4375} = 219.4$$

$$R_{PG} = 1 - 0.0005\frac{42}{12.5}\left(\frac{96}{0.4375} - \frac{760}{\sqrt{21.6}}\right) = 0.906$$

$$F_b' = F_b R_{PG} R_e = 21.6 \times 0.906 \times 1.0 = 19.57\ \text{k/in.}^2 > 19.45\ \text{k/in.}^2$$

(b) *Check the bending stress in a 20-ft panel.* The maximum bending stress is

$$f_b = \frac{2832 \times 12}{1872} = 18.15\ \text{k/in.}^2$$

$$C_b = 1.75 + 1.05\frac{M_1}{M_2} + 0.3\left(\frac{M_1}{M_2}\right)^2$$

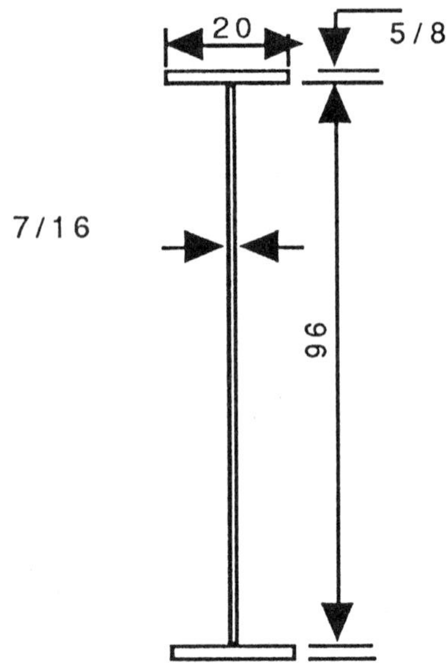

Figure 9.7 Plate girder section.

where $M_1 = 0$; then, $(M_1/M_2) = 0$ and $C_b = 1.75$

$$(A_f + \tfrac{1}{6}A_w) = 12.5 + 7 = 19.5 \text{ in.}^2$$

$$I_{0y} = 416.7 \text{ in.}^4$$

$$r_T = \sqrt{\frac{416.7}{19.5}} = 4.62 \text{ in.}$$

$$\frac{l}{r_T} = \frac{20 \times 12}{4.62} = 51.9 < 53\sqrt{C_b} = 70.1$$

Allowable stress in a 20-ft panel based on lateral buckling criteria is

$$F_b = 0.60F_y = 21.6 \text{ k/in.}^2$$

The reduced allowable bending stress in the compression flange is

$$F_b' = 19.57 \text{ k/in.}^2 > 18.15 \text{ k/in.}^2$$

The trial section conforms to AISC specifications. Use for the web, one plate $\frac{7}{16} \times 96$ in., and for flanges, two plates $\frac{5}{8} \times 20$ in.

4. Stiffener Requirements

Bearing Stiffeners

(a) *Bearing stiffeners are required at both ends of the girder.*

(b) *Check the bearing under concentrated loads.* Assume point bearing and $\frac{1}{4}$-in. web-to-flange welds.

Local web yielding

$$\frac{R}{t_w(N + 5k)} \leq 0.66F_y \quad \text{(AISC Manual, Section K1-2)}$$

$$k = \tfrac{5}{8} + \tfrac{1}{4} = 0.875 \text{ in.}$$

Since point bearing is assumed, $N = 0$.

$$\frac{80}{0.4375(0 + 5 \times 0.875)} = 41.80 \text{ k/in.}^2\text{, which is greater than } 0.66F_y$$

$$= 23.8 \text{ k/in.}^2$$

Provide bearing stiffeners under concentrated loads.

Intermediate Stiffeners

(a) *Check the shear stress in the unstiffened end panel.*

$$\frac{h}{t_w} = \frac{96}{0.4375} = 220$$

$$\frac{a}{h} = \frac{20 \times 12}{96} = 2.5$$

From AISC Table 1-36 and by interpolation, we obtain

$$F_v = 2.0 \text{ k/in.}^2$$

$$f_v = \frac{169.6}{0.4375 \times 96} = 4.04 \text{ k/in.}^2 > 2.0$$

Use intermediate stiffeners.

(b) *End panel stiffener spacing (tension field action not permitted, AISC Manual, Table* 1-36.

$$F_v \geq f_v = 4.05 \text{ k/in.}^2$$

$$\frac{h}{t_w} = 220 \qquad \frac{a}{h} = 0.7, \qquad F_v = 4.8 \text{ k/in.}^2$$

$$a = 0.7 \times 96 = 67.2 \text{ in.}$$

Use $a = 5$ ft, 7 in.

(c) *Check for additional stiffeners.* Shear at the first intermediate stiffener is

$$V = 169.6 - 2.8 \times 5.83 = 154 \text{ k}$$

$$f_v = \frac{154}{0.4375 \times 96} = 3.67 \text{ k/in.}^2$$

Using Table 1-36 and keeping $F_v \geq 3.67$ require a ratio a/h of 0.85 by interpolation

$$a = 0.85 \times 96 = 81.6 \text{ in.}$$

Use $a = 81$ in. (6 ft, 9 in.).

(d) *Check the panel between the second intermediate stiffener and concentrated load.* Shear at the second intermediate stiffener is

$$V = 169.6 - 2.8 \times 12.33 = 135.1 \text{ k}$$

$$f_v = \frac{135}{0.4375 \times 96} = 3.22 \text{ k/in.}^2$$

Keeping $F_v \geq 3.22$ requires a ratio of 0.95

$$a = 0.95 \times 96 = 91.2 \text{ in.} \quad (7 \text{ ft, } 7.2 \text{ in.})$$

The distance between the second stiffener and concentrated load is 7 ft, 8 in. Therefore, no additional stiffener is required.

(e) *Check the middle panel for stiffeners.*

$$f_v = \frac{33.6}{0.4375 \times 96} = 0.80 \text{ k/in.}^2$$

Without a stiffener in the 24-ft panel, the required ratio $a/h = 3$. This means that there is an allowable shear stress F_v of 1.7 k/in.$^2 \geq 0.80$ k/in.^{2}.

Combined Shear and Tension Stress

Check interaction at the concentrated load in the tension field panel

$$f_v = \frac{113.6}{0.4375 \times 96} = 2.70 \text{ k/in.}^2$$

$$f_b = \frac{2832 \times 12}{1872} = 18.2 \text{ k/in.}^2$$

$$F_v = 3.20 \text{ k/in.}^2$$

$$\left(\text{from Table 1-36, } \frac{a}{h} = 0.85 \text{ and } \frac{h}{t_w} = 322\right)$$

The allowable bending stress is

$$F_b = \left(0.825 - 0.375 \times \frac{2.7}{3.20}\right) \times 36 = 18.31 \text{ k/in.}^2 > 18.2$$

A summary of space stiffeners is given below.

Stiffener Size

Intermediate Stiffeners

(a) *Area required using a single plate.*

$$A_{st} = \frac{1 - C_v}{2}\left[\frac{a}{h} - \frac{(a/h)^2}{\sqrt{1 + (a/h)^2}}\right] YDht_w$$

$$C_v = \frac{45{,}000 k_v}{F_y (h/t_w)^2}$$

$$k_v = 4 + \frac{5.34}{(a/h)^2}$$

For the first intermediate stiffener,

$$\frac{a}{h} = 0.7$$

$$k_v = 15.0$$

$$C_v = \frac{45{,}000 \times 15}{36(220)^2} = 0.387$$

$$A_{st} = \frac{1 - 0.387}{2}\left[0.70 - \frac{(0.7)^2}{\sqrt{1 + 0.7^2}}\right]YDht_w$$

$$= 0.0915 \times YDht_w$$

$$Y = 1 \text{ for the same material in web and stiffener}$$

$$D = 2.4 \text{ for single plate stiffener}$$

$$A_{st} = 0.0915 \times 1.0 \times 2.4 \times 96 \times 0.4375 = 9.224 \text{ in.}^2$$

The width of the flange is 20 in. The width of the stiffener is obtained by subtracting the thickness of the web from the width of the flange and then dividing the remainder by 2

$$b_{st} = \tfrac{1}{2}(b - t_w) = \tfrac{1}{2}(20 - 0.4375) = 9.78$$

Use one bar $1\frac{1}{8} \times 9$ in.

$$A_{st} = 1.125 \times 9 = 10.125 \text{ in.}^2 > 9.78 \text{ in.}^2$$

(b) *Check the width-thickness ratio.*

$$\frac{9}{1.125} = 8.0 < 15.8$$

(c) *Check the moment of inertia of the stiffener.*

$$I_{\text{required}} = (96/50)^4 \quad \text{(AISC Manual, Section G4-1)}$$

$$I_{\text{required}} = 13.60 \text{ in.}^4$$

$$I_{\text{furnished}} = \tfrac{1}{3} \times 1.125 \times (9)^3 = 273.4 \text{ in.}^4 > 13.6 \text{ in.}^4$$

(d) *Minimum length required.* The end of the stiffener should not be closer than four times nor more than six times the thickness of the web from the near toe of the web-to-flange weld. With a $\frac{1}{4}$ weld, the stiffener length is

$$96 - \tfrac{1}{4} - 6 \times (\tfrac{7}{16}) = 93.125 \text{ in.} \quad 7 \text{ ft}, 9\tfrac{1}{8} \text{ in.}$$

See Figure 9.8a for final stiffener locations.

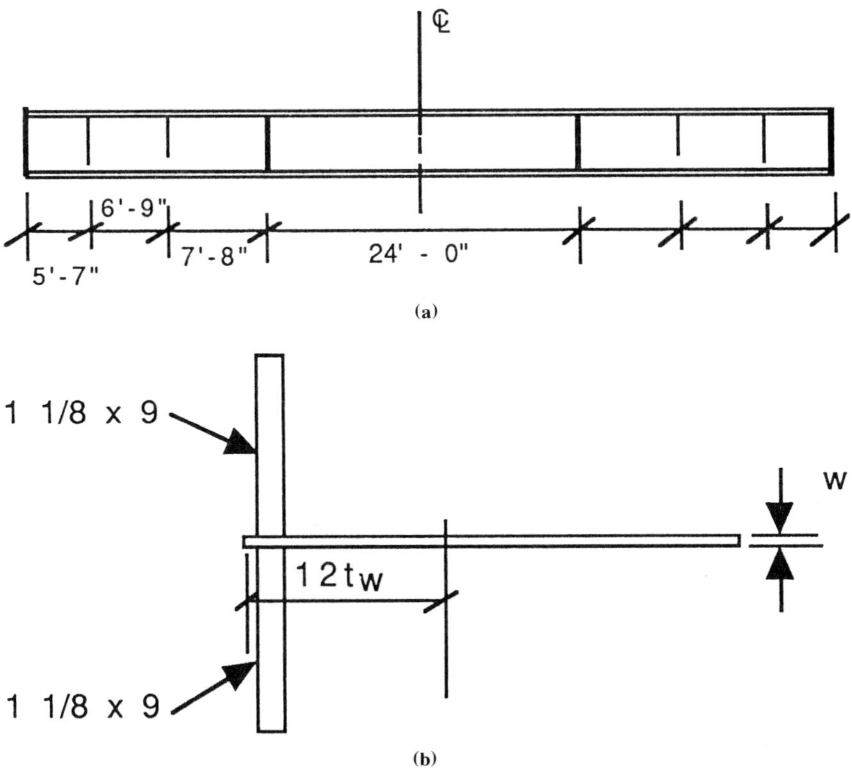

Figure 9.8 End bearing stiffener. (a) Modified dimensions. (b) Stiffener spacing.

Design Bearing Stiffeners

At the end of the girder, design for an end reaction. Try two bars, $1\frac{1}{8} \times$ 9 in.

(a) *Check the width-thickness ratio.*

$$\frac{b}{t_w} = \frac{9}{1.125}8 < 15.8$$

(b) *Check the compressive stress.*

$$I = \frac{1.125}{12} \times (18.4375)^3 = 587.4 \text{ in.}^4$$

$$A_{\text{eff}} = 2 \times 9 \times 1.125 + 6 \times 0.4375 \times 0.4375$$

$$= 22.55 \text{ in.}^2$$

The radius of gyration r is given by

$$r = \sqrt{\frac{I}{A}} = \sqrt{\frac{587.4}{22.55}} = 5.104 \text{ in.}$$

$$Kl = 0.75 \times 96 = 72 \text{ in.}$$

$$\frac{Kl}{r} = \frac{72}{5.104} = 14.1$$

The allowable compressive stress is

$$F_a = 20.9 \text{ k/in.}^2 \quad \text{(AISC Manual, Table C-36)}$$

The compressive stress at the support is

$$f_a = \frac{169.6}{22.55} = 7.52 \text{ k/in.}^2 < 20.9 \text{ k/in}^2$$

Use for bearing stiffeners two plates $1\frac{1}{8} \times 9$ in.$\times$ 7 ft with $9\frac{1}{4}$ in. fillet weld. See Figure 9.8b. Use the same-size stiffeners under concentrated loads.

Example 9.3

Given the frame as shown in Figure 9.9, design a plate girder to carry the building loads above the first floor. The frames are located 30 ft

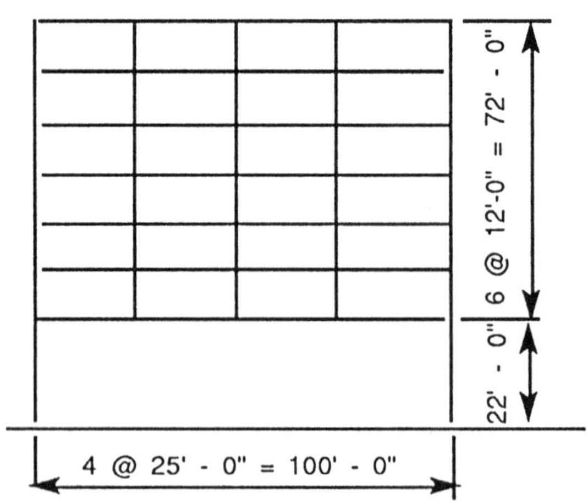

Figure 9.9 Seven-story building with no interior columns on first floor.

on center. The total dead and live load on a typical floor and the roof is 140 $\mathrm{lb/ft^2}$.

Solution

First, determine the load at the bottom of each interior column. The tributary area is

$$A = 30 \times 25$$
$$= 750 \text{ ft}^2$$

The dead load can be assumed to be 55 $\mathrm{lb/ft^2}$, and that leaves a combination of live loads of 85 $\mathrm{lb/ft^2}$. The next step is to determine the reduction factor for the live loads from the equation

$$L = L_0\left(0.25 + \frac{15}{\sqrt{A_I}}\right)$$

where A_I is four times the tributary area of the column, and L and L_0 are the reduced and initial live loads, respectively. L should not be less than 40% of L_0. Since each interior column carries loads from six levels, the tributary area must be multiplied by 6. Hence,

$$L = L_0\left(0.25 + \frac{15}{\sqrt{6 \times 4 \times 750}}\right)$$
$$= 0.362L_0 \quad (\text{use } 0.40L_0)$$
$$= 85 \times 0.40$$
$$= 34 \text{ lb/ft}^2$$

The sum of the dead load and reduced live load is 89 $\mathrm{lb/ft^2}$. The load at the bottom of each interior column in kips is

$$P = 750 \times 0.089 \times 6$$
$$= 400 \text{ k}$$

The uniform load on the plate girder will come from the weight of the girder itself and the loads from the second floor. The reduction factor for a beam is obtained by considering that the influence A_I is twice the tributary area. The tributary area for the girder is 100×30 ft^2. Then,

$$L = L_0\left(0.25 + \frac{15}{\sqrt{2 \times 3000}}\right)$$
$$= 0.444 L_0$$

The uniform load on the girder is

$$w = 55 + 0.444 \times 85$$
$$= 92.7 \text{ lb/ft}^2$$

Find the maximum bending moment. See Figure 9.10. It is obvious from the above that the maximum moment occurs at the middle of the span9.10

$$M_{\text{max}} = 400 \times 1.5 \times 50 - 400 \times 25 + 2.782 \times \frac{100^2}{8}$$
$$= 23{,}480 \text{ k.ft}$$

The required section modulus to resist this moment for A-36 steel is

$$S_x = 11{,}740 \text{ in.}^3$$

Assume the weight of the girder to be on the order of 0.6 k/ft. The maximum moment from the weight of the girder is

$$m_{\text{max}} = 0.6 \times \frac{100^2}{8}$$
$$= 750 \text{ ft k}$$

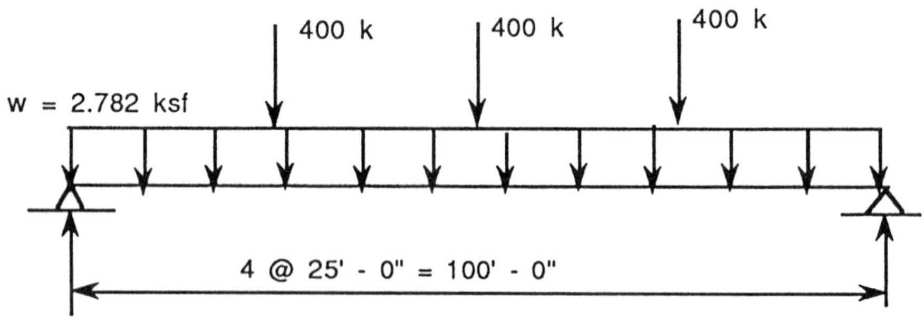

Figure 9.10 Loading on the plate girder for the first floor.

The required section modulus for resisting this moment is

$$S_x = 375 \text{ in.}^3$$

The total required section modulus is the sum of the two

$$S_x = 12{,}115 \text{ in.}^3$$

A girder section can be assumed; it is then checked. Since the height of the girder is limited to 6 ft, the dimensions of the flange and web thickness and the flange width must be relatively large. Begin by assuming a section as shown in Figure 9.11.

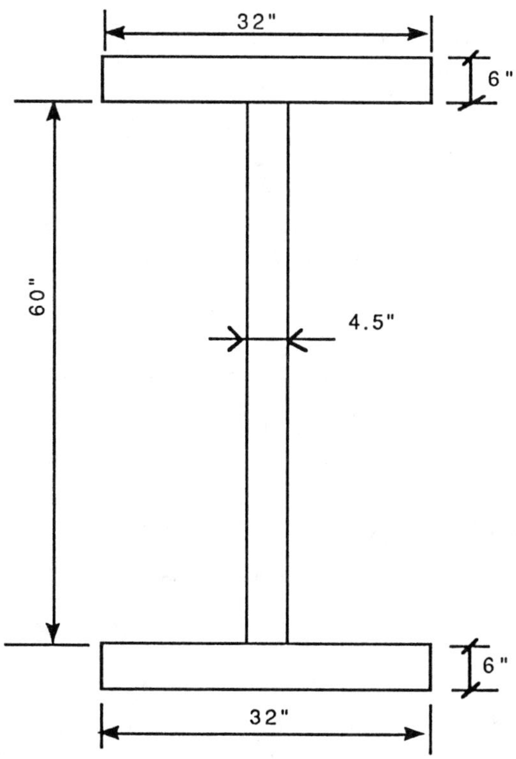

Figure 9.11 Trial dimensions for the plate girder.

The moment of inertia for this section is

$$I = 2 \times 32 \times 6 \times 33^2 + 4.5 \times \frac{60^3}{12}$$

$$= 499{,}176 \text{ in.}^4$$

The section modulus is given by the expression

$$S_x = \frac{I}{h/2}$$

$$= 13{,}866 \text{ in.}^3 > 12{,}115$$

Check the weight of the section.

The area of the section is

$$A = 2 \times 32 \times 6 + 4.5 \times 60$$

$$= 654 \text{ in.}^2$$

Steel weighs 490 lb/ft^3. Then the weight of the section is

$$w = \frac{654}{144} \times 0.490$$

$$= 2.225 \text{ k/ft}$$

Notice that the assumed weight of the girder is 0.6 k/ft. The difference between the actual and assumed weight is

$$\Delta w = 1.625 \text{ k/ft}$$

$$\Delta m = 1.625 \times \frac{100^2}{8}$$

$$= 2030 \text{ ft k}$$

$$\Delta S_x = 1016 \text{ in.}^3$$

The adjusted section modulus is

$$S_x = 1016 + 12{,}115$$

$$= 13{,}131 \text{ in.}^3 < 13{,}866$$

Example 9.4

Given a span of 60 ft and moment due to superimposed loads of 3850 kip ft, design a plate girder in accordance with the AISC specifications using the ASD method. The depth of the girder is limited to 8 ft.

Solution

In addition to the given moment, the weight of girder itself exerts an influence that must be added to the moment. An increment of 5% on top of the given moment will accommodate the contribution due to the weight of the girder. Hence, the total moment for the purpose of selection is

$$M_{\text{total}} = 4042 \text{ k ft}$$

If we assume an allowable stress of 22 k/in.^{2}, the required section modulus is

$$S_{x,\text{required}} = 2205 \text{ in.}^3$$

The general equation for the section modulus can be expressed as in Equation (9.15)

$$S_x = \left(A_f + \frac{A_w}{6}\right)h$$

where h is an average value between the depth of the web and the total depth of the beam. For a plate girder to be noncompact, the range of values for h/t_w is as follows:

$$162 < \frac{h}{t_w} < 322$$

Assume a ratio of 190

$$\frac{h}{t_w} = 190$$

or

$$t_w = \frac{h}{190}$$

For noncompact sections, the ratio for the projected part of the flange and its thickness is obtained from AISC Table B5.1

$$\frac{b}{2t_f} = \frac{95}{\sqrt{\dfrac{F_y}{k_c}}}.$$

where

$$k_c = \frac{4.05}{\left(\frac{h}{t_w}\right)0.46}$$

for all values of h/t_w greater than 70. Then k_c for the above ratio is

$$k_c = 0.362$$

and

$$\frac{b}{2t_f} = 9.5$$

or

$$\frac{b}{t_f} = 19$$

If we assume that $t_f = t_w$ provides a relationship between h and b,

$$\frac{h}{b} = 5$$

or

$$b = \frac{h}{5}$$

The expression for S_x can now be put strictly in terms of h

$$\begin{aligned} S_x &= \left(A_f + \frac{A_w}{6}\right)h \\ &= \left(bt_f + \frac{t_w h}{6}\right)h \\ &= 6.947\sqrt[3]{S_x} \end{aligned}$$

We are given the fact that S_x is required as above to yield a value for h of 90.7 in. Use a value of $h = 90$ in. From this, the values of b, t_f, and t_w are found to be 18, 1, and 0.5 in. The calculated S_x turns out to be 2280 in.^{3}, which is greater than the required value of 2205 in.^{3}

Another method for determining h is to use the required S_x and an $r = h/t_w$ ratio smaller than 200. Examine the graph in Figure 9.12 and note the required depth.

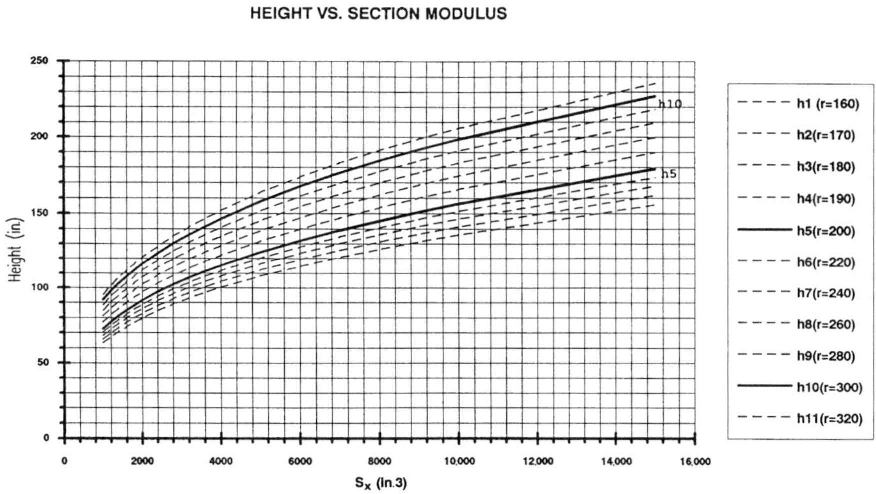

Figure 9.12 Trial height charts for plate girders.

Having determined the size of the section, we can next find the actual weight of the plate girder. The cross-sectional area is

$$\begin{aligned} A &= 18 \times 1 \times 2 + 0.5 \times 89 \\ &= 80.5 \text{ in.}^2 \end{aligned}$$

The weight in kips per foot of the beam is

$$\begin{aligned} w &= \frac{80.5}{144} \times 0.490 \text{ k/ft} \\ &= 0.273 \text{ k/ft} \end{aligned}$$

The moment due to the weight of the beam is

$$\begin{aligned} M &= 0.274 \times \frac{60^2}{8} \\ &= 219 \text{ k/ft} \end{aligned}$$

The actual total moment is

$$\begin{aligned} M_{\text{tot}} &= 3850 + 219 \\ &= 4069 \text{ k/ft} \end{aligned}$$

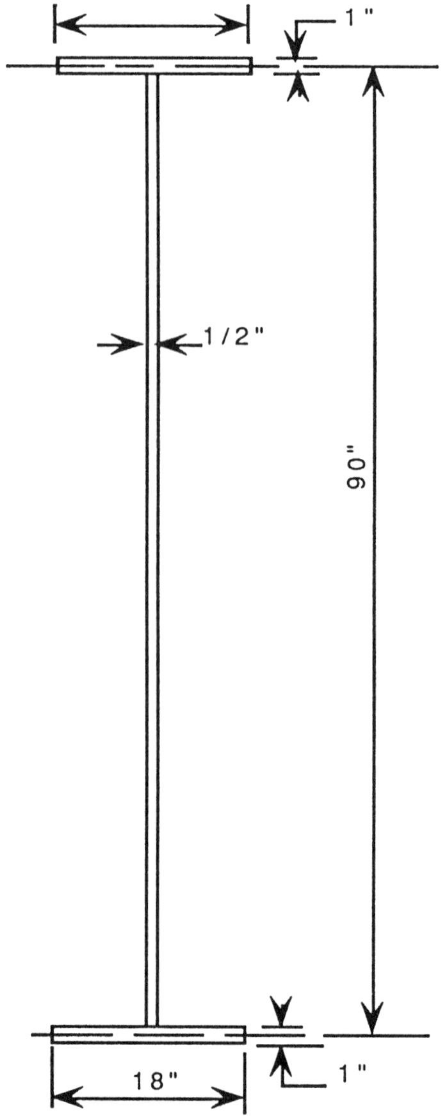

Figure 9.13 Final dimensions for the plate girder.

Check the bending stress in the flange

$$
\begin{aligned}
f_b &= \frac{M}{S_x} \\
&= \frac{4069 \times 12}{2280} \\
&= 21.42 \text{ k/in.}^2
\end{aligned}
$$

A value for the allowable bending stress of 22 k/in.^{2} having been used, it is necessary to check for the reduced value of the allowable stress in the bending of plate girders. From the AISC Manual, Section G2-1, the reduced value of the allowable stress is

$$
\begin{aligned}
F_b' &= F_b R_{PG} R_e \\
R_{PG} &= 1 - 0.0005 \frac{A_w}{A_f}\left(\frac{h}{t_w} - \frac{760}{\sqrt{F_b}}\right) \\
R_e &= 1 \text{ for nonhybrid girder}
\end{aligned}
$$

For the above section.

$$
\begin{aligned}
R_{PG} &= 0.965 \\
F_b' &= 0.965 \times 22 \\
&= 21.24 \text{ k/in.}^2
\end{aligned}
$$

F_b' is slightly lower than the flexural stress calculated in the flange of the girder due to the loads. This slight difference may be ignored, or a slight increase in the size of the section may be effected. The final section is shown in Figure 9.13.

10

Composite Construction

10.1 INTRODUCTION

The slab and beam type of construction is commonly used in bridge and building construction. In recent years, such construction has been improved by letting the slab participate with the steel section in carrying loads. The slab–beam interaction is made possible through the use of shear connectors that are welded to the top flange. Figure 10.1 shows various types of shear connectors.

Composite structures are stronger and stiffer than noncomposite ones in which the steel section carries the given loads independently of the concrete slab. There are a number of advantages to using composite construction:

(**a**) Savings in steel by using lighter sections.
(**b**) Savings in overhead space by using shallower sections.
(**c**) For long spans, savings in the structural steel are significant.

10.2 DESIGN CONCEPTUALIZATION AND ASSUMPTIONS

In composite construction, the slab is mobilized by means of shear connectors to act with the steel section in resisting the given loads. Thus, the steel

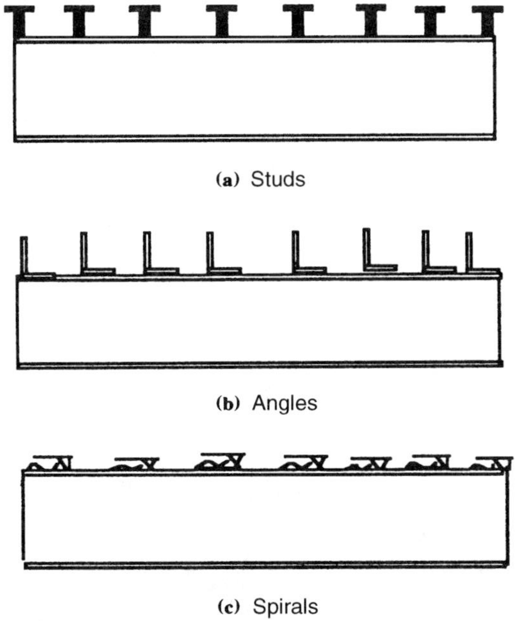

Figure 10.1 Types of shear connectors. (a) Studs. (b) Angles. (c) Spirals.

beam behaves like a T-beam with the slab or part of it acting as a flange in compression. The effective width of the slab that is mobilized to act with the steel beam is obtained in accordance with the ACI code. The effective width b_E is equal to the smallest of the following:

(a) One-fourth the length of the beam
(b) The distance from center to center of the beams
(c) Sixteen times the thickness of the slab plus the width of the flange of the steel beam

When the beam is an exterior one, the slab exists only on one side of it. The effective width is obtained from the following:

(a) One-twelfth the span of the beam
(b) One-half the center-to-center distance between the beams
(c) Six times the thickness of the slab plus the width of the flange of the steel section.

To illustrate the above assumptions, see Figure 10.2a and 10.2b.

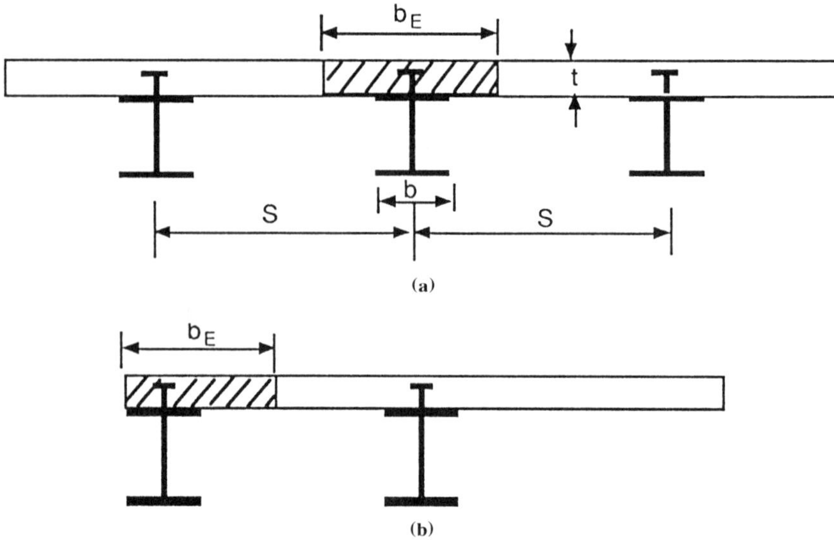

Figure 10.2 Section showing composite construction.

The American Association of State Highway and Transportation Officials (AASHTO) requirements are the same a those specified by the American Concrete Institute (ACI), except for the effective width of an interior beam. The effective width of an interior beam shall not exceed 12 times the thickness of the slab.

A further assumption for composite construction is that bond between the shear connectors and slab is perfect, i.e., no slippage between the top flange of the steel beam and slab is permitted. In determining the section properties, it is more convenient to transform the concrete slab into an equivalent steel section. This is accomplished by dividing the area of the concrete section by the modular ratio n

$$n = \frac{E_s}{E_c} \tag{10.1}$$

where

E_s = modulus of elasticity of steel, 29,000 k/in.^{2}
E_c = modulus of elasticity of concrete, $57\sqrt{f'_c}$ k/in.^{2}
f'_c = compression strength of concrete (lb/in.^{2})

The rest of the analysis is carried out as if the section were made of a homogeneous material. Stresses in the composite beam must not exceed the maximum allowable stresses for each material.

Loading on the composite beam is limited to the same requirements as the homogeneous beam. The loads are:

(**a**) Dead load
(**b**) Live load
(**c**) Deformational loads such as creep, shrinkage, and thermal

The dead load from the concrete slab can be either shored or unshored. In each case, the section properties must reflect the type of construction. All dead loads placed on the structure after the concrete has attained 75% of its strength at 28 days are assumed to be carried by the composite section. For unshored construction, all other dead loads are carried by the steel section.

10.3 DEVELOPMENT OF SECTION PROPERTIES

First, consider the case of a rolled section with a cover plate attached to the bottom flange. See Figure 10.3.

A_s = area of steel section

A_p = area of cover plate

y_s = distance from center of steel section to centroid of the composite section

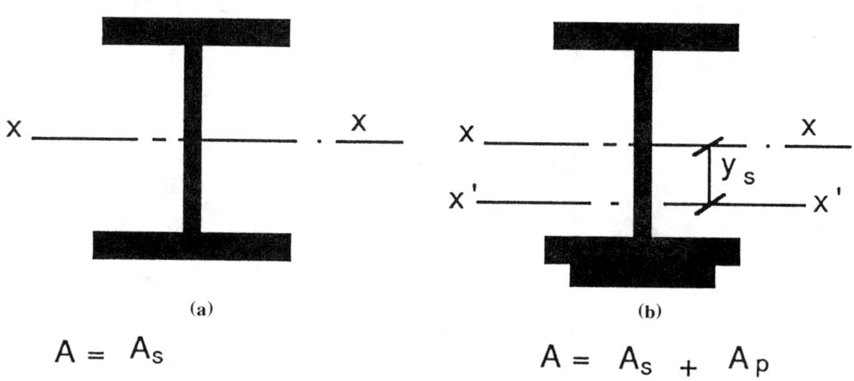

Figure 10.3 Cover plates.

From the mechanics of materials, we determine that the centroid of the section is located a distance y_s from the center of the steel section, as shown in Figure 10.3b

$$y_s = \frac{1}{2} \frac{A_p}{A_s + A_p}(d + t_p) \tag{10.2}$$

where

d = overall depth of the steel section
t_p = thickness of cover plate

The moment of inertia of the composite section is written as

$$I_c = I_s + A_s(y_s)^2 + \tfrac{1}{4}A_p(d + t_p - 2y_s)^2 \tag{10.3}$$

where

I_s = moment of inertia of the steel section
I_c = moment of inertia of the composite section

The section modulus of the beam is given by

$$S_{xct1} = \frac{I_{c1}}{y_{t1}}$$

Referring to Figure 10.4, we can express the positions of the extreme fibers in the composite section as follows:

$$y_{ts} = \frac{d}{2} + y_s \tag{10.4}$$

$$y_{bs} = \tfrac{1}{2}d + t_p - y_{ts} \tag{10.5}$$

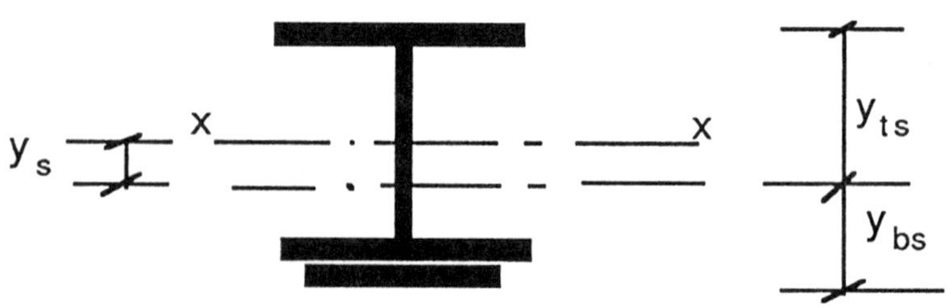

Figure 10.4 Neutral axis of W-flange sections with cover plates.

where

y_{ts} = the fiber at the top of the beam
y_{bs} = the fiber at the bottom of the cover plate

The section modulii of the beam in Figure 10.4 with respect to the top and bottom fibers are

$$S_{xts} = \frac{I_c}{y_{ts}} \tag{10.6}$$

$$S_{xbs} = \frac{I_c}{y_{bs}} \tag{10.7}$$

When concrete is made to participate in the structural function of a floor system by using shear connectors, the cover plate is equivalent to the transformed area of the concrete. The physical properties of a composite section of a steel beam and concrete slab depend on the properties of the steel and concrete components, as will be demonstrated later. In addition to the modular ratio given in Equation (10.1), another parameter k is introduced that will account for the type of loading applied to the composite structure. The parameter $k = 3$ is used with long-term dead loads, whereas $k = 1$ is used with live loads. Long-term dead loads induce creep deformation in concrete structure. The higher value of k compensates for the creep effect.

Consider the composite system shown in Figure 10.5 and introduce the same concept that has been used when a cover plate is added to the steel beam.

When the slab and steel beam act together, the centroidal axis shifts upward an amount y_c. This upward shift depends on the transformed area

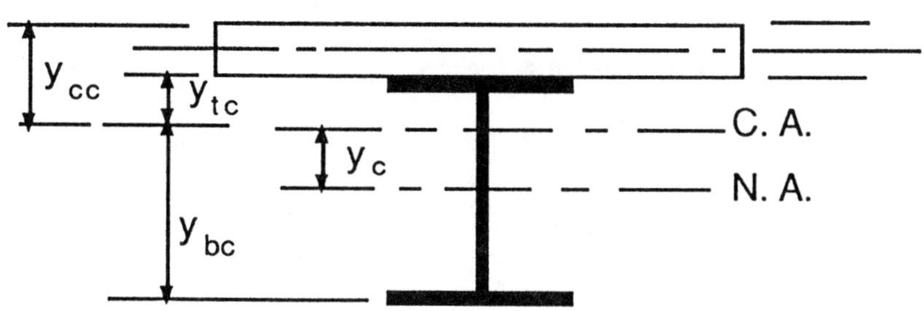

Figure 10.5 Centroidal axis of composite section.

of concrete. The section properties are determined using the step-by-step method outlined below:

1. Determine the effective slab width that acts with the steel beam

(a) $b_E = L/4$
(b) $b_E = S$
(c) $b_E = 16t + b$

where

L = the length of the beam
S = the spacing from center to center of successive beams
t = thickness of the slab
b = width of the flange of the steel beam

Use the smallest of the three values.

2. Determine the effective area of concrete to act with the steel beam

$$A_c = b_E \times t$$

3. Transform the concrete section into steel

$$A_t = \frac{A_c}{nk} \tag{10.8}$$

where

k = a load factor to account for creep in concrete
n = modular ratio as defined in Equation (10.1)

(a) For live loads, $k = 1$

$$A_{t1} = \frac{A_c}{n}$$

(b) For long-term dead loads, $k = 3$

$$A_{t2} = \frac{A_c}{3n}$$

4. Sum the areas of the various components in the composite section

$$A_1 = A_s + A_{t1} \tag{10.9a}$$

$$A_2 = A_s + A_{t2} \tag{10.9b}$$

5. Determine the location of the centroidal axis for the composite section

$$y_{c1} = \frac{1}{2}\frac{A_{t1}}{A_{t1} + A_s} \times (d + t) \tag{10.10a}$$

$$y_{c2} = \frac{1}{2}\frac{A_{t2}}{A_{t2} + A_s} \times (d + t) \tag{10.10b}$$

where

d = depth of the steel beam

6. Determine the moment of inertia for the composite section

$$I_{c1} = I_s + A_s y_{c1}^2 + \frac{b'_1 t^3}{12} + \frac{b'_1 t}{4}(d + ty_{c1})^2 \tag{10.11a}$$

$$I_{c2} = I_s + A_s y_{c2}^2 + \frac{b'_2 t^3}{12} + \frac{b'_2 t}{4}(d + ty_{c2})^2 \tag{10.11b}$$

where

I_s = moment of inertia of steel section about its centroidal axis
$b'_{1,2}$ = the transformed width of the slab for cases 1 and 2, respectively
$y_{c1,2}$ = the shift in the centroidal axis for cases 1 and 2, respectively

7. Determine the distances from the centroidal axis for the composite section to all critical points in the section

$$y_{b1} = \frac{d}{2} + y_{c1} \tag{10.12}$$

$$y_{t1} = \frac{d}{2} - y_{c1} \tag{10.13}$$

$$y_{cc1} = \frac{d}{2} - y_{c1} + t \tag{10.14}$$

$$y_{b2} = \frac{d}{2} + y_{c2} \tag{10.15}$$

$$y_{t2} = \frac{d}{2} - y_{c2} \tag{10.16}$$

$$y_{cc2} = \frac{d}{2} - y_{c2} + t \tag{10.17}$$

8. Determine the section modulus for each component of the composite section

$$S_{xcb1} = \frac{I_{c1}}{y_{b1}} \tag{10.18}$$

$$S_{xct1} = \frac{I_{c1}}{y_{t1}} \tag{10.19}$$

$$S_{xcc1} = \frac{I_{c1}}{y_{c1}} \tag{10.20}$$

$$S_{xcb2} = \frac{I_{c2}}{y_{b2}} \tag{10.21}$$

$$S_{xct2} = \frac{I_{c2}}{y_{t2}} \tag{10.22}$$

$$S_{xcc2} = \frac{I_{c2}}{y_{c2}} \tag{10.23}$$

9. Determine the fiber stresses in the composite section

The final stress in any fiber across the cross section of the beam is the sum of all stress components. Thus,

$$f_b = f_{bs} + f_{bc1} + f_{bc2} \tag{10.24}$$

where

f_{bs} = the stress in the bottom flange carried by the steel section before composite action takes place

f_{bc1} = the stress in the bottom flange due to live loads carried by the composite action of the section

f_{bc2} = the stress in the bottom flange due to long dead loads carried by the composite of the section

Given the various moment components in the beam due to different loadings, the expressions for the fiber stresses are as follows:

$$f_{bs} = \frac{M_s}{S_{xs}} \tag{10.25}$$

$$f_{bc2} = \frac{M_D}{S_{xc2}} \tag{10.26}$$

$$f_{bc1} = \frac{M_l}{S_{xc1}} \tag{10.27}$$

Example 10.1

Given the floor plan as shown in Figure 10.6, calculate the section properties for composite action.

Solution

Use the step-by-step method outlined in the text above.

Step 1. Determine the width b_E

(a) $L/4 = 72$ in.
(b) $S = 72$ in. (center-to-center spacing of beams)
(c) $b + 16t = 62$ in. controls

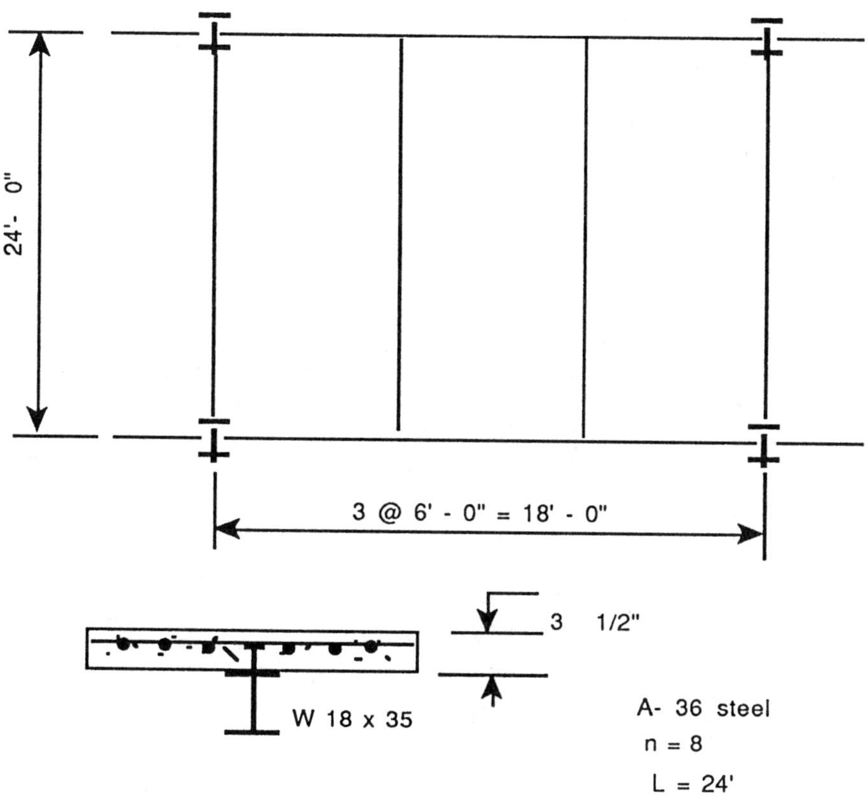

Figure 10.6 Typical floor frame: Interior bay.

Step 2. Determine the effective area of the slab that will act with the steel section

$$\begin{aligned} A_c &= 62 \times 3.5 \\ &= 217 \\ &= in.^2 \end{aligned}$$

Step 3. Transform the concrete section into steel

$$\begin{aligned} A_{t1} &= \frac{A_c}{nk}, \qquad k = 1 \\ &= \frac{62 \times 3.5}{8} \\ &= 27.1 \text{ in.}^2 \\ A_{t2} &= \frac{62 \times 3.5}{8 \times 3}, \qquad k = 3 \\ &= 9.0 \text{ in.}^2 \end{aligned}$$

Step 4. Sum the areas of the various components in the composite section

$$\begin{aligned} A_1 &= 27.1 + 10.3 \\ &= 37.40 \text{ in.}^2 \\ A_2 &= 9.04 + 10.3 \\ &= 19.34 \text{ in.}^2 \end{aligned}$$

Step 5. Determine the location of the centroid for the composite section

$$\begin{aligned} y_{c1} &= \frac{A_1}{A_1 + A_s} \tfrac{1}{2}(d + t) \\ &= \frac{27.1}{37.40} \times \tfrac{1}{2}(3.5 + 17.7) \\ &= 7.682 \text{ in.} \\ y_{c2} &= \frac{A_2}{A_2 = A_s} \tfrac{1}{2}(d + t) \\ &= \frac{9.04}{19.34} \times \tfrac{1}{2}(3.5 + 17.7) \\ &= 4.954 \text{ in.} \end{aligned}$$

Step 6. Determine the moment of inertia for the composite section

$$I_{c1} = I_s + A_s y_{c1}^2 + \frac{b'_1 t^2}{12} + \frac{b'_1 t}{4}(d + t - 2y_{c1})^2$$

$$= 510 + 10.3(7.682)^2 + \frac{7.75 \times 3.5^3}{12} + \frac{27.12}{4}(17.7 + 3.5 - 2 \times 7.682)^2$$

$$= 1376 \text{ in.}^4$$

$$I_{c2} = I_s + A_s y_{c2}^2 + \frac{b'_2 t^3}{12} + \frac{b'_2 t}{4}(d + t - 2y_{c2})^2$$

$$= 510 + 10.3(4.954)^2 + \frac{2.583 \times 3.5^3}{12} + \frac{9.04}{4}(17.7 + 3.5 - 2 \times 4.954)^2$$

$$= 1061 \text{ in.}^4$$

Step 7. Determine the distances from the centroidal axis for the composite section to all critical points in the section

$$y_{b1} = \frac{d}{2} + y_{c1}$$

$$= \frac{17.7}{2} + 7.682$$

$$= 16.532 \text{ in.}$$

$$y_{t1} = \frac{d}{2} - y_{c1}$$

$$= \frac{17.7}{2} - 7.682$$

$$= 1.168 \text{ in.}$$

$$y_{cc1} = \frac{d}{2} + t - y_{c1}$$

$$= \frac{17.7}{2} + 3.5 - 7.682$$

$$= 4.668 \text{ in.}$$

$$y_{b2} = \frac{d}{2} + y_{c2}$$

$$= \frac{17.7}{2} + 4.954$$

$$= 13.804 \text{ in.}$$

$$y_{t2} = \frac{d}{2} - y_{c2}$$

$$= \frac{17.7}{2} - 4.954$$

$$= 3.646 \text{ in.}$$

$$y_{cc2} = \frac{d}{2} + t - y_{c2}$$

$$= \frac{17.7}{2} + 3.5 - 4.954$$

$$= 7.396 \text{ in.}$$

Step 8. Determine the section modulus for each component of the composite section

$$S_{xcb1} = \frac{I_{c1}}{y_{b1}}$$

$$= \frac{1376}{16.532}$$

$$= 83.2 \text{ in.}^3$$

$$S_{xct1} = \frac{I_{c1}}{y_{t1}}$$

$$= \frac{1376}{1.168}$$

$$= 1178 \text{ in.}^3$$

$$S_{xcc1} = \frac{1376}{4.668}$$

$$= 294.8 \text{ in.}^3$$

$$S_{xcb2} = \frac{I_{c2}}{y_{b2}}$$

$$= \frac{1061}{13.804}$$

$$= 76.9 \text{ in.}^3$$

$$S_{xct2} = \frac{I_{c2}}{y_{t2}}$$

$$= \frac{1061}{3.646}$$

$$= 291 \text{ in.}^3$$

$$S_{xcc2} = \frac{1061}{7.396}$$

$$= 143.5 \text{ in.}^3$$

Example 10.2

Given the following loads on the composite section of Example 10.1, check the bending stresses in the composite section and make recommendations for change if needed. The construction is shored.

Loads

Weight of steel beam	= 0.035 k/ft
Weight of slab (3.5 × 12.5 × 6)/1000	= 0.262 k/ft
Partition load 20 lb/ft^2 × 6/1000	= 0.120 k/ft
Ceiling load 5 lb/ft^2 × 6/1000	= 0.030 k/ft
Total long-range dead load	= 0.412 k/ft
Live load 50 lb/ft^2 × 6/1000	= 0.300 k/ft

Corresponding bending moments

$$M_s \text{ (due to steel section)} = \frac{0.035 \times 24^2}{8}$$

$$= 2.52 \text{ ft k}$$

$$M_D \text{ (due to long-range dead loads)} = \frac{0.412 \times 24^2}{8}$$

$$= 29.66 \text{ ft k}$$

$$M_L \text{ (due to live loads)} = \frac{0.300 \times 24^2}{8}$$

$$= 21.6 \text{ ft k}$$

Solution

Since the construction is shored, the slab and other long-range dead loads will be carried by the composite section. Bending stresses are obtained as follows:

f_{bs} = the bending stress in the lower flange due to the weight of the steel section

f_{bc1} = the bending stress in the bottom flange of the composite section due to the live load

f_{bc2} = the bending stress in the bottom flange of the composite section due to long-range dead loads

f_b = the bending stress in the bottom flange of the composite section due to all the loads on the beam

$$= f_{bs} + f_{bc1} + f_{bc2}$$

$$= \frac{2.52 \times 12}{57.9} + 29.66 \times \frac{12}{76.9} + 21.6 \times \frac{12}{83.2}$$

$$= 8.263 \text{ k/in.}^2 \ll 24 \text{ k/in.}^2$$

Based on the above result in the bending stress of the bottom flange of the composite section, it is recommended that a smaller steel section be used.

If the steel beam were to carry these loads singly with composite action, the required section modulus would be

$$S_x = \frac{\sum_{n=1}^{J} M_n}{F_b}$$

$$j = 3 \text{ in this case}$$

$$S_x = \frac{53.8 \times 12}{24}$$

$$= 26.9 \text{ in.}^3$$

The section modulus of the steel beam that would meet the bending requirement of the loads need not not be more than the one shown above, but it can be smaller. Two such sections meet these requirements: W12 × 19 and M14 × 18. Try the latter section. Its physical properties are listed below:

$$A_s = 5.10 \ in.^2$$

$$d = 14.00 \text{ in.}$$

$$b = 4.00 \text{ in.}$$

$$S_{xs} = 21.1 \text{ in.}^3$$

$$I_s = 148 \text{ in.}^4$$

Step 1. Determine the effective width of the slab:

(a) $b_E = \dfrac{24 \times 12}{4}$

$= 72$ in.

(b) $b_E = 6 \times 12$

$= 72$ in.

(c) $b_E = 16 \times 3.5 + 4$

$= 60$ in. controls

Step 2. Determine the area of the concrete

$$A_c = 60 \times 3.5$$

$$= 210 \text{ in.}^2$$

Step 3. Determine the transformed area of concrete

$$A_1 = \frac{210}{1 \times 8} \quad \text{(for live loads)}$$

$$= 26.2 \text{ in.}^2 \quad \text{(for long-range dead loads)}$$

$$A_2 = \frac{210}{3 \times 8}$$

$$= 8.75 \text{ in.}^2$$

Step 4. Calculate the total composite area for the composite section

$$A_1 = 26.2 + 5.10$$

$$= 31.3 \text{ in.}^2$$

$$A_2 = 8.75 + 5.1$$

$$= 13.85 \text{ in.}^2$$

Step 5. Determine the position of the centroidal axis for the composite section

$$y_{c1} = \frac{1}{2} \times \frac{26.25}{31.3}(14.0 + 3.5)$$

$$= 7.234 \text{ in.}$$

$$y_{c2} = \frac{1}{2} \times \frac{8.75}{13.85}(14.0 + 3.5)$$

$$= 5.528 \text{ in.}$$

Step 6. Determine the moment of inertia for the composite section

$$I_{c1} = 148 + 5.1(7.237)^2 + 26.2$$

$$\times \tfrac{1}{4}(14.0 + 3.50 - 2 \times 7.237)^2 + \frac{7.5 \times 3.5^3}{12}$$

$$= 497.5 \text{ in.}^4$$

$$I_{c2} = 148 + 5.1(5.528)^2 + 8.75$$

$$\times \tfrac{1}{2}(14.0 + 3.50 - 2 \times 5.528)^2 + \frac{2.5 \times 3.5^3}{12}$$

$$= 403.6 \text{ in.}^4$$

Step 7. Determine the distances from the centroidal axis for the composite section to all critical points in the section

$$y_{b1} = \frac{d}{2} + y_{c1}$$

$$= \frac{14.0}{2} + 7.237$$

$$= 14.237 \text{ in.}$$

$$y_{t1} = \frac{d}{2} - y_{c1}$$

$$= \frac{14.0}{2} - 7.237$$

$$= -0.237 \text{ in}$$

$$y_{cc1} = \frac{d}{2} + t - y_{c1}$$

$$= \frac{14.0}{2} + 3.5 - 7.237$$

$$= 3.263 \text{ in.}$$

$$y_{b2} = \frac{d}{2} + y_{c2}$$

$$= \frac{14.0}{2} + 5.528$$

$$= 12.528 \text{ in.}$$

$$y_{t2} = \frac{d}{2} - y_{c2}$$

$$= \frac{14.0}{2} - 5.528$$

$$= 1.472 \text{ in.}$$

$$y_{cc2} = \frac{d}{2} + t - y_{c2}$$

$$= \frac{14.0}{2} + 3.5 - 5.528$$

$$= 4.972 \text{ in.}$$

Step 8. Determine the section modulus for each component of the composite section

$$S_{xcb1} = \frac{I_{c1}}{y_{b1}}$$

$$= \frac{497.5}{14.237}$$

$$= 34.94 \text{ in.}^3$$

$$S_{xct1} = \frac{I_{c1}}{y_{t1}}$$

$$= \frac{497.5}{-0.263}$$

$$= -2099 \text{ in.}^3$$

$$S_{xcc1} = \frac{I_{c1}}{y_{c1}}$$

$$S_{xcc1} = \frac{497.5}{3.263}$$

$$= 152.5 \text{ in.}^3$$

$$S_{xcb2} = \frac{I_{c2}}{y_{b2}}$$

$$= \frac{403.6}{12.528}$$

$$= 32.2 \text{ in.}^3$$

$$S_{xct2} = \frac{I_{c2}}{y_{t2}}$$
$$= \frac{403.6}{1.472}$$
$$= 274.2 \text{ in.}^3$$
$$S_{xcc2} = \frac{403.6}{4.972}$$
$$= 81.2 \text{ in.}^3$$

Step 9. Determine the bending stresses.
Lower flange

$$f_b = \frac{M_s}{S_{xs}} + \frac{M_D}{S_{xbc2}} + \frac{M_L}{S_{xbc1}}$$
$$= \frac{2.52 \times 12}{21.1} + \frac{29.66 \times 12}{32.22} + \frac{21.6 \times 12}{34.94}$$
$$= 20.0 \text{ k/in.}^2 < 24$$

Top of concrete

$$f_{cc} = \frac{M_D}{S_{xcc2} \times k \times n} + \frac{M_L}{S_{xcc1} \times n \times k}$$
$$= \frac{29.66 \times 12}{81.2 \times 3 \times 8} + \frac{21.6 \times 12}{152.5 \times 1 \times 8}$$
$$= 0.45 \text{ k/in.}^2 < 1.8$$

The section M12 × 18 satisfies the required conditions.

10.4 SHORT-CUT METHOD FOR DETERMINING S_{xbc}

In a composite section, the sum of the various stress components should not exceed the maximum allowable bending stress for steel and concrete. The bending stress in a composite beam is

$$f_b = \frac{M_{D1}}{S_B} + \frac{M_{D2}}{S_{bc3}} + \frac{M_L}{S_{bc1}} \tag{10.28}$$

where

S_B = section modulus of the steel beam
M_{D1} = moment due to dead loads carried by steel section alone before composite action takes place

M_{D2} = moment due to long-range dead loads carried by composite action
M_L = moment due to live loads carried by composite action

Since at all times $F_b \geq f_b$,

$$F_b \geq \frac{M_{D1}}{S_B} + \frac{M_{D2}}{S_{bc3}} + \frac{M_L}{S_{bc1}} \tag{10.29}$$

The section modulii S_{bc3} and S_{bc1} signify the use of $k = 3$ and $k = 1$, respectively, in the calculations of the section properties for the composite beam. Let

$$S_{bc3} = S_{bc1} = S_{xbc}$$

Then,

$$F_b \geq \frac{M_{D1}}{S_B} + \frac{M_{D2} + M_L}{S_{xbc}} \tag{10.30}$$

Equation (10.30) may be expressed as follows:

$$S_{bxc}E_b \geq \frac{M_{D1}}{S_B}\frac{S_{bxc}}{F_b} + \frac{(M_{D2} + M_L)}{F_b} \tag{10.31}$$

Let

$$f_1 = \frac{M_{D1}}{S_B E_b}$$

and

$$f_2 = \frac{(M_{D2} + M_L)}{F_b}$$

Then,

$$S_{xbc} \geq \frac{f_2}{1 - f_1} \tag{10.32}$$

This procedure consists of calculating f_2 based on the given moments and the allowable steel bending stress, and selecting a steel section that yields an f_1 of about 2 to 4% for shored construction. For unshored construction, f_1 is approximately equal to the ratio $M_{D1}/(M_{D2} + M_L)$. Then solve Equation (10.32) and find a section in Table 10.1 that has a section modulus S_{xbc} equal to or greater than that calculated above.

TABLE 10.1 Section Properties of Composite Beams

Steel Section	Area of Steel A_s	Depth of Section d (in.)	Moment of Inertia I_x	Slab Thickness 3 1/2 in.									
				$b/n = 4$		$b/n = 6$		$b/n = 8$		$b/n = 10$		$b/n = 12$	
				S_{xbc}	Y_{bc} (in.)	S_{xbc}	Y_{bc} (in.)	S_{xbc}	Y_{bc} (in.)	S_{xbc}	Y_{bc} (in.)	S_{xbc}	Y_{bc} (in.)
W36 × 160	47	36.01	9750	620.0	22.54	640.3	24.11	855.1	25.38	666.3	26.44	675.1	27.33
W36 × 150	44.2	35.85	9040	581.3	22.56	600.6	24.26	614.5	25.56	625.0	26.62	633.2	27.51
W36 × 135	39.7	35.55	7800	514.3	22.87	532.3	24.53	545.0	25.85	554.4	26.92	561.7	27.81
W33 × 141	41.6	33.3	7450	517.3	21.28	534.4	22.82	546.5	24.05	555.6	25.06	562.8	25.89
W33 × 130	38.3	33.09	6710	473.6	21.44	489.5	23.02	500.7	24.27	509.0	25.28	515.4	26.11
W33 × 118	34.7	32.86	5900	425.3	21.66	440.0	23.28	450.1	24.55	457.6	25.56	463.3	26.39
W30 × 116	34.2	30.01	4930	389.1	19.87	402.4	21.38	411.6	22.55	418.3	23.48	423.5	24.24
W30 × 108	31.7	29.83	4470	358.7	20.02	371.1	21.56	379.6	22.73	385.7	23.66	390.4	24.41
W30 × 99	29.1	29.65	3990	326.6	20.21	338.2	21.77	345.9	22.95	351.5	23.88	355.7	24.62
W27 × 94	27.7	26.92	3270	292.8	18.57	302.5	20.02	308.9	21.11	313.6	21.95	317.1	22.63
W27 × 84	24.8	26.71	2850	260.9	18.81	269.5	20.28	275.2	21.37	279.2	22.20	282.3	22.85
W24 × 94	27.7	24.31	2700	266.2	16.82	277.2	18.15	283.1	19.14	287.5	19.92	290.7	20.53
W24 × 84	24.7	24.1	2370	239.8	17.04	247.6	18.39	252.7	19.38	256.4	20.14	259.2	20.74
W24 × 76	22.4	23.92	2100	216.7	17.23	223.7	18.59	228.2	19.58	231.5	20.32	233.9	20.90
W24 × 68	20.1	23.73	1830	193.3	17.45	199.5	18.82	203.5	19.79	206.3	20.51	208.5	21.07
W24 × 62	18.2	23.74	1550	170.4	17.79	176.4	19.17	180.1	20.12	182.8	20.83	184.7	21.37
W24 × 55	16.22	23.57	1350	151.7	18.06	156.8	19.43	160.1	20.36	162.3	21.04	164.0	21.55
W21 × 68	20	21.13	1480	175.4	15.64	181.1	16.87	184.7	17.75	187.3	18.40	189.2	18.91
W21 × 62	18.3	20.99	1330	160.3	15.80	165.4	17.04	168.6	17.90	170.9	18.54	172.6	19.02
W21 × 57	16.7	21.06	1170	144.6	16.13	149.3	17.37	152.3	18.22	154.5	18.84	156.1	19.32
W21 × 50	14.7	20.83	984	126.0	16.35	130.0	17.57	132.6	18.39	134.4	18.98	135.8	19.43
W21 × 44	13	20.66	843	110.9	16.59	114.4	17.79	116.6	18.58	118.2	19.14	119.4	19.55

W18 × 55	16.2	18.11	890	126.6	14.06	130.6	15.15	133.1	15.90	134.9	16.44	136.3	16.85
W18 × 50	14.7	17.99	800	115.4	14.24	118.8	15.32	121.0	16.04	122.6	16.56	123.9	16.95
W18 × 46	13.5	18.06	712	105.0	14.52	108.3	15.59	110.3	16.30	111.8	16.81	113.0	17.19
W18 × 40	11.8	17.9	612	92.1	14.76	94.8	15.80	96.6	16.48	97.8	16.95	98.8	17.30
W18 × 35	10.3	17.7	510	79.6	14.96	81.9	15.96	83.4	16.60	84.5	17.04	85.4	17.36
W16 × 50	14.7	16.26	659	106.0	12.95	109.3	13.94	111.5	14.61	113.0	15.09	114.2	15.45
W16 × 45	13.3	16.13	586	96.0	13.10	98.9	14.07	100.8	14.72	102.1	15.18	103.2	15.52
W16 × 40	11.8	16.01	518	85.8	13.30	88.3	14.25	89.9	14.87	91.1	15.30	92.0	15.62
W16 × 31	9.12	15.88	375	65.7	13.81	67.6	14.70	68.8	15.25	69.8	15.63	70.5	15.90
W16 × 26	7.68	15.69	301	55.0	14.04	56.5	14.87	57.5	15.37	58.3	15.71	59.0	15.96
W14 × 38	11.2	14.1	385	73.8	11.94	76.0	12.79	77.5	13.34	78.6	13.72	79.5	14.00
W14 × 34	10	13.98	340	66.2	12.09	68.1	12.91	69.4	13.43	70.3	13.79	71.2	14.05
W14 × 30	8.85	13.84	291	58.3	12.23	59.9	13.02	61.1	13.51	62.0	13.84	62.7	14.08
W14 × 26	7.69	13.91	245	50.5	12.57	52.0	13.33	53.0	13.78	53.8	14.09	54.5	14.31
W14 × 22	6.49	13.74	199	42.5	12.76	43.8	13.45	44.6	13.87	45.4	14.14	46.0	14.34
W12 × 30	8.79	12.34	238	53.6	11.04	55.1	11.75	56.3	12.20	57.2	12.50	57.9	12.72
W12 × 26	7.65	12.22	204	46.8	11.19	48.2	11.87	49.2	12.28	50.0	12.56	50.7	12.76
W12 × 22	6.48	12.31	156	38.7	11.56	39.9	12.20	40.8	12.57	41.6	12.83	42.3	13.00
W12 × 19	5.57	12.16	130	33.3	11.68	34.3	12.27	35.2	12.61	35.9	12.83	36.5	12.99
W12 × 16	4.71	11.99	103	27.9	11.79	28.8	12.32	29.6	12.62	30.2	12.82	30.9	12.96
W10 × 26	7.61	10.33	144	40.9	9.64	42.2	10.24	43.3	10.60	44.1	10.85	44.9	11.02
W10 × 22	6.49	10.17	118	34.8	9.76	36.0	10.31	36.9	10.63	37.7	10.85	38.5	11.01
W10 × 19	5.62	10.24	96.3	29.9	10.02	31.0	10.54	31.9	10.84	32.7	11.04	33.4	11.18
W10 × 15	4.41	9.99	68.9	23.3	10.12	24.2	10.57	25.0	10.82	25.7	10.99	26.4	11.10
W8 × 21	6.16	8.28	75.3	28.9	8.23	30.1	8.69	31.1	8.97	32.0	9.15	32.8	9.28
W8 × 18	5.26	8.14	61.9	24.8	8.30	25.9	8.72	26.8	8.97	27.7	9.13	28.5	9.24
W8 × 15	4.44	8.11	48	20.8	8.46	21.8	8.85	22.7	9.07	23.5	9.21	24.3	9.31
W8 × 13	3.84	7.99	39.6	18.0	8.50	19.0	8.85	19.9	9.05	20.7	9.17	21.4	9.26

(cont'd.)

TABLE 10.1 (*cont'd.*)

Steel Section	Area of Steel A_s	Depth of Section d (in.)	Moment of Inertia I_x	Slab Thickness 4 in.									
				$b/n = 4$		$b/n = 6$		$b/n = 8$		$b/n = 10$		$b/n = 12$	
				S_{xbc}	Y_{bc} (in.)	S_{xbc}	Y_{bc} (in.)	S_{xbc}	Y_{bc} (in.)	S_{xbc}	Y_{bc} (in.)	S_{xbc}	Y_{bc} (in.)
W36 × 160	47	36.01	8750	630.2	23.09	651.7	24.77	666.9	26.11	678.3	27.20	687.1	28.11
W36 × 150	44.2	35.85	9040	591.1	23.22	611.4	24.94	625.7	26.29	836.3	27.39	644.5	28.30
W36 × 135	39.7	35.55	7800	523.6	23.46	542.4	25.23	555.3	26.60	564.8	27.70	572.1	28.60
W33 × 141	41.8	33.3	7450	526.4	21.83	544.3	23.47	556.7	24.76	565.9	25.79	573.0	26.64
W33 × 130	38.3	33.09	6710	482.2	22.01	498.8	23.69	510.2	24.99	518.6	26.02	525.0	26.88
W33 × 118	34.7	32.86	5900	433.4	22.25	448.6	23.97	458.9	25.27	466.4	26.30	472.0	27.13
W30 × 116	34.2	30.01	4930	396.7	20.42	410.6	22.02	419.9	23.22	426.7	24.17	431.9	24.93
W30 × 108	31.7	29.83	4470	385.9	20.59	378.8	22.20	387.4	23.41	393.5	24.35	398.2	25.10
W30 × 99	29.1	29.65	3990	333.5	20.79	345.3	22.43	353.1	23.64	358.7	24.56	362.9	25.30
W27 × 94	27.7	26.92	3270	298.9	19.12	308.9	20.64	315.5	21.75	320.2	22.59	323.8	23.26
W27 × 84	24.8	26.71	2850	266.5	19.38	275.4	20.91	281.1	22.01	285.2	22.83	288.3	23.48
W24 × 94	27.7	24.31	2700	274.2	17.34	283.5	18.73	289.6	19.74	294.0	20.52	297.4	21.13
W24 × 84	24.7	24.1	2370	245.2	17.57	253.2	18.97	258.5	19.98	262.2	20.74	265.1	21.33
W24 × 76	22.4	23.92	2100	221.7	17.78	228.9	19.18	233.5	20.17	236.8	20.91	239.3	21.48
W24 × 68	20.1	23.73	1830	197.9	18.01	204.3	19.41	208.3	20.38	211.2	21.09	213.4	21.64
W24 × 62	18.2	23.74	1550	174.8	18.36	180.8	19.76	184.7	20.71	187.3	21.40	189.4	21.93
W24 × 55	16.2	23.57	1350	155.7	18.83	160.9	20.01	164.1	20.94	166.4	21.60	168.2	22.09
W21 × 68	20	21.13	1480	179.9	16.15	185.7	17.42	189.4	18.30	192.1	18.94	194.1	19.43
W21 × 62	18.3	20.99	1330	164.4	16.32	169.6	17.58	173.0	18.44	175.4	19.07	177.2	19.54
W21 × 57	16.7	21.06	1170	148.5	16.66	153.4	17.92	156.4	18.76	158.7	19.37	160.4	19.83
W21 × 50	14.7	20.83	984	129.5	16.89	133.7	18.11	136.3	18.92	138.2	19.49	139.7	19.92
W21 × 44	13	20.66	843	114.1	17.13	117.7	18.33	120.0	19.10	121.6	19.64	122.9	20.03

W18 × 55	16.2	18.11	890	130.3	14.55	134.4	15.66	137.1	16.39	139.0	16.92	140.5	17.32
W18 × 50	14.7	17.99	800	118.7	14.73	122.3	15.81	124.7	16.53	126.4	17.04	127.8	17.41
W18 × 46	13.5	18.06	712	108.2	15.01	111.6	16.09	113.8	16.79	115.4	17.28	116.7	17.64
W18 × 40	11.8	17.9	612	94.9	15.25	97.8	16.29	99.6	16.95	101.0	17.41	102.1	17.74
W18 × 35	10.3	17.7	510	82.1	15.45	84.6	16.44	86.2	17.06	87.4	17.48	88.4	17.78
W16 × 50	14.7	16.26	659	109.4	13.41	112.9	14.41	115.1	15.07	116.8	15.54	118.2	15.89
W18 × 45	13.3	16.13	586	99.0	13.56	102.1	14.54	104.1	15.17	105.7	15.62	106.9	15.95
W16 × 40	11.8	16.01	518	88.6	13.76	91.2	14.71	93.0	15.31	94.3	15.73	95.4	16.04
W16 × 31	9.12	15.88	375	68.0	14.27	70.0	15.14	71.4	15.68	72.5	16.03	73.4	16.29
W16 × 26	7.88	15.69	301	56.9	14.50	58.6	15.30	59.8	15.78	60.8	16.10	61.8	16.33
W14 × 38	11.2	14.1	385	76.4	12.37	78.9	13.22	80.5	13.75	81.8	14.12	82.9	14.39
W14 × 34	10	13.98	340	68.6	12.52	70.7	13.34	72.1	13.84	73.3	14.18	74.3	14.43
W14 × 30	8.85	13.84	291	60.5	12.66	62.3	13.44	63.7	13.91	64.7	14.22	65.7	14.45
W14 × 26	7.69	13.91	245	52.5	13.00	54.2	13.74	55.4	14.17	56.4	14.47	57.3	14.67
W14 × 22	8.49	13.74	199	44.3	13.18	45.7	13.85	46.8	14.24	47.7	14.50	48.5	14.68
W12 × 30	8.79	12.34	238	55.8	11.44	57.6	12.15	58.9	12.58	60.0	12.87	61.0	13.08
W12 × 26	7.65	12.22	204	48.8	11.60	50.4	12.26	51.6	12.66	52.6	12.92	53.6	13.11
W12 × 22	6.48	12.31	156	40.5	11.96	41.9	12.58	43.1	12.94	44.0	13.17	45.0	13.34
W12 × 19	5.57	12.16	130	34.9	12.07	36.2	12.64	37.2	12.96	38.2	13.17	39.0	13.32
W12 × 16	4.71	11.99	103	29.3	12.17	30.5	12.68	31.5	12.96	32.4	13.15	33.2	13.28
W10 × 26	7.61	10.33	144	42.9	10.02	44.6	10.61	45.9	10.95	47.0	11.18	48.0	11.35
W10 × 22	8.49	10.17	118	36.7	10.13	38.1	10.66	39.3	10.98	40.4	11.18	41.4	11.33
W10 × 19	5.62	10.24	96.3	31.6	10.39	33.0	10.89	34.1	11.18	35.2	11.36	36.1	11.49
W10 × 15	4.41	9.99	68.9	24.8	10.48	26.0	10.90	27.0	11.14	28.0	11.30	29.0	11.40
W8 × 21	6.16	8.28	75.3	30.8	8.57	32.4	9.03	33.7	9.29	34.9	9.46	36.0	9.58
W8 × 18	5.26	8.14	61.9	26.5	8.64	27.9	9.05	29.2	9.28	30.4	9.43	31.5	9.54
W8 × 15	4.44	8.11	48	22.4	8.79	23.7	9.16	24.9	9.37	26.1	9.51	27.2	9.60
W8 × 13	3.84	7.99	39.6	19.5	8.83	20.8	9.16	22.0	9.35	23.1	9.46	24.2	9.55

(cont'd.)

TABLE 10.1 (*cont'd.*)

Steel Section	Area of Steel A_s	Depth of Section d (in.)	Moment of Inertia I_x	Slab Thickness 4 1/2 in.									
				$b/n = 4$		$b/n = 6$		$b/n = 8$		$b/n = 10$		$b/n = 12$	
				S_{xbc}	Y_{bc} (in.)	S_{xbc}	Y_{bc} (in.)	S_{xbc}	Y_{bc} (in.)	S_{xbc}	Y_{bc} (in.)	S_{xbc}	Y_{bc} (in.)
W36 × 160	47	36.01	9750	640.3	23.61	662.8	25.40	678.4	26.79	689.9	27.91	698.8	28.83
W36 × 150	44.2	35.85	9040	600.8	23.76	622.0	25.58	636.6	26.98	647.3	28.10	655.6	29.02
W36 × 135	39.7	35.55	7800	532.7	24.02	552.1	25.88	565.3	27.30	574.9	28.41	582.1	29.32
W33 × 141	41.6	33.3	7450	535.3	22.36	544.0	24.09	566.7	25.42	576.0	26.47	583.1	27.33
W33 × 130	38.3	33.09	6710	490.6	22.55	507.9	24.32	519.5	25.65	527.9	26.70	534.4	27.54
W33 × 118	34.7	32.86	5900	441.3	22.81	457.0	24.60	467.4	25.94	474.9	26.98	480.6	27.80
W30 × 116	34.2	30.01	4930	404.3	20.96	418.6	22.62	428.1	23.85	435.0	24.81	440.2	25.57
W30 × 108	31.7	29.83	4470	373.0	21.13	366.3	22.81	395.0	24.04	401.3	24.99	406.0	25.73
W30 × 99	29.1	29.85	3990	340.2	21.35	352.3	23.04	360.3	24.27	365.9	25.19	370.1	25.92
W27 × 94	27.7	26.92	3270	305.0	19.65	315.4	21.21	322.1	22.34	326.8	23.18	330.5	23.84
W27 × 84	24.8	26.71	2850	272.1	19.92	281.2	21.49	287.1	22.59	291.2	23.42	294.4	24.05
W24 × 94	27.7	24.31	2700	280.1	17.83	289.8	19.27	296.1	20.30	300.6	21.07	304.0	21.88
W24 × 84	24.7	24.1	2370	250.6	18.08	258.9	19.52	264.3	20.53	268.2	21.28	271.1	21.86
W24 × 76	22.4	23.92	2100	226.6	18.29	234.1	19.73	238.9	20.72	242.3	21.45	244.9	22.00
W24 × 68	20.1	23.73	1830	202.5	18.53	209.0	19.96	213.2	20.92	216.2	21.62	218.5	22.15
W24 × 62	18.2	23.74	1550	179.2	18.89	185.3	20.30	189.2	21.25	192.0	21.92	194.2	22.43
W24 × 55	16.2	23.57	1350	159.6	19.17	164.9	20.56	168.3	21.46	170.7	22.10	172.5	22.58
W21 × 68	20	21.13	1480	184.3	16.64	190.4	17.93	194.2	18.80	197.0	19.44	199.2	19.92
W21 × 62	18.3	20.99	1330	168.6	16.81	174.0	18.09	177.4	18.94	180.0	19.56	181.9	20.01
W21 × 57	16.7	21.06	1170	152.4	17.16	157.4	18.43	160.6	19.26	163.0	19.85	164.8	20.29
W21 × 50	14.7	20.83	984	133.0	17.39	137.3	18.62	140.1	19.41	142.1	19.96	143.8	20.37
W21 × 44	13	20.66	843	117.3	17.63	121.0	18.82	123.4	19.57	125.2	20.09	126.6	20.47

W18 × 55	16.2	18.11	890	134.0	15.01	138.3	16.12	141.2	16.85	143.3	17.37	145.0	17.75
W18 × 50	14.7	17.99	800	122.1	15.18	125.9	16.28	128.4	16.98	130.3	17.47	131.9	17.83
W18 × 46	13.5	18.06	712	111.4	15.48	115.0	16.55	117.3	17.23	119.1	17.71	120.6	18.05
W18 × 40	11.8	17.9	612	97.8	15.72	100.8	16.74	102.8	17.39	104.4	17.82	105.7	18.14
W18 × 35	10.3	17.7	510	84.7	15.91	87.3	16.88	89.1	17.48	90.5	17.88	91.7	18.17
W16 × 50	14.7	16.26	659	112.8	13.84	116.5	14.85	119.0	15.50	120.9	15.95	122.5	16.29
W16 × 45	13.3	16.13	586	102.2	14.00	105.5	14.98	107.7	15.60	109.5	16.03	110.9	16.34
W16 × 40	11.8	16.01	518	91.4	14.20	94.3	15.14	96.2	15.73	97.8	16.13	99.1	16.42
W16 × 31	9.12	15.88	375	70.3	14.70	72.5	15.56	74.1	16.07	75.4	16.41	76.6	16.66
W16 × 26	7.68	15.69	301	59.0	14.92	60.9	15.70	62.3	16.17	63.5	16.47	64.6	16.68
W14 × 38	11.2	14.1	385	79.2	12.78	81.9	13.62	83.8	14.14	85.3	14.50	86.7	14.75
W14 × 34	10	13.98	340	71.1	12.93	73.4	13.73	75.2	14.22	76.6	14.55	77.9	14.79
W14 × 30	8.85	13.84	291	62.8	13.07	64.9	13.83	66.5	14.28	87.8	14.58	69.0	14.80
W14 × 26	7.69	13.91	245	54.6	13.40	56.5	14.12	58.0	14.54	59.2	14.82	60.4	15.01
W12 × 22	6.49	13.74	199	46.1	13.57	47.8	14.22	49.1	14.60	50.3	14.84	51.4	15.01
W12 × 30	8.79	12.34	238	58.1	11.83	60.2	12.52	61.8	12.94	63.2	13.21	64.5	13.41
W12 × 26	7.65	12.22	204	50.9	11.98	52.8	12.62	54.3	13.00	55.6	13.26	56.8	13.43
W12 × 22	6.48	12.31	156	42.4	12.34	44.1	12.93	45.5	13.28	46.8	13.50	48.0	13.66
W12 × 19	5.57	12.16	130	36.6	12.44	38.2	12.99	39.5	13.29	40.8	13.49	41.9	13.63
W12 × 16	4.71	11.99	103	30.9	12.53	32.4	13.02	33.6	13.29	34.8	13.46	36.0	13.58
W10 × 26	7.61	10.33	144	45.1	10.38	47.1	10.95	48.7	11.29	50.2	11.51	51.6	11.66
W10 × 22	6.49	10.17	118	38.7	10.48	40.5	11.00	42.0	11.30	43.4	11.50	44.8	11.63
W10 × 19	5.62	10.24	96.3	33.5	10.74	35.2	11.22	36.6	11.49	38.0	11.67	39.3	11.80
W10 × 15	4.41	9.99	68.9	26.4	10.81	27.9	11.22	29.3	11.45	30.7	11.59	32.0	11.69
W8 × 21	6.16	8.28	75.3	32.9	8.90	34.9	9.34	36.6	9.60	38.2	9.76	39.7	9.88
W8 × 18	5.26	8.14	61.9	28.4	8.96	30.3	9.36	31.9	9.58	33.5	9.73	35.0	9.83
W8 × 15	4.44	8.11	48	24.1	9.11	25.9	9.47	27.5	9.67	29.1	9.79	30.6	9.88
W8 × 13	3.84	7.99	39.6	21.2	9.14	22.9	9.46	24.5	9.64	26.0	9.75	27.5	9.83

(*cont'd.*)

TABLE 10.1 (*cont'd.*)

Steel Section	Area of Steel A_s	Depth of Section d (in.)	Moment of Inertia I_x	Slab Thickness 5 in.									
				$b/n = 4$		$b/n = 6$		$b/n = 8$		$b/n = 10$		$b/n = 12$	
				S_{xbc}	Y_{bc} (in.)	S_{xbc}	Y_{bc} (in.)	S_{xbc}	Y_{bc} (in.)	S_{xbc}	Y_{bc} (in.)	S_{xbc}	Y_{bc} (in.)
W36 × 160	47	36.01	9750	650.4	24.13	673.7	25.99	689.7	27.43	701.3	28.57	710.3	29.50
W36 × 150	44.2	35.85	9040	610.4	24.29	632.4	26.18	647.3	27.63	658.1	28.77	666.4	29.69
W38 × 135	39.7	35.55	7800	541.7	24.57	561.7	26.50	575.1	27.95	584.7	29.08	592.0	29.98
W33 × 141	41.6	33.3	7450	544.2	22.87	563.5	24.67	576.5	26.04	586.0	27.10	593.2	27.96
W33 × 130	38.3	33.09	6710	499.1	23.08	516.8	24.91	528.7	26.27	537.2	27.33	543.6	28.17
W33 × 118	34.7	32.86	5900	449.1	23.35	465.3	25.21	475.9	26.57	483.4	27.60	489.1	28.42
W30 × 116	34.2	30.01	4930	411.8	21.46	426.6	23.18	436.3	24.44	433.2	25.40	448.5	26.15
W30 × 108	31.7	29.83	4470	380.1	21.65	393.8	23.38	402.6	24.83	408.9	25.57	413.7	26.31
W30 × 99	29.1	29.85	3990	346.8	21.88	359.3	23.62	367.3	24.85	373.0	25.78	377.4	26.49
W27 × 94	27.7	26.92	3270	311.1	20.15	321.8	21.76	328.6	22.89	333.5	23.73	337.3	24.36
W27 × 84	24.8	26.71	2850	277.7	20.43	287.1	22.03	293.0	23.14	297.3	23.95	300.6	24.57
W24 × 94	27.7	24.31	2700	286.1	18.30	296.1	19.77	302.6	20.81	307.3	21.59	310.9	22.18
W24 × 84	24.7	24.1	2370	256.0	18.56	264.6	20.03	270.2	21.05	274.2	21.79	277.3	22.36
W24 × 76	22.4	23.92	2100	231.7	18.78	239.3	20.24	244.3	21.23	247.8	21.95	250.6	22.49
W24 × 68	20.1	23.73	1830	207.1	19.03	213.8	20.47	218.1	21.43	221.3	22.11	223.7	22.63
W24 × 62	18.2	23.74	1550	183.5	19.39	189.9	20.81	193.9	21.75	196.8	22.41	199.1	22.90
W24 × 55	16.2	23.57	1350	163.5	19.68	169.0	21.06	172.5	21.95	175.0	22.57	177.0	23.03
W21 × 68	20	21.13	1480	188.8	17.10	195.1	18.40	199.2	19.28	202.2	19.90	204.6	20.36
W21 × 62	18.3	20.99	1330	172.8	17.28	178.4	18.57	182.0	19.41	184.7	20.01	186.9	20.45
W21 × 57	16.7	21.06	1170	156.4	17.63	161.6	18.90	165.0	19.72	167.5	20.30	169.5	20.72
W21 × 50	14.7	20.83	984	136.6	17.86	141.1	19.08	144.0	19.86	146.3	20.40	148.1	20.79
W21 × 44	13	20.66	843	120.5	18.11	124.4	19.28	127.0	20.01	129.0	20.51	130.6	20.88

W18 × 55	16.2	18.11	890	137.7	15.44	142.3	16.56	145.4	17.28	147.8	17.78	149.7	18.15
W18 × 50	14.7	17.99	800	125.6	15.62	129.6	16.71	132.4	17.40	134.5	17.88	136.3	18.23
W18 × 46	13.5	18.06	712	114.7	15.91	118.5	16.98	121.1	17.65	123.1	18.11	124.8	18.44
W18 × 40	11.8	17.9	612	100.7	16.15	104.0	17.17	106.2	17.79	108.0	18.21	109.6	18.52
W18 × 35	10.3	17.7	510	87.4	16.34	90.2	17.30	92.2	17.88	93.9	18.26	95.3	18.54
W16 × 50	14.7	16.26	659	116.3	14.26	120.3	15.26	123.1	15.90	125.2	16.34	127.1	16.67
W16 × 45	13.3	16.13	586	105.4	14.41	109.0	15.38	111.5	15.99	113.5	16.41	115.2	16.71
W16 × 40	11.8	16.01	518	94.4	14.61	97.5	15.54	99.7	16.12	101.5	16.50	103.1	16.78
W16 × 31	9.12	15.88	375	72.8	15.11	75.2	15.95	77.1	16.44	78.7	16.77	80.2	17.00
W16 × 26	7.68	15.69	301	61.1	15.32	63.3	16.08	65.0	16.52	66.5	16.81	67.9	17.02
W14 × 38	11.2	14.1	385	82.1	13.17	85.1	14.00	87.3	14.51	89.1	14.85	90.8	15.10
W14 × 34	10	13.98	340	73.7	13.32	76.4	14.11	78.4	14.58	80.2	14.90	81.8	15.12
W14 × 30	8.85	13.84	291	65.2	13.45	67.6	14.19	69.5	14.63	71.2	14.92	72.7	15.13
W14 × 26	7.69	13.91	245	56.8	13.78	59.0	14.48	60.8	14.89	62.4	15.15	63.9	15.34
W14 × 22	6.49	13.74	199	48.1	13.94	50.1	14.57	51.7	14.93	53.3	15.16	54.7	15.33
W12 × 30	8.79	12.34	238	60.6	12.19	63.0	12.88	65.0	13.28	66.8	13.54	68.4	13.73
W12 × 26	7.65	12.22	204	53.2	12.34	55.4	12.97	57.2	13.34	58.9	13.58	60.5	13.75
W12 × 22	6.48	12.31	156	44.5	12.69	46.5	13.27	48.3	13.60	49.9	13.82	51.5	13.97
W12 × 19	5.57	12.16	130	38.5	12.79	40.4	13.32	42.1	13.61	43.7	13.80	45.2	13.93
W12 × 16	4.71	11.99	103	32.6	12.87	34.4	13.34	36.1	13.60	37.6	13.76	39.2	13.87
W10 × 26	7.61	10.33	144	47.5	10.72	49.9	11.28	52.0	11.60	53.8	11.82	55.6	11.97
W10 × 22	6.49	10.17	118	40.8	10.81	43.1	11.32	45.0	11.61	46.8	11.80	48.6	11.93
W10 × 19	5.62	10.24	96.3	35.5	11.07	37.6	11.54	39.5	11.80	41.3	11.97	43.0	12.09
W10 × 15	4.41	9.99	68.9	28.2	11.14	30.1	11.53	32.0	11.75	33.7	11.88	35.5	11.98
W8 × 21	6.16	8.28	75.3	35.2	9.22	37.6	9.65	39.8	9.89	41.9	10.05	43.9	10.16
W8 × 18	5.26	8.14	61.9	30.6	9.27	32.9	9.66	35.0	9.88	37.1	10.01	39.1	10.11
W8 × 15	4.44	8.11	48	26.1	9.42	28.3	9.76	30.4	9.96	32.5	10.08	34.5	10.16
W8 × 13	3.84	7.99	39.6	23.0	9.44	25.2	9.75	27.3	9.92	29.3	10.03	31.4	10.10

(cont'd.)

TABLE 10.1 (*cont'd.*)

Steel Section	Area of Steel A_s	Depth of Section d (in.)	Moment of Inertia I_x	Slab Thickness 6 in.									
				$b/n = 4$		$b/n = 6$		$b/n = 8$		$b/n = 10$		$b/n = 12$	
				S_{xbc}	Y_{bc} (in.)	S_{xbc}	Y_{bc} (in.)	S_{xbc}	Y_{bc} (in.)	S_{xbc}	Y_{bc} (in.)	S_{xbc}	Y_{bc} (in.)
W36 × 160	47	36.01	9750	670.4	25.11	695.3	27.12	711.8	26.62	723.8	29.78	733.0	30.71
W36 × 150	44.2	35.85	9040	629.6	25.29	652.9	27.32	668.3	28.82	679.4	29.97	687.8	30.89
W36 × 135	39.7	35.55	7800	559.6	25.60	580.6	27.65	594.3	29.15	604.1	30.28	611.6	31.17
W33 × 141	41.6	33.3	7450	562.0	23.84	582.5	25.77	596.1	27.18	605.8	28.25	613.2	29.10
W33 × 130	38.3	33.09	6710	515.8	24.07	534.6	26.01	546.8	27.42	555.6	28.47	562.3	29.30
W33 × 118	34.7	32.86	5900	464.8	24.37	481.6	26.32	492.6	27.71	500.3	28.74	506.3	29.54
W30 × 116	34.2	30.01	4930	426.8	22.43	442.4	24.24	452.5	25.52	459.8	26.47	465.3	27.21
W30 × 108	31.7	29.83	4470	394.3	22.63	408.7	24.44	417.9	25.70	424.5	26.64	429.6	27.35
W30 × 99	29.1	29.65	3990	260.2	22.88	373.2	24.68	381.5	25.92	387.5	26.83	392.1	27.52
W27 × 94	27.7	26.92	3270	323.5	21.10	334.7	22.76	342.0	23.90	347.3	24.72	351.4	25.35
W25 × 84	24.8	26.71	2850	289.0	21.40	298.9	23.04	305.2	24.14	309.9	24.93	313.5	25.52
W24 × 94	27.7	24.31	2700	298.3	19.19	309.0	20.72	316.0	21.76	321.1	22.52	325.1	23.10
W24 × 84	24.7	24.1	2370	267.1	19.47	276.3	20.98	282.3	21.99	286.8	22.71	290.3	23.26
W24 × 76	22.4	23.92	2100	241.9	19.70	250.1	21.18	255.5	22.16	259.5	22.85	262.7	23.37
W24 × 68	20.1	23.73	1830	216.4	19.95	223.7	21.40	228.5	22.34	232.0	23.00	235.0	23.49
W24 × 62	18.2	23.74	1550	192.4	20.33	199.2	21.75	203.6	22.65	207.0	23.28	209.7	23.74
W24 × 55	16.2	23.57	1350	171.8	20.61	177.4	21.98	181.3	22.84	184.3	23.43	186.8	23.85
W21 × 68	20	21.13	1480	198.1	17.96	205.0	19.29	209.6	20.14	213.1	20.74	216.0	21.18
W21 × 62	18.3	20.99	1330	181.4	18.15	187.6	19.44	191.8	20.27	195.0	20.84	197.8	21.26
W21 × 57	16.7	21.06	1170	164.5	18.51	170.3	19.77	174.2	20.57	177.2	21.11	179.8	21.51
W21 × 50	14.7	20.83	984	143.9	18.73	149.0	19.94	152.4	20.68	155.2	21.19	157.6	21.56
W21 × 44	13	20.66	843	127.2	19.98	131.6	20.12	134.7	20.82	137.3	21.29	139.5	21.62

W18 × 55	16.2	18.11	890	145.7	16.25	150.9	17.37	154.6	18.07	157.6	18.55	160.2	18.90
W18 × 50	14.7	17.99	800	132.9	16.43	137.6	17.51	141.0	18.18	143.8	18.63	146.3	18.96
W18 × 46	13.5	18.06	712	121.6	16.73	126.0	17.78	129.3	18.42	131.9	18.85	134.3	19.16
W18 × 40	11.8	17.9	612	106.9	16.96	110.8	17.95	113.7	18.54	116.2	18.94	118.4	19.22
W18 × 35	10.3	17.7	510	93.0	17.4	96.5	18.06	99.1	18.61	101.5	18.96	103.6	19.22
W16 × 50	14.7	16.26	659	123.8	15.03	128.5	16.03	132.0	16.65	134.8	17.07	137.4	17.37
W16 × 45	13.3	16.13	586	112.3	15.18	116.6	16.14	119.9	16.73	122.6	17.12	125.1	17.40
W16 × 40	11.8	16.01	518	100.6	15.38	104.5	16.29	107.4	16.84	110.0	17.20	112.4	17.46
W16 × 31	9.12	15.88	375	78.0	15.87	81.2	16.67	83.8	17.13	86.2	17.44	88.4	17.65
W16 × 26	7.68	15.69	301	65.8	16.06	68.7	16.78	71.2	17.19	73.4	17.46	75.6	17.64
W14 × 38	11.2	14.1	385	88.4	13.90	92.1	14.72	95.2	15.20	97.8	15.52	100.3	15.75
W14 × 34	10	13.98	340	79.5	14.04	83.0	14.81	85.9	15.26	88.4	15.55	90.9	15.76
W14 × 30	8.85	13.84	291	70.5	14.17	73.8	14.88	76.5	15.30	79.0	15.56	81.4	15.75
W14 × 26	7.69	13.91	245	61.7	14.49	64.7	15.16	67.3	15.54	69.7	15.78	72.1	15.95
W14 × 22	6.49	13.74	199	52.5	14.64	55.3	15.23	57.8	15.56	60.2	15.78	62.5	15.92
W12 × 30	8.79	12.34	238	66.1	12.88	69.4	13.54	72.3	13.92	75.0	14.17	77.6	14.34
W12 × 26	7.85	12.22	204	58.2	13.02	61.3	13.62	64.1	13.97	66.7	14.19	69.3	14.35
W12 × 22	6.48	12.31	156	49.1	13.36	52.1	13.91	54.7	14.22	57.3	14.42	59.8	14.55
W12 × 19	5.57	12.16	130	42.7	13.45	45.6	13.94	48.2	14.22	50.7	14.39	53.2	14.51
W12 × 16	4.71	11.99	103	36.5	13.51	39.3	13.95	41.9	14.19	44.4	14.34	46.9	14.44
W10 × 26	7.61	10.33	144	52.9	11.36	56.3	11.91	59.4	12.21	62.4	12.41	65.2	12.55
W10 × 22	6.49	10.17	118	45.8	11.45	49.1	11.94	52.1	12.21	55.0	12.38	57.8	12.50
W10 × 19	5.62	10.24	96.3	40.0	11.70	43.2	12.14	46.2	12.39	49.0	12.54	51.9	12.65
W10 × 15	4.41	9.99	68.9	32.3	11.75	35.3	12.12	38.2	12.32	41.1	12.44	43.9	12.53
W8 × 21	6.16	8.28	75.3	40.4	9.82	44.1	10.24	47.5	10.47	50.9	10.62	54.2	10.72
W8 × 18	5.26	8.14	61.9	35.4	9.87	39.0	10.24	42.4	10.44	45.8	10.57	49.1	10.66
W8 × 15	4.44	8.11	48	30.6	10.01	34.1	10.34	37.5	10.51	40.8	10.62	44.1	10.70
W8 × 13	3.84	7.99	39.6	27.3	10.03	30.8	10.32	34.1	10.47	37.5	10.57	40.8	10.64

(cont'd.)

TABLE 10.1 (*cont'd.*)

Steel Section	Area of Steel A_s	Depth of Section d (in.)	Moment of Inertia I_x	Slab Thickness 7 in.									
				$b/n = 4$		$b/n = 6$		$b/n = 8$		$b/n = 10$		$b/n = 12$	
				S_{xbc}	Y_{bc} (in.)	S_{xbc}	Y_{bc} (in.)	S_{xbc}	Y_{bc} (in.)	S_{xbc}	Y_{bc} (in.)	S_{xbc}	Y_{bc} (in.)
W36 × 160	47	36.01	9750	690.6	26.03	716.7	28.15	734.0	29.70	746.3	30.87	755.8	31.79
W36 × 150	44.2	35.85	9040	648.9	26.23	673.3	28.36	689.2	29.90	700.7	31.06	709.5	31.96
W36 × 135	39.7	35.55	7800	577.5	26.57	599.4	28.71	613.5	30.22	623.7	31.35	631.4	32.22
W33 × 141	41.6	33.3	7450	580.0	24.76	601.6	26.77	615.7	28.21	625.8	29.29	633.6	30.13
W33 × 130	38.3	33.09	6710	532.7	25.01	552.4	27.03	565.1	28.45	574.3	29.50	581.4	30.31
W33 × 118	34.7	32.86	5900	480.4	25.33	498.1	27.34	509.4	28.74	517.6	29.75	523.9	30.53
W30 × 116	34.2	30.01	4930	442.1	23.34	458.5	25.20	469.2	26.49	476.9	27.44	482.9	28.16
W30 × 108	31.7	29.83	4470	408.7	23.55	423.8	25.41	433.5	26.67	440.6	27.59	446.1	28.28
W30 × 99	29.1	29.65	3990	373.6	23.81	387.3	25.65	396.1	26.88	402.5	27.77	407.8	28.44
W27 × 94	27.7	26.92	3270	336.1	21.99	348.1	23.68	355.9	24.81	361.7	25.61	366.4	26.21
W27 × 84	24.8	26.71	2850	300.6	22.29	311.1	23.95	318.0	25.04	323.2	25.80	327.4	26.37
W24 × 94	27.7	24.31	2700	311.0	20.02	322.5	21.59	330.1	22.63	335.9	23.37	340.5	23.93
W24 × 84	24.7	24.1	2370	278.5	20.31	288.5	21.84	295.2	22.84	300.3	23.54	304.5	24.07
W24 × 76	22.4	23.92	2100	252.5	20.55	261.5	22.04	267.5	23.00	272.1	23.67	276.0	24.17
W24 × 68	20.1	23.73	1830	226.2	20.81	234.1	22.26	239.5	23.17	243.8	23.80	247.4	24.26
W24 × 62	18.2	23.74	1550	201.6	21.19	209.0	22.59	214.0	23.47	218.1	24.07	221.5	24.50
W24 × 55	16.2	23.57	1350	179.9	21.47	186.4	22.82	191.0	23.64	194.6	24.20	197.8	24.60
W21 × 68	20	21.13	1480	207.9	18.77	215.6	20.09	220.9	20.93	225.2	21.50	228.9	21.93
W21 × 62	18.3	20.99	1330	190.5	18.96	197.5	20.24	202.4	21.04	206.5	21.59	210.0	21.99
W21 × 57	16.7	21.06	1170	173.1	19.32	179.6	20.57	184.2	21.34	188.0	21.86	191.4	22.23
W21 × 50	14.7	20.83	984	151.7	19.54	157.5	20.72	161.7	21.44	165.3	21.92	168.5	22.26
W21 × 44	13	20.66	843	134.3	19.77	139.5	20.89	143.3	21.55	146.7	21.99	149.7	22.31

W18 × 55	16.2	18.11	890	154.2	17.01	160.3	18.12	164.9	18.79	168.8	19.25	172.3	19.58
W18 × 50	14.7	17.99	800	140.8	17.19	146.4	18.25	150.7	18.89	154.4	19.32	157.7	19.63
W18 × 46	13.5	18.06	712	129.1	17.48	134.4	18.51	138.5	19.13	142.0	19.53	145.3	19.83
W18 × 40	11.8	17.9	612	113.7	17.71	118.4	18.67	122.3	19.23	125.6	19.60	128.8	19.87
W18 × 35	10.3	17.7	510	99.2	17.88	103.5	18.77	107.1	19.28	110.4	19.62	113.5	19.85
W16 × 50	14.7	16.26	659	131.8	15.76	137.6	16.74	142.0	17.34	145.9	17.74	149.4	18.03
W16 × 45	13.3	16.13	586	119.9	15.91	125.1	16.85	129.4	17.41	133.1	17.78	136.6	18.05
W16 × 40	11.8	16.01	518	107.5	16.10	112.4	16.99	116.3	17.51	119.9	17.85	123.3	18.09
W16 × 31	9.12	15.88	375	83.9	16.57	88.1	17.34	91.7	17.78	95.1	18.06	98.3	18.26
W16 × 26	7.68	15.69	301	71.1	16.75	75.0	17.44	78.5	17.82	81.8	18.07	85.0	18.24
W14 × 38	11.2	14.1	385	95.3	14.59	100.2	15.38	104.3	15.84	108.1	16.14	11.7	16.36
W14 × 34	10	13.98	340	86.0	14.72	90.6	15.46	94.5	15.89	98.3	16.17	101.8	16.36
W14 × 30	8.85	13.84	291	76.5	14.84	80.9	15.53	84.8	15.92	88.4	16.17	92.0	16.35
W14 × 26	7.69	13.91	245	67.2	15.16	71.4	15.79	75.1	16.15	78.7	16.38	82.1	16.53
W14 × 22	6.49	13.74	199	57.6	15.29	61.5	15.85	65.2	16.16	68.7	16.36	72.1	16.50
W12 × 30	8.79	12.34	238	72.3	13.53	76.9	14.17	81.0	14.53	85.0	14.76	88.8	14.92
W12 × 26	7.65	12.22	204	63.9	13.66	68.3	14.24	72.4	14.56	76.3	14.77	80.1	14.92
W12 × 22	6.46	12.31	156	54.4	14.00	58.6	14.52	62.5	14.81	66.3	14.99	70.1	15.12
W12 × 19	5.57	12.16	130	47.7	140.7	51.8	14.54	55.7	14.79	59.5	14.95	63.2	15.06
W12 × 16	4.71	11.99	103	41.1	14.12	45.2	14.53	49.0	14.75	52.8	14.89	56.6	14.99
W10 × 26	7.61	10.33	144	59.1	11.98	63.9	12.50	68.4	12.79	72.8	12.98	77.1	13.11
W10 × 22	6.49	10.17	118	51.5	12.05	56.2	12.52	60.7	12.78	65.0	12.94	69.3	13.05
W10 × 19	5.62	10.24	96.3	45.4	12.30	50.0	12.72	54.4	12.95	58.7	13.10	62.9	13.20
W10 × 15	4.41	9.99	68.9	37.1	12.33	41.7	12.68	46.0	12.87	50.4	12.99	54.7	13.07
W8 × 21	6.16	8.28	75.3	46.6	10.40	51.9	10.80	57.0	11.02	62.0	11.16	66.9	11.26
W8 × 18	5.26	8.14	61.9	41.2	10.44	46.4	10.80	51.5	10.99	56.5	11.11	61.5	11.19
W8 × 15	4.44	8.11	48	36.0	10.58	41.2	10.89	46.3	11.06	51.3	11.16	56.2	11.23
W8 × 13	3.84	7.99	39.6	32.5	10.59	37.6	10.86	42.7	11.01	47.7	11.10	52.8	11.16

(cont'd.)

TABLE 10.1 (*cont'd.*)

Steel Section	Area of Steel A_s	Depth of Section d (in.)	Moment of Inertia I_x	Slab Thickness 8 in.									
				$b/n = 4$		$b/n = 6$		$b/n = 8$		$b/n = 10$		$b/n = 12$	
				S_{xbc}	Y_{bc} (in.)	S_{xbc}	Y_{bc} (in.)	S_{xbc}	Y_{bc} (in.)	S_{xbc}	Y_{bc} (in.)	S_{xbc}	Y_{bc} (in.)
W36 × 160	47	36.01	9750	711.0	26.92	738.4	29.12	756.3	30.69	769.2	31.87	779.2	32.78
W36 × 150	44.2	35.85	9040	668.3	27.13	693.9	29.34	710.5	30.89	722.4	32.05	731.7	32.94
W36 × 135	39.7	35.55	7800	595.5	27.49	618.3	29.69	633.0	31.21	643.6	32.22	651.9	33.18
W33 × 141	41.6	33.3	7450	598.3	25.63	621.0	27.71	635.8	29.17	648.5	30.24	654.9	31.06
W33 × 130	38.3	33.09	6710	549.9	25.90	570.5	29.97	583.9	29.40	593.6	30.44	601.3	31.23
W33 × 118	34.7	32.86	5900	496.3	26.23	514.8	28.29	526.8	29.68	535.5	30.68	542.4	31.44
W30 × 116	34.2	30.01	4930	457.7	24.19	475.0	26.10	486.4	27.39	494.7	26.32	501.4	29.02
W30 × 108	31.7	29.83	4470	423.4	24.42	439.2	26.31	449.7	27.56	457.4	28.46	483.6	29.13
W30 × 99	29.1	29.65	3990	387.4	24.68	401.8	26.54	411.3	27.77	418.4	28.63	424.2	29.27
W27 × 94	27.7	26.92	3270	349.1	22.82	362.0	24.53	370.6	25.65	377.2	26.43	382.6	27.01
W27 × 84	24.8	26.71	2850	312.5	23.13	323.9	24.80	331.5	25.86	337.5	26.60	342.5	27.15
W24 × 94	27.7	24.31	2700	324.1	20.81	336.6	22.40	345.1	23.43	351.8	24.16	357.3	24.69
W24 × 84	24.7	24.1	2370	290.5	21.11	301.5	22.65	309.0	23.63	315.0	24.31	320.1	24.82
W24 × 76	22.4	23.92	2100	263.6	21.35	273.5	22.84	280.4	23.78	285.9	24.43	290.7	24.90
W24 × 68	20.1	23.73	1830	236.4	21.61	245.2	23.05	251.5	23.94	256.7	24.54	261.2	24.98
W24 × 62	18.2	23.74	1550	211.2	21.99	219.4	23.38	225.4	24.23	230.3	24.80	234.6	25.21
W24 × 55	16.2	23.57	1350	188.7	22.26	196.0	23.59	201.5	24.38	206.1	24.91	210.2	25.29
W21 × 68	20	21.13	1480	218.2	19.53	226.9	20.85	233.3	21.66	238.6	22.22	243.3	22.62
W21 × 62	18.3	20.99	1330	200.2	19.72	208.2	20.99	214.1	21.77	219.2	22.29	223.7	22.67
W21 × 57	16.7	21.06	1170	182.2	20.08	189.7	21.31	195.3	22.05	200.1	22.55	204.5	22.91
W21 × 50	14.7	20.83	984	160.0	20.29	166.8	21.45	172.1	22.14	176.7	22.59	180.9	22.92
W21 × 44	13	20.66	843	141.9	20.52	148.1	21.61	153.0	22.24	157.4	22.66	161.5	22.95

W18 × 55	16.2	18.11	890	163.3	17.72	170.6	18.82	176.4	19.47	181.4	19.91	186.1	20.23
W18 × 50	14.7	17.99	800	149.3	17.90	156.1	18.94	161.5	19.56	166.4	19.97	171.0	20.26
W18 × 46	13.5	18.06	712	137.1	18.19	143.6	19.20	148.9	19.79	153.6	20.18	158.1	20.45
W18 × 40	11.8	17.9	612	121.0	18.41	127.0	19.34	132.0	19.88	136.6	20.24	140.9	20.48
W18 × 35	10.3	17.7	510	105.9	18.57	111.5	19.43	116.3	19.92	120.8	20.23	125.1	20.45
W16 × 50	14.7	16.26	659	140.6	16.44	147.6	17.42	153.3	17.99	158.5	18.38	163.4	18.65
W16 × 45	13.3	16.13	586	128.1	16.59	134.6	17.51	140.1	18.05	145.2	18.41	150.0	18.66
W16 × 40	11.8	16.01	518	115.1	16.78	121.3	17.64	126.5	18.14	131.4	18.47	136.1	18.70
W16 × 31	9.12	15.88	375	90.4	17.23	95.9	17.97	100.8	18.39	105.5	18.66	110.1	18.84
W16 × 26	7.68	15.69	301	77.1	17.40	82.3	18.06	87.1	18.42	91.7	18.65	96.3	18.81
W14 × 38	11.2	14.1	385	103.0	15.24	109.3	16.01	114.9	16.45	120.1	16.74	125.2	16.95
W14 × 34	10	13.98	340	93.1	15.36	99.2	16.09	104.6	16.49	109.8	16.76	114.8	16.94
W14 × 30	8.85	13.84	291	83.3	15.47	89.1	16.14	94.4	16.51	99.6	16.75	104.6	16.92
W14 × 26	7.69	13.91	245	73.5	15.79	79.1	16.40	84.3	16.73	89.3	16.95	94.3	17.10
W14 × 22	6.49	13.74	199	63.3	15.91	68.7	16.45	73.9	16.74	78.9	16.92	83.8	17.05
W12 × 30	8.79	12.34	238	79.3	14.15	85.5	14.77	91.2	15.11	96.8	15.33	102.2	15.49
W12 × 26	7.65	12.22	204	70.5	14.27	76.5	14.83	82.1	15.14	87.7	15.34	93.1	15.47
W12 × 22	6.48	12.31	156	60.4	14.60	66.3	15.10	71.8	15.38	77.2	15.55	82.6	15.67
W12 × 19	5.57	12.16	130	53.4	14.67	59.1	15.11	64.6	15.35	70.0	15.50	75.4	15.61
W12 × 18	4.71	11.99	103	46.5	14.71	52.2	15.10	57.7	15.30	63.1	15.43	68.5	15.52
W10 × 26	7.61	10.33	144	66.1	12.57	72.8	13.08	79.1	13.36	85.3	13.53	91.4	13.66
W10 × 22	6.49	10.17	118	58.1	12.64	64.6	13.09	70.9	13.33	77.1	13.49	83.2	13.59
W10 × 19	5.62	10.24	96.3	51.6	12.88	58.0	13.28	64.2	13.50	70.4	13.64	76.4	13.74
W10 × 15	4.41	9.99	68.9	42.9	12.90	49.2	13.23	55.5	13.41	61.7	13.52	67.8	13.59
W8 × 21	6.16	8.28	75.3	53.6	10.97	61.0	11.35	68.2	11.57	75.3	11.70	82.4	11.79
W8 × 18	5.26	8.14	61.9	47.9	11.00	55.2	11.34	62.4	11.53	69.6	11.64	76.7	11.72
W8 × 15	4.44	8.11	48	42.4	11.13	49.7	11.43	56.8	11.59	64.0	11.69	71.1	11.75
W8 × 13	3.84	7.99	39.6	38.6	11.13	45.9	11.40	53.1	11.54	60.3	11.62	67.4	11.68

Source: Tables produced by Howard Reginald Lee.

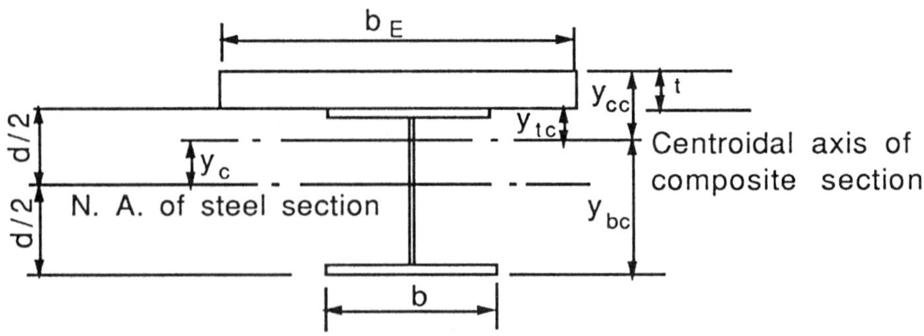

Figure 10.7 Location of neutral axis of a composite section.

Example 10.3

The floor system in Figure 10.8 is for a stackroom in a public library. The slab is 4.5 in. thick. The live load in the stackroom is 150 lb/ft^2 and the ceiling load 10 lb/ft^2. The construction is shored. Select a steel beam that will carry these loads economically using composite construction and the ASD method.

Solution

(a) *Moment calculation.* Assume a beam load of 35 lb/ft. The moment due to the weight of the steel section is

$$M_s = 0.035 \times \frac{24^2}{8}$$
$$= 2.52 \text{ k/ft}$$

The moment due to the weight of the slab and the ceiling loads is

$$M_D = \left(\frac{4.5}{12 \times 1000} \times 150 + 0.010\right) \times 6 \times \frac{24^2}{8}$$
$$= 28.6 \text{ k/ft}$$

The moment due to the live load is

$$M_L = \frac{150}{1000} \times 6 \times \frac{24^2}{8}$$
$$= 64.8 \text{ k/ft}$$

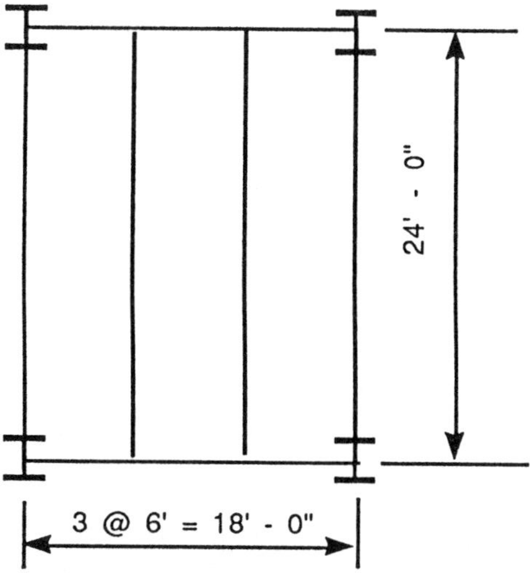

Figure 10.8 Floor frame: Typical Bay.

(b) *Determine b_E/n.*

$$L/4 = 72 \text{ in.}$$
$$S = 72 \text{ in.}$$
$$16t + b = 72 + b$$
$$b_E = 72 \text{ in.}$$

For a concrete strength of 4000 lb/in.^{2}, $n = 8$. Then, $b_E/8 = 9$.

(c) *Use the short-cut method.*

$$f_2 = \frac{M_D + M_L}{F_b}$$
$$= \left(\frac{28.6 + 64.8}{24}\right) \times 12$$
$$= 46.7 \text{ in.}^3$$

Since it is a shored construction, the assumed value of f_1 should be about 2 to 4%. Using the lower limit yields

$$S_{xbc} \geq \frac{f_2}{1 - f_1}$$
$$\geq 46.6 \text{ in.}^3$$

From Table 10.1 and by using 9 for b_E/n, W14 $\times$ 22 is observed to have a section modulus of 49.0 in.^{3}. The section modulus of the steel beam is 29.0 in.^{3}. For a trial section, W14 $\times$ 22 will be used.

The step-by-step method calculations yield the following results:

$$S_{xcb1} = 49.5 \text{ in.}^3$$

$$S_{xcb2} = 44.8 \text{ in.}^3$$

$$S_{xcc1} = 208.3 \text{ in.}^3$$

$$S_{xcc2} = 112.3 \text{ in.}^3$$

The final stresses are calculated as follows:

$$f_b = \frac{2.52 \times 12}{29} + \frac{26.5 \times 12}{44.8} + \frac{64.8 \times 12}{49.5}$$

$$= 23.44 \text{ k/in.}^2 < 24$$

The stress at the top of the slab is

$$f_{cc} = \frac{M_D}{nkS_{xcc2}} + \frac{M_L}{nkS_{xcc1}}$$

$$= \frac{26.5 \times 12}{8 \times 3112.3} + \frac{64.8 \times 12}{1 \times 8 \times 208.3}$$

$$= 0.585 \text{ k/in.}^2 < 0.45f_c' = 1.8 \text{ k/in.}^2$$

Use W14 × 22.

10.5 SHEAR CONNECTORS

The basic assumption in the design of composite structures is that no slippage occurs between concrete and steel when subjected to external or internal loading. One way of obtaining this is by providing shear connectors at the top flange of the steel section. See Figure 10.9. The function of shear connectors is to transfer horizontal shear, thus preventing the relative movement between concrete and the steel section. Shear connectors must be designed to transfer the stresses in both the horizontal and vertical directions.

The capacity of a shear connector at working stress for studs is

$$Q = 370d_s^2\sqrt{f_c'} \tag{10.33}$$

and for channels

$$Q = 220(t_f + 0.5t_w)c\sqrt{f_c'} \tag{10.34}$$

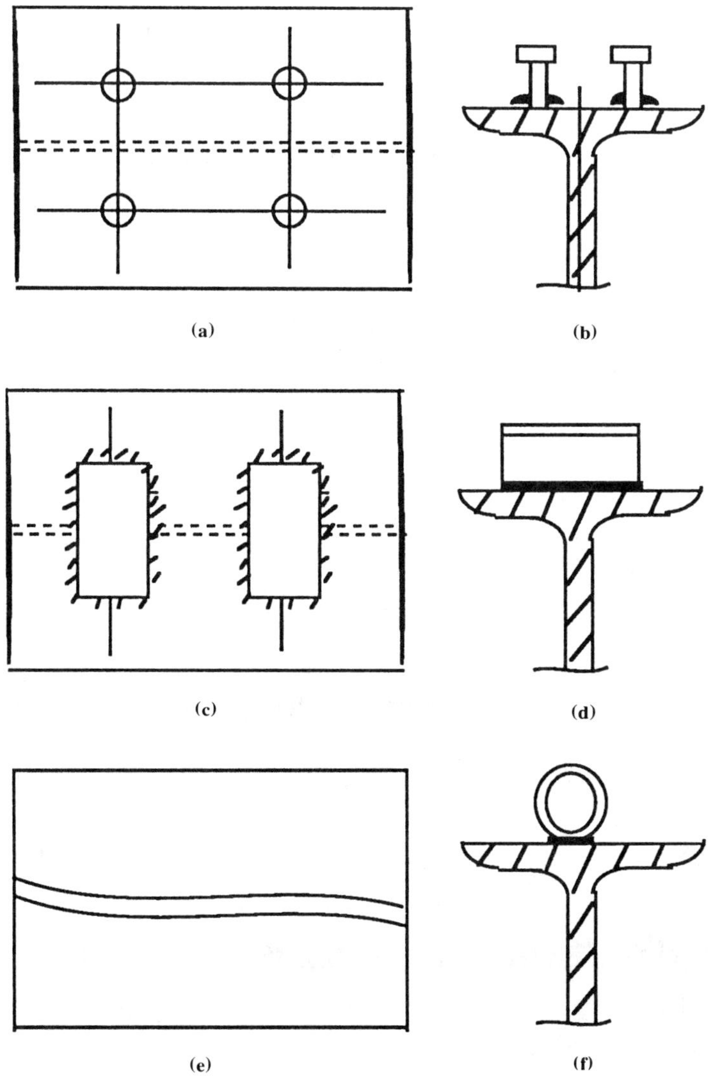

Figure 10.9 Types of shear connectors.

where

d_s = stud diameter (in.)
h_s = stud height (in.)
f'_c = compressive strength of concrete (lb/in.^{2})
t_f = thickness of the flange of the channel
t_w = thickness of the web of the channel
c = the length of the channel

The allowable shear capacity of connectors may be obtained from Table 14.1 of the AISC specifications.

The total horizontal shear to be resisted between the points of maximum and zero moments must be equal to the smaller of the values given by the two equations that follow. These equations are based on the ultimate capacity for concrete and steel.

$$V_h = \frac{0.85 f_c' A_c}{2} \tag{10.35}$$

and

$$V_h = \frac{F_y A_s}{2} \tag{10.36}$$

where

A_c = actual effective concrete area (in.^{2})
A_s = area of steel beam (in.^{2})
f_c' = concrete strength (k/in.^{2})

The number of shear connectors over one-half the beam for full composite action is given by

$$N = \frac{V}{Q} \tag{10.37}$$

10.6 LRFD METHOD: DESIGN ASSUMPTIONS

With the LRFD method, several assumptions are made to simplify the design procedure:

1. *Force determination.* Forces in a composite member are calculated on the basis of the effective area.
2. *Elastic analysis.* In the elastic analysis of prismatic members, the stiffness of the member is considered uniform throughout its length. The stiffness is calculated with a moment of inertia based on the transformed section.
3. *Plastic stress distribution.* For a slab in the positive moment region, the concrete stress is $0.85f_c'$ distributed uniformly in the compression block. The steel stress is F_y throughout the steel section.

4. *Fully composite beam*. Either the tensile yield strength of the steel section or compressive strength of the slab governs the maximum flexural strength of a fully composite beam subjected to a positive moment. Shear connectors are provided in sufficient numbers to develop the maximum flexural composite beam capacity. No slip is allowed between concrete and steel.

10.7 LRFD FLEXURAL MEMBERS

The considerations for the effective width of the slab are the same as those with the ASD method. The strength of the composite beam with shear connectors is

$$M_u = \phi_b M_n \tag{10.38}$$

where

M_n = the moment determined from the plastic stress distribution

ϕ_b = 0.85, with M_n determined from the plastic stress distribution on the composite section for $h_c/t_w \leq 640/\sqrt{F_{yf}}$

ϕ_b = 0.90, with M_n determined from the superposition of elastic stresses, considering the effect of shoring or nonshoring for $h_c/t_w > 640/\sqrt{F_{yf}}$

The plastic stress distribution for the composite beam is shown in Figure 10.10. The design flexural strength of a composite beam depends on the position of the neutral axis.

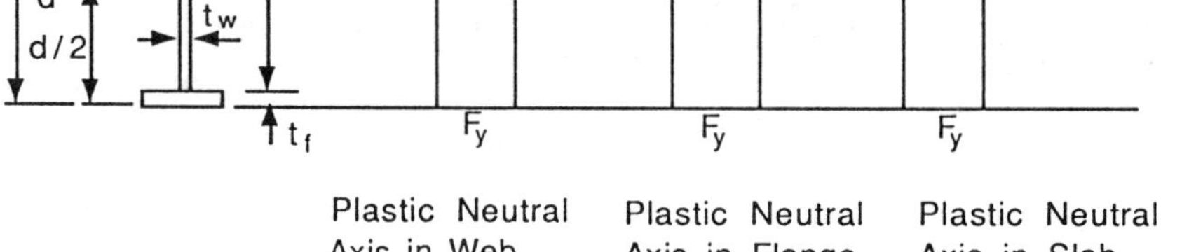

Figure 10.10 Use of plastic stress method in composite beams.

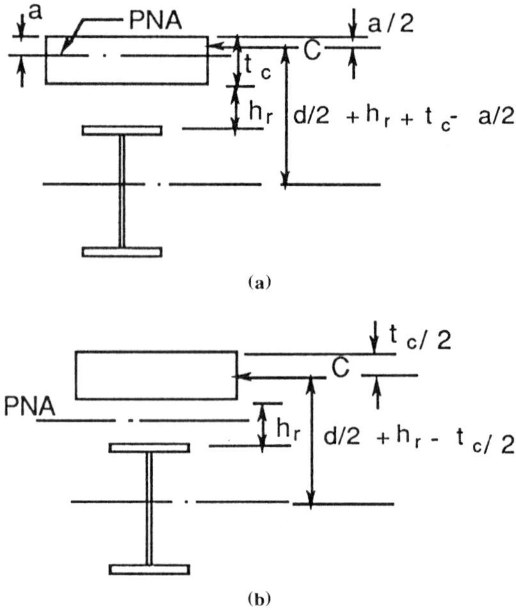

Figure 10.11 Neutral axis of composite sections using the plastic method.

Case 1. *Plastic neutral axis (PNA) falls in the slab (fully plastic).* For an analysis of this case, refer to Figure 10.11a

$$C = 0.85 f'_c b_E a$$

$$T = C$$

$$T = A_s F_y$$

$$a = \frac{A_s F_y}{0.85 f'_c b_E} \tag{10.39}$$

$$M_n = A_s F_y \left(\frac{d}{2} + h_r + t_c - \frac{a}{2} \right) \tag{10.40}$$

Case 2. *The neutral axis falls between the slab and top of the steel section.* Referring to Figure 10.11b, we observe that the nominal plastic moment is

$$M_n = A_s F_y \left(\frac{d}{2} + h_r - \frac{t_c}{2} \right) \tag{10.41}$$

Other possibilities for the PNA are in the flange and/or web of the steel beam. The above two cases represent the upper limit for the nominal plastic moment of the composite beam.

10.8 SHEAR CONNECTORS: LRFD METHOD

The compressive force developed in the slab of a fully plastic composite section is the smallest of

$$C = A_{sw}F_{yw} + 2A_{sf}F_{yf} \tag{10.42}$$

$$C = 0.85f'_c A_c \tag{10.43}$$

$$C = Q_n \tag{10.44}$$

where

A_c = area of concrete slab within the effective width (in.^{2})
A_{sw} = area of steel web (in.^{2})
A_{sf} = area of steel flange (in.^{2})
F_{yw} = minimum specified yield stress of web steel (k/in.^{2})
F_{yf} = minimum specified yield stress of flange steel (k/in.^{2})

The nominal capacity of shear connectors between the point of maximum moment and zero moment is

$$Q_n = 0.50A_{sc}\sqrt{f'_c E_c} \leq A_{sc}F_u \tag{10.45}$$

where

A_{sc} = cross-section of a stud shear connector (in.^{2})
f'_c = specified compressive strength of concrete (k/in.^{2})
F_u = minimum specified tensile strength of a stud shear connector (k/in.^{2})
E_c = modulus of elasticity of concrete (k/in.^{2})
= $(w^{1.5})\sqrt{f'_c}$
w = unit weight of concrete (lb/ft^3)
f'_c = concrete strength (k/in.^{2})

The nominal shear strengths of $\frac{3}{4}$-in. headed studs embedded in concrete slab are given in Table. 10.2

TABLE 10.2 Nominal Stud Shear Strength Q_n (k) for $\frac{3}{4}$-in. headed studs

f'_c (k/in.^{2})	w (lb/ft^3)	Q_n (k)
3.0	115	17.7
3.0	145	21.0
3.5	115	19.8
3.5	145	23.6
4.0	115	21.9
4.0	145	26.1

Source: Courtesy of AISC, Manual of Steel Construction, LRFD, First Edition, Section 4, p. 7, Table. 4.1.

For stud shear connectors embedded in a slab on a formed steel deck, a reduction factor is applied to the nominal shar strength of the stud. The factor is applicable only to the strength calculated on the basis of the concrete capacity. The reduction factor is

$$R = \left(\frac{0.85}{N_r}\right)\frac{w_r}{h_r}\left(\frac{H_s}{h_r} - 1\right) \leq 1.0 \tag{10.46}$$

where

N_r = number of stud connectors in one rib ≤ 3
w_r = average width of rib (in.)
h_r = nominal rib height (in.)
H_s = length of stud connector after welding limited to $h_r + 3$ in.

To use the LRFD composite design selection table as provided by the AISC, reference to Figure 10.12 is required. A preliminary section selection formula is

$$WT = \left[\frac{M_u(12)}{(d/2 + Y_{\text{con}} - a/2)\phi F_y}\right] \times 3.4 \tag{10.47}$$

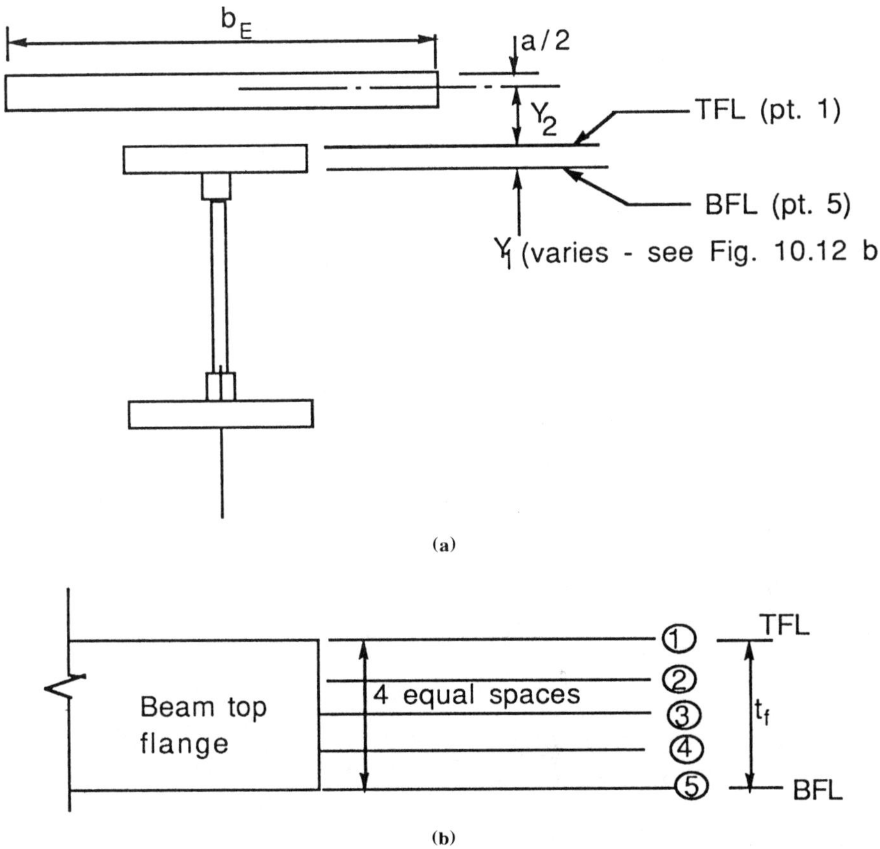

Figure 10.12 Composite sections by LRFD. (Figure 4.3 from LRFD, Manual of Steel Construction, first edition, 1986, by American Institute of Steel Construction p.5.)

where

$$WT = \text{weight of steel beam} \quad (\text{lb/ft}^3)$$

Example 10.5

A floorplan shows bays of 25 × 30 ft as depicted in Figure 10.13. Typical beams run at 10 ft on centers in the short direction. The slab is 4 in. thick supported by a steel deck 3 in. high and 6 in. wide ribs. The ribs are placed perpendicular to the beams. The concrete strength is 4 k/in.^{2}. The

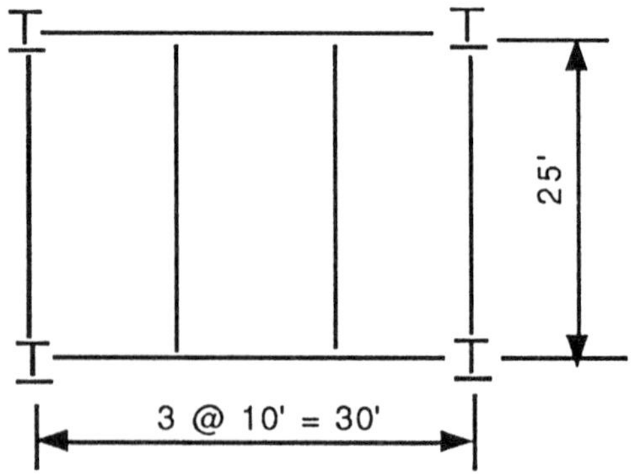

Figure 10.13 Typical floor bay.

construction is unshored. Live load on the floor is 85 lb/ft^2. Select a beam size using the LRFD method. A-36 steel is used. For shear connectors, use $\frac{3}{4}$-in. headed studs

Solution

Load Calculations

Slab load $\frac{4}{12} \times 0.15 \times 10$	$= 0.50$ k/ft
Steel deck (assumed) 0.003×10	$= 0.03$ k/ft
Steel beam (assumed) 0.05	$= 0.05$ k/ft
Total dead load	$= 0.58$ k/ft
Live load 0.085×10	$= 0.85$ k/ft

Factored Load Calculation

$$\begin{aligned} u &= 1.2D + 1.6L \\ &= 1.2 \times 0.58 + 1.6 \times 0.85 \\ &= 2.06 \text{ k/ft} \end{aligned}$$

Moment Calculation

$$\begin{aligned} M_u &= 2.06 \times \frac{25^2}{8} \\ &= 170 \text{ k/ft} \end{aligned}$$

TABLE 10.3 Preliminary Section Selection

d	$\dfrac{M_u(12)(3.4)}{\phi E_y}$	$d/2$	$Y_{con} - a/2$	WT
16	226.7	8	6	16.2
18	226.7	9	6	15.1
14	226.7	7	6	17.4
12	226.7	6	6	18.9

Preliminary Selection of Section

$$WT = \left[\frac{M_u(12)}{(d/2 + Y_{con} - a/2)\phi F_y}\right] \times 3.4$$

where

$$Y_{con} = 3 + 4 = 7 \text{ in.}$$

$$a/2 = 1 \text{ in.} \quad \text{(assumed)}$$

$$\phi = 0.85$$

$$F_y = 36 \text{ k/in.}^2$$

From the above results and Table 10.3, W12 × 19 seems to be the more likely section.

Calculations of Shear Connectors

The nominal strength of one stud is

$$Q_n = 0.5 \times 0.44\sqrt{4 \times 145^{1.5}\sqrt{4}} \quad \text{[Equation (10.45)]}$$

$$= 26.1 \text{ k}$$

Let the total shear provided by the connectors be

$$\sum Q_n = A_s F_y$$

$$= 5.57 \times 36$$

$$= 200.5 \text{ k}$$

Since C is the compression force in the slab and it is transferred to the steel section by the connectors, then

$$\sum Q_n = 0.85f'_c A_c$$
$$= 0.85f'_c b_E a$$
$$a = \frac{200.5}{0.85 b_E f'_c}$$
$$b_E = \frac{25 \times 12}{4}$$
$$= 75 \text{ in.} \quad \text{controls}$$
$$b_E = 10 \times 12$$
$$= 120 \text{ in.}$$

Then, the depth of the compression zone is

$$a = \frac{200.5}{0.085 \times 75 \times 4}$$
$$= 0.786 \text{ in.}$$
$$Y_2 = 7 - \frac{0.786}{2}$$
$$= 6.60 \text{ in.}$$

By interpolation from the composite selection table in AISC, LRFD Manual, First Edition for W14 × 22 and a value of 6.60 in. for Y_2

$$\phi M_n = 179 + (0.1/0.5)(186 - 179)$$
$$= 180.4 \text{ k/ft} > M_u = 170 \text{ k/ft}$$

Computation of the Number of Studs

Using a stud connector that protudes 2.5 in. above the top of the rib, we obtain

$$R = \frac{0.85}{\sqrt{1}} \times \frac{6}{3}\left(\frac{5.5}{3} - 1\right)$$
$$\times 1.416$$

No reduction is required.

From Table 10.2, we see that the nominal strength of one stud connector is 26.1 k. The number of stud connectors required is

$$N = \frac{\sum Q_n}{Q_n}$$

$$= \frac{200.5}{26.1}$$

$$= 7.7$$

Use 8 studs over one-half the span or a total of 18 studs for the entire length of the beam. The studs should be spaced uniformly.

11

Plastic Analysis and Design of Structures

11.1 INTRODUCTION

This chapter deals with the design of structural members based on their behavior beyond the elastic limit, with a particular emphasis on collapse loads. In treating the subject of the plastic design of structures, resistance to bending action is the primary means by which the loads are supported. The application of this method is best suited to the analysis of continuous beams and rigid frames.

Except for very tall buildings, the most significant stresses are bending stresses when deformations are relatively small. The effect of shear stresses, except for column members under special conditions, can be neglected in the determination of the collapse load without loss of accuracy. The behavior of the material beyond the elastic range is very highly variable. For structural steel, because of its high ductility, collapse of the structure occurs with very large deformation. The plastic design approach takes advantage of this unique property. Evidence of the ductility of steel is made clear by examining the stress–strain diagram obtained from a simple tension or compression test shown in Figure 11.1.

Ordinarily, the strain in the plastic range is more than 15 times higher than that of the elastic one. To simplify the analysis, the stress–strain diagram of structural steel is idealized as shown in Figure 11.1b.

11.2 BENDING OF BEAMS

In the elastic range, the analysis of beams in bending is based on simplified assumptions and conditions. These same assumptions are used in the

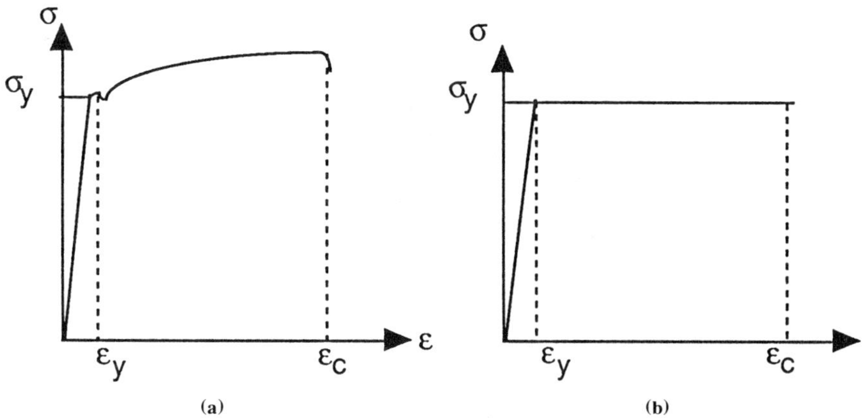

Figure 11.1 Stress vs. strain curve for structural steel. (a) Real. (b) Ideal.

analysis of bending of beams in the plastic range. The assumptions are as follows:

1. Strain is proportional to the distance from the neutral axis.
2. Stress–strain is idealized to consist of two straight lines as shown in Figure 11.1b

$$\sigma = E\varepsilon, \qquad 0 < \varepsilon < \varepsilon_y$$

$$\sigma = \sigma_y, \qquad \varepsilon > \varepsilon_y$$

3. Strains are small enough so that ϕ (angle of curvature) is equal to tan ϕ.
4. The equilibrium conditions stipulates that the internal forces and moments are obtainable as explained below.

The normal force is given by

$$F = \int_{\text{Area}} \sigma \, dA$$

and the bending moment by

$$M = \int_{\text{Area}} y \cdot \sigma \, dA$$

Consider the bending of a beam with a rectangular cross section as shown in Figure 11.2. For relatively small loads, the maximum fiber stress

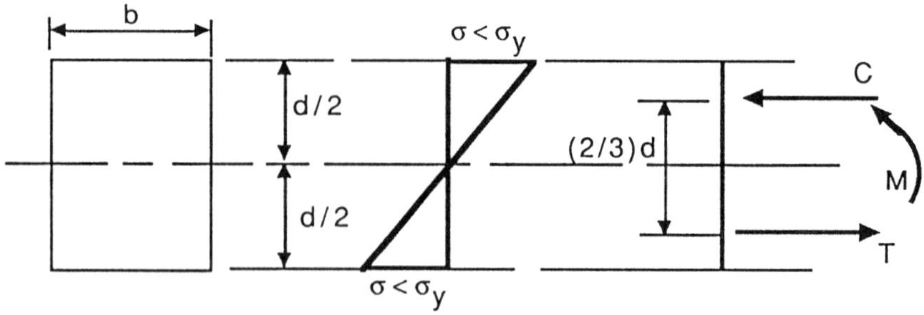

Figure 11.2 Stress distribution diagram: Elastic range.

is below the yield. Additional loads leading to an extreme fiber stress of $\sigma = \sigma y$ maintain the elastic stress condition in Figure 11.3.

The internal bending moment for this level of stress condition is denoted by M_y. From the condition of equilibrium, the resisting moment can be expressed as follows:

$$M_y = 2bt_f \times \frac{1}{2} \times (\sigma_y + \sigma_f) \times \left(\frac{d}{2} - \frac{t_f}{2}\right) + 2t_w \times \frac{\sigma_f}{2} \times \left(\frac{d}{2} - t_f\right) \times \frac{2}{3} \times \left(\frac{d}{2} - t_f\right) \tag{11.1}$$

With assumptions 1 and 2, the expression for the stress at the bottom of the flange σ_f is expressed as

$$\sigma_f = \frac{\dfrac{d - 2t_f}{2}}{\dfrac{d}{2}} \sigma_y \tag{11.2a}$$

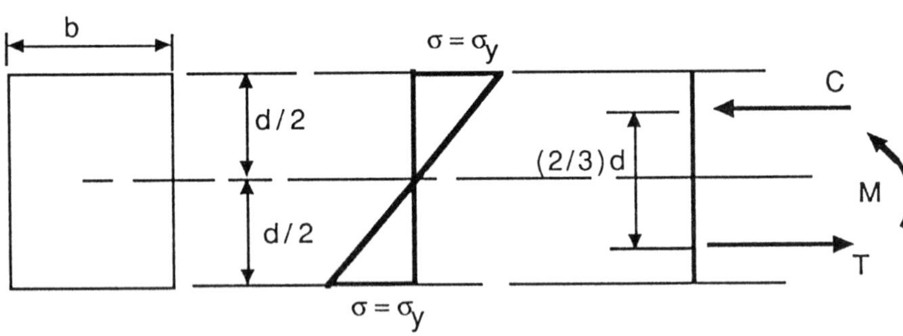

Figure 11.3 Stress distribution: Yield limit.

In simplified form,

$$\sigma_f = \frac{d - 2t_f}{d}\sigma_y \tag{11.2b}$$

Substituting Equation (11.2b) in Equation (11.1) and simplifying the result yield

$$M_y = \frac{1}{2}bt_f\sigma_y\left(1 + \frac{d - 2t_f}{d}\right)(d - t_f) + \frac{2}{3}t_w\sigma_y\left(\frac{d - 2t_f}{d}\right)\left(\frac{d}{2} - t_f\right)^2 \tag{11.3}$$

Further simplification reduces Equation (11.3)

$$M_y = \sigma_y\left[\frac{bt_f}{d}(d - t_f)^2 + \frac{t_w}{6d}(d - 2t_f)^3\right] \tag{11.4}$$

Note that Equation (11.4) is valid for both wide-flange sections and rectangular ones. To use it for a rectangular section, let t_w be replaced by b and t_f made equal to zero. Then,

$$M_y = \sigma_y\frac{bd^2}{6} \tag{11.5}$$

This is the same equation for a rectangular section. Let the section modulus about the x axis be denoted by S_x; the bending moment expression becomes

$$M_y = \sigma_y S_x \tag{11.6}$$

where

$$S_x = \frac{bd^2}{6} \tag{11.7}$$

for a rectangular section.

By a similar procedure, the moment of inertia for both wide-flange and rectangular sections can be expressed as

$$I_x = \frac{1}{2}bt_f(d - t_f)^2 + \frac{t_w}{12}(d - t_f)^3 \tag{11.8}$$

The section modulus can be obtained by dividing Equation (11.8) by $d/2$ or by substitution between Equations (11.4) and (11.6) to yield

$$S_x = \frac{bt_f}{d}(d - t_f)^2 + \frac{t_w}{6d}(d - 2t_f)^3 \tag{11.9}$$

As the loading continues beyond the state of yield in the first fibers, the yield condition penetrates deeper and deeper into the section. See Figure 11.4.

In the fully plastic state, all fibers throughout the entire cross section reach the yield condition. See Figure 11.5.

Similarly, the equations for the plastic moment and plastic section modulus are written as follows:

$$M_p = \left[bt_f(d - t_f) + t_w\left(\frac{d}{2} - t_f\right)^2\right]\sigma_y \tag{11.10}$$

and

$$Z_p = bt_f(d - t_f) + t_w\left(\frac{d}{2} - t_f\right)^2 \tag{11.11}$$

For a rectangular section, substituting $t_f = 0$ and $t_w = b$ yields

$$M_p = \frac{bd^2}{4}\sigma_y \tag{11.12}$$

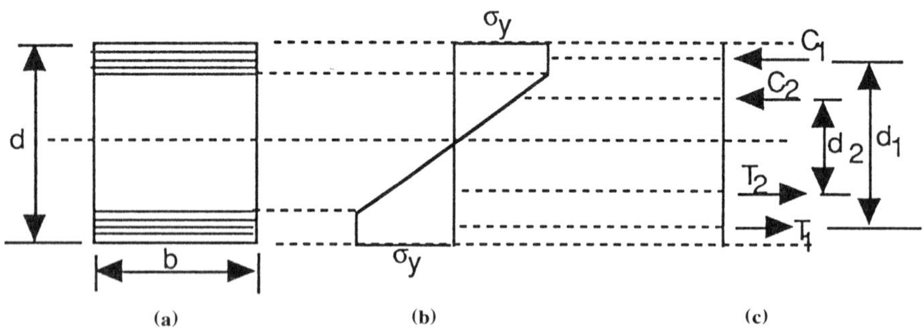

Figure 11.4 Stress distribution: Plastic-elastic range.

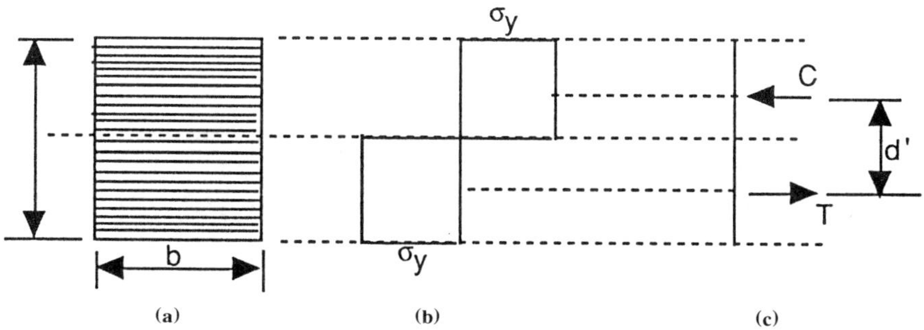

Figure 11.5 Stress distribution: Fully plastic state.

and

$$Z_p = \frac{bd^2}{4} \tag{11.13}$$

For a ductile material like structural steel, a member reaching yield at the extreme fibers retains a reserve of strength that varies with the shape factor. The shape factor is defined as the ratio of the plastic moment M_p to the maximum elastic moment M_y.

For a rectangular section, the shape factor is

$$f = \frac{\sigma_y \dfrac{bd^2}{4}}{\sigma_y \dfrac{bd^2}{6}} = 1.5 \tag{11.14}$$

In general, the shape factor is determined from the ratio of the plastic section modulus to the elastic section modulus. Using Equations (11.11) and (11.9) results in the following expression for the shape factor:

$$f = \frac{bt_f(d - t_f) + t_w\left(\dfrac{d}{2} - t_f\right)^2}{\dfrac{bt_f}{d}(d - t_f)^2 + \dfrac{t_w}{6d}(d - 2t_f)^3} \tag{11.15}$$

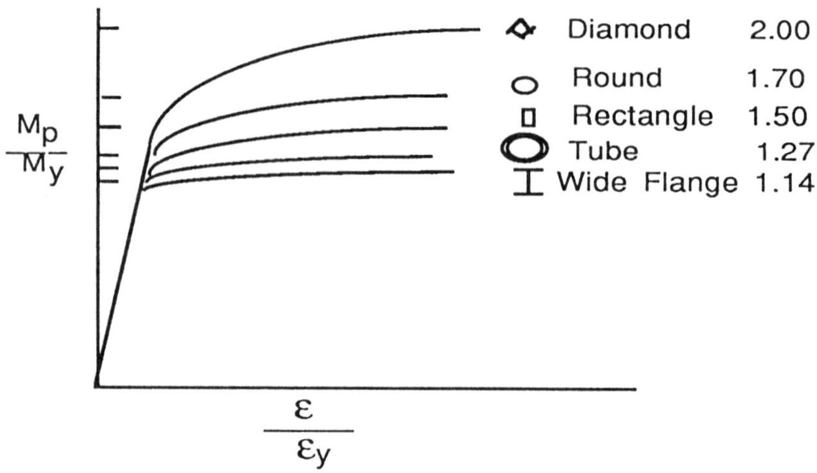

Figure 11.6 Shape factors for various sections.

It is determined from Equation (11.15) that the shape factor for wide-flange sections varies from a low of 1.08 to a high of 1.23, with 1.14 being the most common value. The same procedure is applied to determine the shape factors of other shapes. Figure 11.6 illustrates these variations.

Example 11.1

Find the shape factor for a circular tube as shown in Figure 11.7.

Solution

The element of force is given by

$$dF_h = trd\beta\sigma_h$$

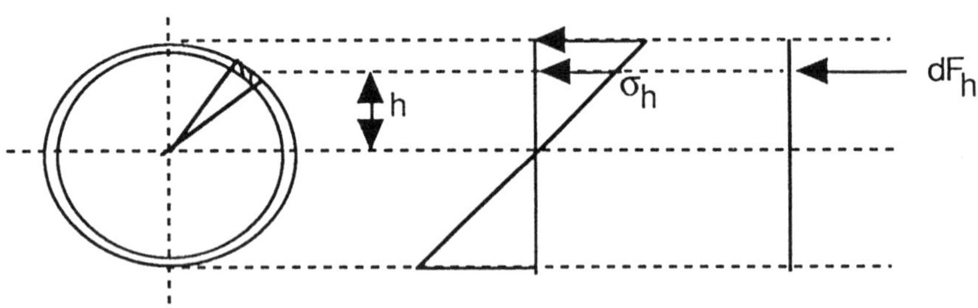

Figure 11.7 Hollow tube section.

where

$$t = \text{thickness of tube}$$

$$\beta = \text{angle from the zero axis}$$

$$r = \text{radius of the tube}$$

$$\sigma = \frac{h}{r}\sigma_y$$

$$dF_h = th\sigma_y \, d\beta$$

$$dM_y = th^2\sigma_y \, d\beta$$

$$h = r \sin \beta$$

$$dM_y = tr^2\sigma_y \sin^2 \beta \, d\beta$$

$$M_y = 4tr^2\sigma_y \int_{\beta=0}^{\pi/2} \sin^2 \beta \, d\beta$$

$$M_y = tr^2\sigma_y\pi$$

$$dF_p = tr\sigma_y \, d\beta$$

$$dM_p = trh\sigma_y \, d\beta$$

$$M_p = 4tr^2\sigma_y \int_{\beta=0}^{\pi/2} \sin \beta \, d\beta$$

$$M_p = 4tr^2\sigma_y$$

The shape factor is

$$f = \frac{M_p}{M_y}$$

$$= \frac{4tr^2\sigma_y}{tr^2\sigma_y\pi}$$

$$= 1.27$$

Example 11.2

Find the shape factor of a solid circular section as shown in Figure 11.8. Let yield take place at the extreme fibers. The moment about a centroidal axis can be written in terms of the yield stress σ_y and the parameters r

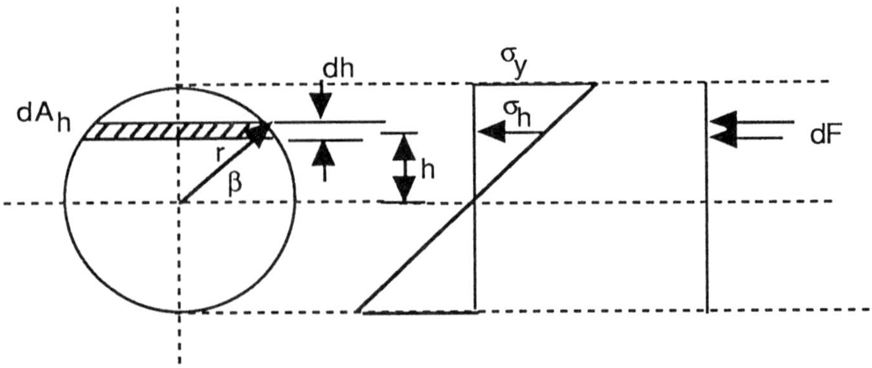

Figure 11.8 Circular section.

and h. Consider the area dA_h and the stress σ_h. The equation for the element of moment dM_y is written as follows.

Solution

$$dM_y = 2r \cos \beta \, dh \sigma_h h$$

$$\sigma_h = \frac{h}{r} \sigma_y$$

$$dM_y = 2h^2 \sigma_y \cos \beta \, dh$$

$$h = r \sin \beta$$

$$dh = r \cos \beta \, d\beta$$

$$dM_y = 2r^3 \sigma_y \sin^2 \beta \cos^2 \beta \, d\beta$$

$$My = \frac{r^3 \sigma_y \pi}{4}$$

Similarly, with $\sigma = \sigma_y$ for all fibers,

$$dM_p = 2r \cos \beta \, dh \sigma_y h$$

When all appropriate substitutions are made,

$$dM_p = 2r^3 \sigma_y \sin \beta \cos^2 \beta \, d\beta$$

and

$$M_p = \frac{4r^3\sigma_y}{3}$$

$$f = \frac{M_p}{M_y}$$

$$= \frac{\frac{4r^3\sigma_y}{3}}{\frac{r^3\sigma_y\pi}{4}}$$

$$= 1.70$$

Example 11.3

Find the shape factor of a diamond shaped cross section as shown in Figure 11.9.

Solution

$$dM_y = 2 \times dh\,\sigma_h h$$

$$\frac{x}{b} = \frac{H-h}{H}$$

$$x = \frac{b}{H}(H-h)$$

$$\sigma_h = \frac{h}{H}\sigma_y$$

$$dM_y = \frac{2b}{H^2}(H-h)h^2\sigma_y\,dh$$

$$M_y = \frac{bH^2}{3}\sigma_y$$

By a similar procedure,

$$M_p = \frac{4bH^2}{6}\sigma_y$$

and

$$f = \frac{\frac{4bH^2}{6}\sigma_y}{\frac{bH^2}{3}\sigma_y} = 2.00$$

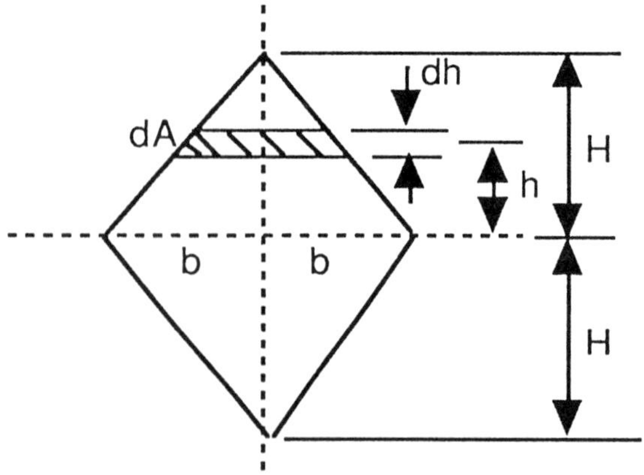

Figure 11.9 Diamond section.

11.3 DESIGN OF BEAMS: FAILURE MECHANISM APPROACH

Designing structural members with the ASD approach, we consider the member to have reached its limit whenever the extreme fibers have reached the maximum allowable stress

$$f_b = 0.66F_y \tag{11.16}$$

Consider a simply supported beam having a rectangular section and carrying a concentrated load at the center of the span. See Figure 11.10. As the extreme fibers at the center of the span reach yield, any further loading will be carried by the fibers below. A plastic node is formed that radiates downward and to the sides as loading progresses to produce the partial plastic condition as shown in Figure 11.10d. Additional loading can be applied to the beam up to the point of failure as shown in Figure 11.10f. That load will constitute the collapse load, and the service loading condition can be obtained by dividing the collapse load by an appropriate factor of safety.

Example 11.4

Use a simply supported beam with a rectangular section carrying a concentrated load at midspan. See Figure 11.11. The section has 4 in. × 8

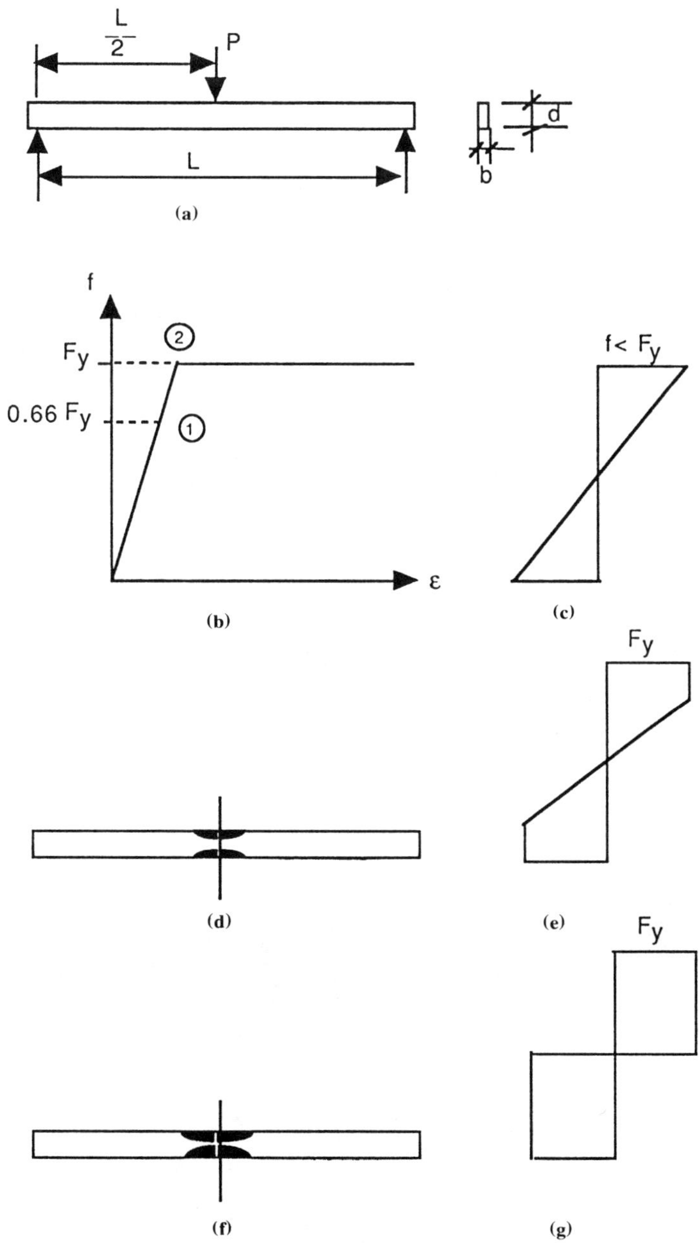

Figure 11.10 Development of the plastic hinge.

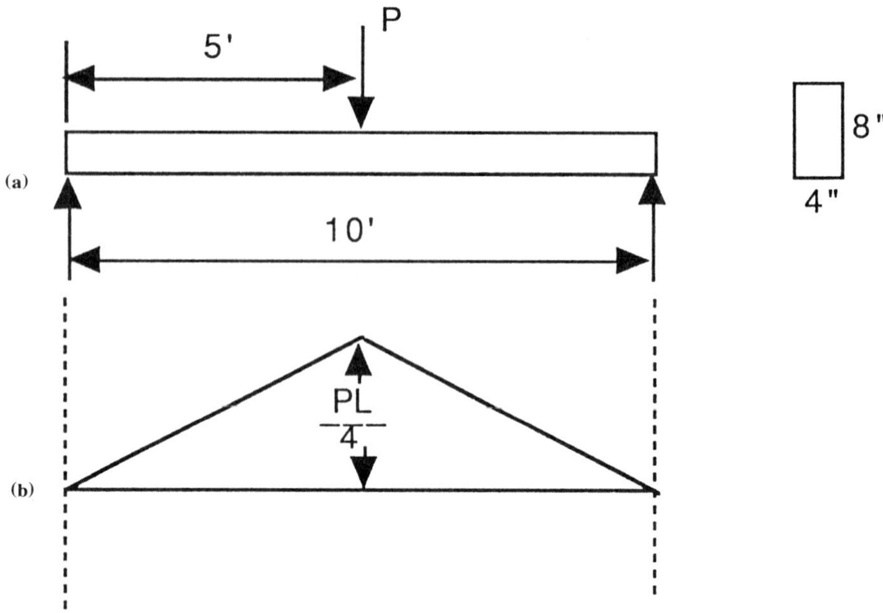

Figure 11.11 Simply supported beam with a concentrated load at midspan.

in. dimensions. Find the load-carrying capacity of the beam using the:

(**a**) Elastic method
(**b**) Plastic method

Solution

With the ASD method, the maximum allowable stress for A-36 steel is 24 k/in.^{2}

$$M_{\max} = \frac{PL}{4}$$

$$M = S_x F_b$$

The load is expressed as follows:

$$P = \frac{4S_x F_b}{L}$$

$$S_x = \frac{bd^2}{6}$$

$$= \frac{4 \times 8^2}{6}$$

$$= 42.7 \text{ in.}^3$$

$$P = \frac{4 \times 42.7 \times 24}{10 \times 12}$$

$$= 34.2 \text{ k}, \quad \text{(maximum allowable load)}$$

Use the plastic method. The moment diagram is assumed to be the same as in the elastic method

$$P_c = \frac{4Z_x F_y}{L}$$

$$Z_x = \frac{bd^2}{4} = 64 \text{ in.}^3$$

$$P_c = \frac{4 \times 64 \times 36}{4}$$

$$= 76.8 \text{ k}$$

Using a factor of safety of 1.5 yields a maximum allowable load of 50.7 k that the beam can carry. The ratio of the loads obtained by the two methods for rectangular sections is 1.5, which is the same as the shape factor. It should be noticed that using the plastic method in the design of rectangular sections in steel provides a 50% advantage. On the average, the advantage of using the plastic method with wide-flange sections is about 14%.

The analysis and design of beams using the plastic method are based on the following assumptions:

1. A plastic hinge forms at a point of maximum moment where the moment is M_p.

2. The plastic moment is the product of the plastic section modulus and yield stress F_y.

3. The shape factor ($f = Z_x/S_x$) determines the reserve strength that can be utilized beyond the elastic limit.

11.4 FIXED END BEAM

The example of a fixed end beam with a constant section and uniform gravity load serves as a good illustration of the concepts of plastic analysis and design of structural members. Figure 11.12 provides a diagrammatic explanation of how moment redistribution takes place during loading beyond the elastic limit. In addition to the modest increase of loading that results from the formation of plastic hinges at the support, redistribution of the moments will enhance the capacity of the beam.

As the load is increased, the moment redistribution will go through different stages as indicated by Figure 11.12. A pattern of deflection is associated with each of the moment conditions. Figure 11.12b represents a

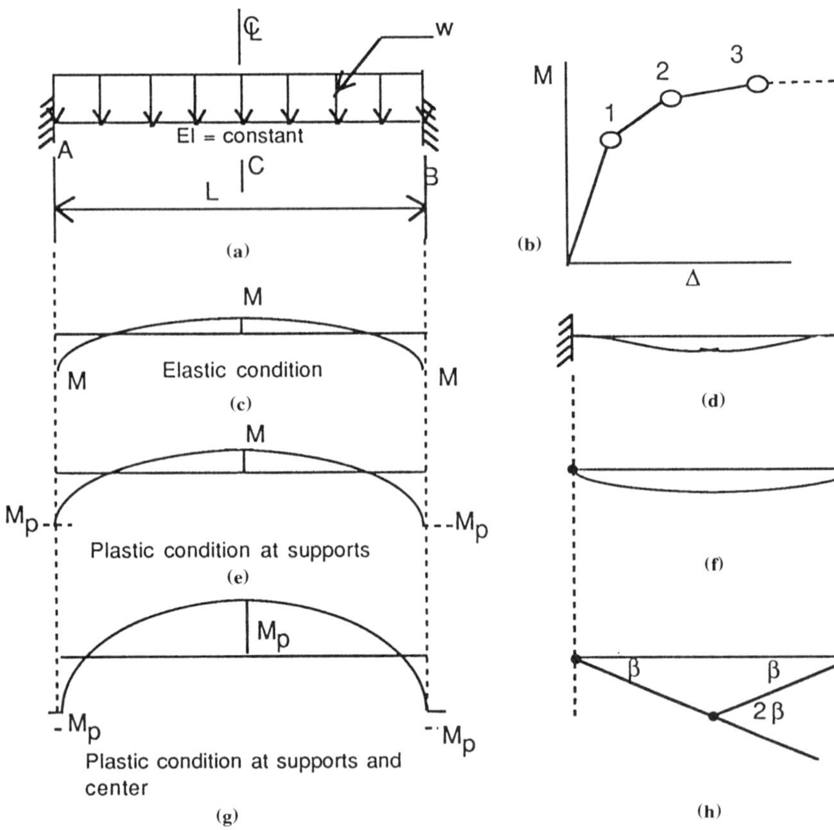

Figure 11.12 Fixed end beam with a uniform load on entire span.

pictorial conceptualization of the relationship between moment and deflection. Note that the region between 1 and 2 describes the moment state when plastic hinges have formed at the supports. This condition transforms the beam into a simply supported one for the remainder of the loading condition.

It is of interest to compare the load-carrying capacity of the fixed end beam using the ASD and plastic methods. Let w_a and w_p represent the allowable and collapse loads, respectively. The maximum moment occurs at the supports, and it is given by

$$M = \frac{w_a L^2}{12} \tag{11.17}$$

The allowable uniform load is

$$w_a = \frac{12M}{L^2} \tag{11.18}$$

From Figure 11.12g, the sum of the negative and positive moments is equal to the moment of a simply supported beam with a uniform load that induces collapse in the beam. Hence,

$$2M_p = \frac{w_c L^2}{8} \tag{11.19}$$

$$M_p = \frac{w_c L^2}{16} \tag{11.20}$$

and

$$w_c = \frac{16M_p}{L^2} \tag{11.21}$$

With the ASD method, the maximum allowable moment is

$$M = S_x F_b \tag{11.22}$$

where

$$F_b = 0.66F_y$$

For the plastic condition, the moment capacity is

$$M_p = Z_x F_y \tag{11.23}$$

The ratio of the collapse and allowable loads is

$$\frac{w_c}{w_a} = \frac{\dfrac{16Z_xF_y}{L^2}}{\dfrac{12S_x0.66F_y}{L^2}} = 2f \tag{11.24}$$

where

$$\frac{Z_x}{S_x} = f \quad \text{(the shape factor)}$$

For a wide-flange beam, the average shape factor is 1.14. Then, the ratio of the collapse load to the service load is

$$\frac{w_c}{w_a} = 2.28 \tag{11.25}$$

Designing a fixed end beam by the plastic method and employing a factor of safety of 1.5 provide a 52% advantage for using a wide-flange section.

11.5 PLASTIC HINGES: MECHANISM OF FAILURE

A structure is considered to be stable and hence functional until a sufficient number of plastic hinges have been formed to render the structure unstable. As soon as the structure reaches an unstable condition, it is considered to have failed. Segments of the beam between the plastic hinges are able to move excessively without additional loads. This condition in a member is called a mechanism. The concepts of mechanism formation in a structure due to loading beyond the elastic limit and virtual work are used in the analysis and design of steel structures.

Example 11.5

The fixed end beam shown in Figure 11.13 is loaded with a concentrated load at midspan. Find the moment that produces a failure mechanism in the beam. Use the concepts of virtual work.

Solution

The principle of virtual work states that for small displacements, internal work is identically equal and opposite to external work for a structure.

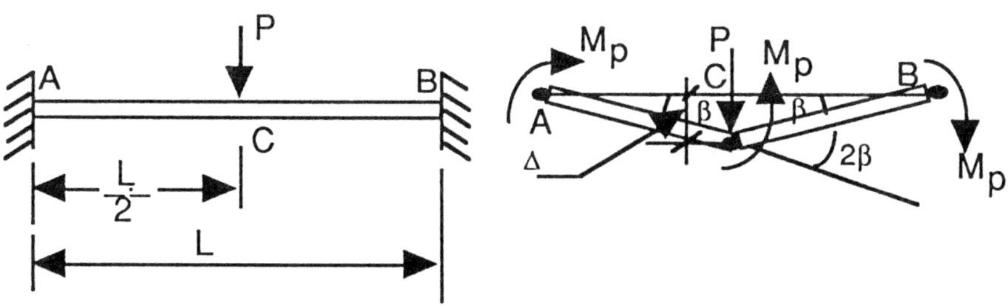

Figure 11.13 Fixed end beam with a concentrated load at midspan.

Internal work is defined as the product of all internal forces mobilized within the structure to resist the action of all external forces acting on the structure. Since work is defined as the product of the force and its displacement in the direction of the force, it is reasonable to use this concept in mapping out all the external and internal work in the above example.

Let external work be denoted by W_E and internal work by W_I. Then,

$$W_E = P_c \Delta \tag{11.26}$$

The internal work W_I is the result of the moments M_p at the support and midspan. Similarly,

$$W_I = M_p \beta + M_p 2\beta + M_p \beta \tag{11.27}$$

Note that between points A and C, the member is a straight line. Hence, for a small displacement, the tangent of the angle is equal to the angle itself in radians. Thus,

$$\Delta = \beta \times \frac{L}{2} \tag{11.28}$$

Substituting Equation (11.28) into (11.26) yields

$$W_E = P_c \frac{L}{2} \beta \tag{11.29}$$

Equating the right-hand side of equations (11.27) and (11.29) results in the following:

$$M_p = P_c \frac{L}{8} \tag{11.30}$$

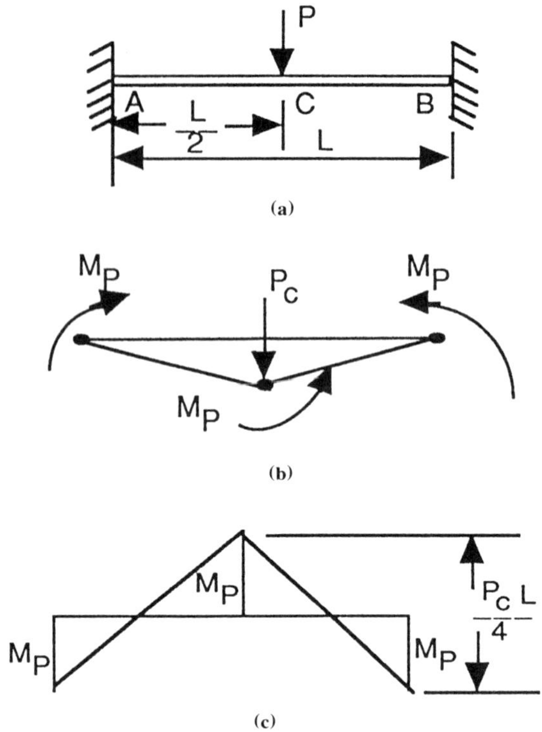

Figure 11.14 Failure mechanism of a fixed end beam with a concentrated load at midspan.

The same results may be obtained by using the moment equilibrium in the beam at the collapse condition. See Figure 11.14. It is noticed from Figure 11.14c that the sum of positive and negative moments is equal to that of a simply supported beam with a concentrated load applied at midspan

$$2M_p = P_c \frac{L}{4} \tag{11.31}$$

and

$$M_p = P_c \frac{L}{8} \tag{11.32}$$

Note that the same results have been obtained with both methods. Deciding which of the methods to use in a particular case will depend on the simplicity of one compared to the other in each individual case.

Example 11.6

Use a uniform load applied to a fixed end beam as shown in Figure 11.15. Find the moment capacity of the beam by plastic analysis using the failure mechanism and moment equilibrium approaches.

Solution

With the failure mechanism, the expressions for the external and internal work are written as follows:

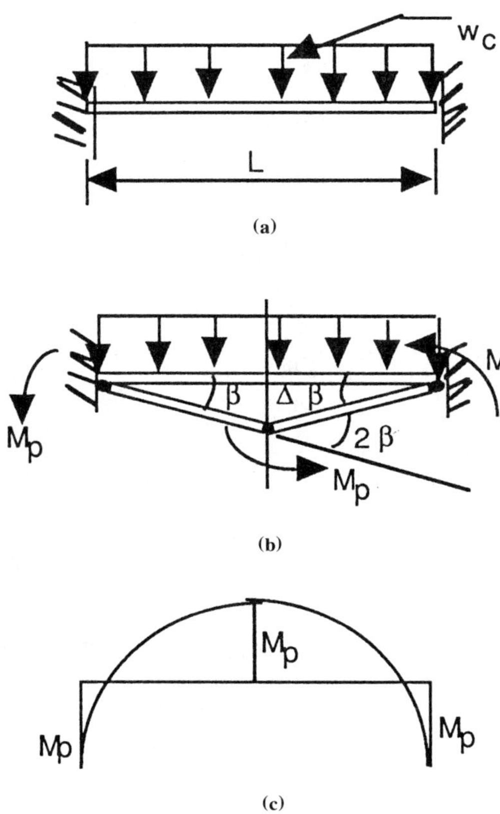

Figure 11.15 Failure mechanism of a fixed end beam with a uniformly distributed load.

External work W_E

$$W_E = w_c \times L \times \frac{\Delta}{2} \tag{11.33}$$

Note that the total load is multiplied by the average displacement in this case. The displacement is

$$\Delta = \beta \times \frac{L}{2} \tag{11.34}$$

Then,

$$W_E = w_c \times L^2 \times \frac{\beta}{4} \tag{11.35}$$

Internal work

$$W_I = 4M_p \tag{11.36}$$

Combining Equations (11.35) and (11.36) yields

$$M_p = w_c \times \frac{L^2}{16} \tag{11.37}$$

From Figure 11.15c, it is noticed that

$$2M_p = w_c \times \frac{L^2}{8}$$

from which the same result is obtained for M_p.

Example 11.7

Place a concentrated load at any point kL from the left support on a fixed end beam as shown in Figure 11.16. Write the general expression for the moment capacity M_p using the failure mechanism concept.

Solution

The assumption that plastic hinges are developed at points of maximum moment is used in locating the mechanisms of failure in the above

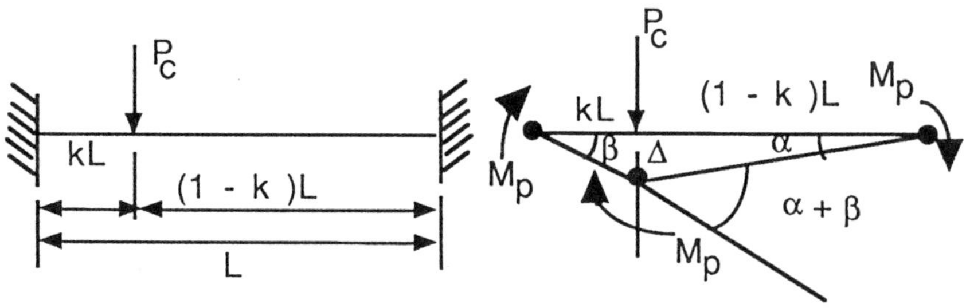

Figure 11.16 Fixed end beam with one single concentrated load at any point.

structure. Next, we develop the geometric relationship between the angles of rotation at the plastic hinges:

$$\alpha = \frac{k}{1-k}\beta \tag{11.38}$$

and

$$\alpha + \beta = \frac{1}{1-k}\beta \tag{11.39}$$

The virtual work equation is written as follows:

$$P_c kL\beta = M_p\beta + M_p\frac{1}{1-k}\beta + M_p\frac{k}{1-k}\beta \tag{11.40}$$

Equation (11.40) yields

$$M_p = \frac{k(1-k)L}{2}P_c \tag{11.41}$$

When the load is located in the middle of the span, $k = \frac{1}{2}$. The value of the plastic moment is reduced to $M_p = P_cL/8$, which is the same as in Equation (11.30).

Example 11.8

Design a fixed end beam with a concentrated collapse load placed at one-third of the span from the left support. The beam length is 27 ft and carries a load of 30 k. The load factors are as follows:

Gravity loads

$$F = 1.70$$

Lateral loads

$$F = 1.30$$

Solution

From Equation (11.41),

$$M_p = \frac{k(1-k)L}{2} P_c$$

where $k = \frac{1}{3}$, $P_c = 1.70 \times 30$ k, with

$$\begin{aligned} M_p &= \frac{\frac{1}{3}(1-\frac{1}{3})27}{2} \times 30 \times 1.70 \\ &= 153 \text{ ft k} \\ Z_x &= \frac{M_p}{F_y} \\ &= \frac{153}{36} \times 12 \\ &= 51.0 \text{ in.}^3 \end{aligned}$$

From the AISC selection table, it is clear that W16 × 31 with a plastic section modulus of 54 in.^{3} is the most economical section.

With the ASD method, the maximum moment occurs at the left support

$$M = \frac{ab^2}{L^2} \times P$$

where

$$\begin{aligned} a &= kL \\ b &= (1-k)L \\ M &= -\frac{9 \times 18^2}{27^2} \times 30 \\ &= -120 \text{ ft k} \end{aligned}$$

The section modulus is

$$\begin{aligned} S_x &= \frac{120 \times 12}{0.66 F_y} \\ &= 60 \text{ in.}^3 \end{aligned}$$

The ASD selection table indicates W16 × 40 or W18 × 40 is the most economical section that can be used. A savings of about 22% results with the plastic method.

Example 11.9

Design a beam having the same length and support conditions as in Example 11.8 with a uniform load of 2.0 k/ft.

Solution

From Equation (11.37), the plastic moment is

$$M_p = \frac{w_c L^2}{16}$$

$$= \frac{1.7 \times 2 \times 27^2}{16}$$

$$= 154.9 \text{ ft k}$$

$$Z_x = \frac{154.9 \times 12}{36}$$

$$= 51.6 \text{ in.}^3$$

Use W16 × 31 with a plastic section modulus of 54.0 in.^{3}. The maximum moment obtained from the elastic analysis is

$$M = \frac{wL^2}{12}$$

$$= \frac{2 \times 27^2}{12}$$

$$= 121.5 \text{ ft k}$$

W16 × 40 or W18 × 40 will satisfy the requirements.

11.6 FIXED END BEAM WITH MULTIPLE CONCENTRATED LOADS

In the following section, the general equation is derived for the plastic moment of a fixed end beam with two concentrated loads P_1 and P_2 at k_1L and k_2L, respectively, from the left support. See Figure 11.17.

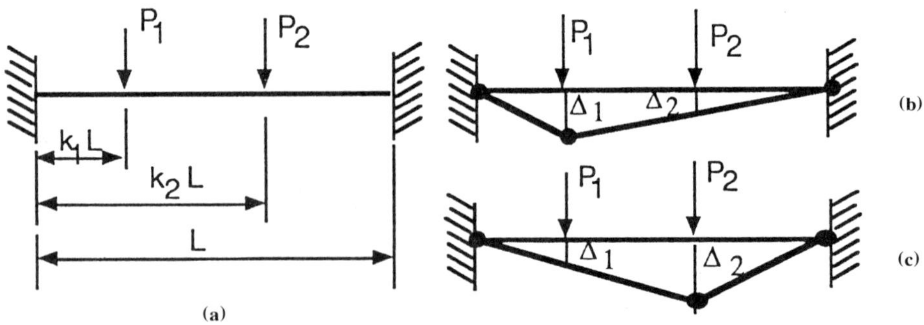

Figure 11.17 Fixed end beam with two concentrated loads.

The loading condition illustrated in Figure 11.17a leads to two possible failure patterns, as shown in Figures 11.17b and 11.17c. The analysis is performed for each case, and the one that produces the larger plastic moment is considered for the purpose of design.

The external work for the failure depicted in Figure 11.17b is

$$W_E = P_1\Delta_1 + P_2\Delta_2 \tag{11.42}$$

where

$$\Delta_1 = \beta k_1 L$$

$$\Delta_2 = \alpha(1 - k_2)L$$

and

$$W_E = P_1\,\beta k_1 L + P_2\,\alpha(1 - k_2)L \tag{11.43}$$

$$\alpha = \frac{k_1}{1 - k_1}\beta \tag{11.44}$$

Then,

$$W_E = P_1 k_1 L\beta + P_2\frac{k_1(1 - k_2)}{1 - k_1}L\beta \tag{11.45}$$

The expression for the internal work is

$$W_I = M_p\beta + M_p(\beta + \alpha) + M_p\alpha \tag{11.46}$$

Simplified, Equation (11.45) is written as follows:

$$W_I = 2M_p(\beta + \alpha) \tag{11.47}$$

Substituting Equation (11.44) in Equation (11.47) yields

$$M_p = \tfrac{1}{2}k_1 L[P_1(1 - k_1) + P_2(1 - k_2)] \tag{11.48}$$

A similar procedure for the failure pattern shown in Figure 11.17c yields the following expression for the plastic moment:

$$M_p = \tfrac{1}{2}(1 - k_2)(P_1 k_1 + P_2 k_2)L \tag{11.49}$$

Example 11.10

A fixed end beam is loaded with two concentrated loads as shown in Figure 11.17. We are given the following:

$$k_1 = 0.30, \qquad P_1 = 15 \text{ k}$$
$$k_2 = 0.60, \qquad P_2 = 20 \text{ k}$$

Find the design plastic moment.

Solution

From Equation (11.48),

$$M_p = \tfrac{1}{2}(0.30)[15(1 - 0.3) + 20(1 - 0.6)]30$$
$$= 83.2 \text{ ft k}$$

From Equation (11.49),

$$M_p = \tfrac{1}{2}(0.40)[15 \times 0.3) + 20 \times 0.6)]30$$
$$= 99 \text{ ft k}$$

Note that the pattern failure shown in Figure 11.17c controls selection and/or design. Use W14 × 22; M_p = 100 ft k.

11.7 CONTINUOUS BEAMS

In analyzing continuous beams by the plastic method, certain assumptions are necessary to simplify the solution. These assumptions are as follows:

1. Deformations are sufficiently small so that an equilibrium condition can be formulated for the undeformed structure.
2. The structure remains stable up to the point of attainment of the ultimate load.

3. The connections remain functional to provide full continuity.
4. Shear and normal forces are small enough to have a negligible effect on the plastic moment M_p.
5. Loading on the structure can increase equally for all the nodes or individual members as long as no local failure occurs.

There are two methods for the plastic analysis of structures: the statical (moment equilibrium) and failure mechanism methods. In the statical methods of analysis, the objective is to find an equilibrium moment condition for which $M < M_p$, such that a mechanism of failure is formed. The procedure is outlined step by step as follows:

1. Select the redundant or redundants if more than one exists in the structure.
2. Draw the moment diagram for the determinate structure.
3. Draw the moment diagram for the structure loaded with the redundants.
4. Superimpose the diagram from Step 3 on that from Step 2.
5. Compute the value of the ultimate load from the equilibrium equation constructed in Step 4.
6. Check that the moment $M \leq M_p$ for all positions in the structure.

The method of failure mechanism is based on the same set of assumptions as the former method of analysis. With the assumption of a small deformation, the principle of virtual work is utilized to establish the relationship between the ultimate load and plastic moment. It can be applied to continuous beams with constant or variable sections. If the load is uniform along the beam, the correct value of the plastic moment is obtained by determining the location of the point of maximum moment. Consider the beam in Figure 11.18.

For complicated frames, the statical method of analysis is more difficult, and finding the correct equilibrium equation becomes illusive. In these cases, the mechanism method is more practical. In a continuous frame, more than one mechanism is considered. The mechanism pattern with the largest plastic moment requirement represents the correct failure mechanism. The moment will not exceed the plastic moment anywhere in the frame. The procedure is outlined as follows:

1. Determine the location of the plastic hinge. It falls at the point of maximum moment.
2. Select possible independent and combined mechanisms.
3. Solve the equilibrium equation using virtual work.
4. $M \leq M_p$ at all locations in the frame.

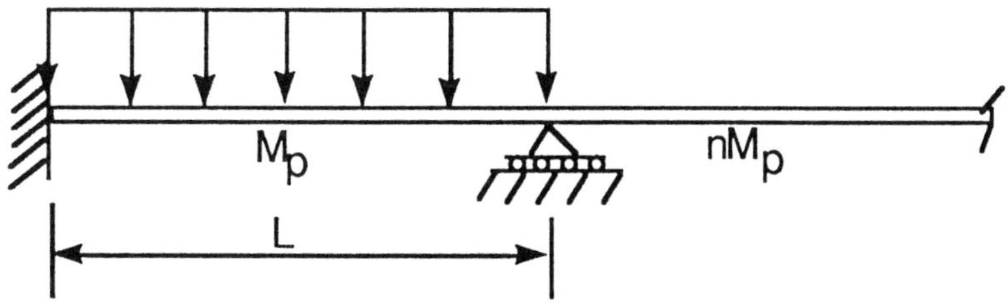

(a)

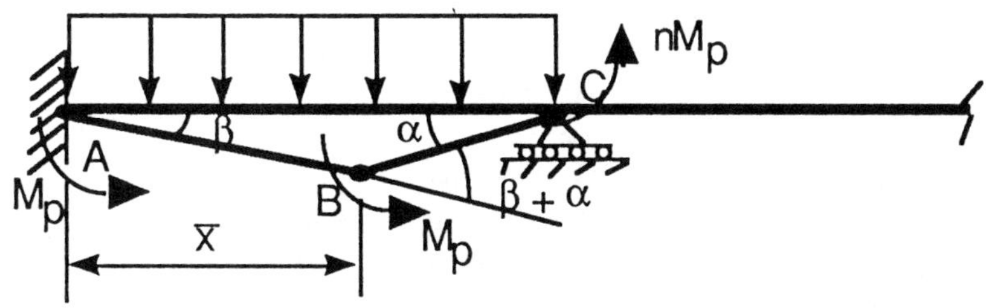

(b)

Figure 11.18 Continuous beam at one end and fixed at the other with variable sections and a uniformly distributed load.

There are two possibilities depending on the value of n:

1. $n = 1$ and $n > 1$: The plastic hinge falls to the left of C.
2. $n < 1$: The plastic hinge falls to the right of c.

Let the plastic hinge between the supports fall a distance $\bar{x}$ from the left support as shown in Figure 11.18b. For the case $n \geq 1$, the internal work is

$$W_I = 2M_p(\beta + \alpha) \tag{11.50}$$

and

$$\alpha = \frac{\bar{x}}{L - \bar{x}}\beta \tag{11.51}$$

$$(\beta + \alpha) = \frac{L}{L - \bar{x}}\beta \tag{11.52}$$

Then,

$$W_I = 2M_p\frac{L}{L - \bar{x}}\beta \tag{11.53}$$

$$W_E = \tfrac{1}{2}wL\bar{x}\beta \tag{11.54}$$

from which

$$\begin{aligned} M_p &= \tfrac{1}{4}w\bar{x}(L - \bar{x}) \\ \frac{dM_p}{dx} &= 0 = L - 2\bar{x} \\ \bar{x} &= 0.50L \end{aligned} \tag{11.55}$$

This is the same result that was obtained for a fixed end beam with a uniform load.

For $n < 1$, the internal work is given by

$$\begin{aligned} W_I &= M_p\beta + M_p(\beta + \alpha) + nM_p\alpha \\ &= M_p[2L + (n - 1)\bar{x}]\frac{1}{L - \bar{x}}\beta \end{aligned} \tag{11.56}$$

The expression for the external work is unchanged. Then,

$$\tfrac{1}{2}wL\bar{x}\beta = M_p[2L + (n - 1)\bar{x}]\frac{1}{L - \bar{x}}\beta \tag{11.57}$$

Simplifying the above equation and solving for the moment yield

$$M_p = \frac{wL}{2}\,\frac{\bar{x}(L - \bar{x})}{2L + \bar{x}(n - 1)}$$

When the equation is differentiated with respect to $\bar{x}$ and the differential is equated to zero, the following quadratic equation is obtained:

$$2L^2 - 4L\bar{x} - (n - 1)\bar{x}^2 = 0 \tag{11.58}$$

The solution for $\bar{x}$ is

$$\bar{x} = \frac{1}{n - 1}[\sqrt{4 + 2(n - 1)} - 2]L \tag{11.59}$$

Equation (11.59) is the general solution for the distance from the left support at which the maximum moment occurs.

For $n = 1$, $\bar{x} = 0/0$. This is an indeterminate case. Use L'Hopital's rule to determine the value of $\bar{x}$

$$\begin{aligned} \bar{x} &= \lim_{n=1} \frac{d}{dn}\sqrt{(4 + 2(n - 1))} \bigg/ \frac{d}{dn}(n - 1)L \\ &= 0.50L \end{aligned} \tag{11.60}$$

This corresponds to the same results for a fixed end beam. For a value of $n = 0.75$, $\bar{x} = 0.52L$.

Consider the case of a continuous beam with a simple support and uniform load as shown in Figure 11.19. The expression for internal work is

$$W_I = M_p(\beta + \alpha) + nM_p\alpha \tag{11.61}$$

The external work is

$$W_E = \tfrac{1}{2}wL\bar{x}\beta \tag{11.62}$$

$$(\beta + \alpha) = \frac{L}{L - \bar{x}}\beta \tag{11.63}$$

$$\alpha = \frac{\bar{x}}{L - \bar{x}}\beta \tag{11.64}$$

Substituting Equations (11.63) and (11.64) in Equation (11.61) and simplifying yield

$$W_I = \frac{M_p\beta}{L - \bar{x}}(L + n\bar{x}) \tag{11.65}$$

Combining Equations (11.62) and (11.65) gives

$$M_p = \tfrac{1}{2}wL\frac{\bar{x}(L - \bar{x})}{(L + n\bar{x})} \tag{11.66}$$

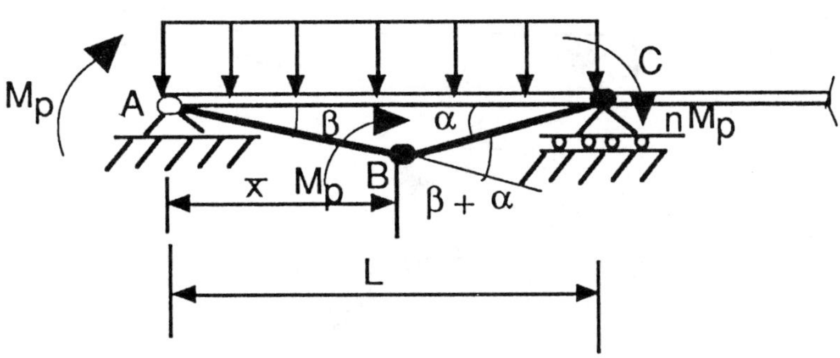

Figure 11.19 Continuous beam at one end and simply supported at the other with variable sections and a uniformly distributed load.

Differentiate the moment with respect to x and solve for $\bar{x}$ to obtain

$$\bar{x} = \frac{1}{n}(-1 + \sqrt{(1+n)})L \qquad (11.67)$$

For $n = 1$, $\bar{x} = 0.414L$. For $n = 0$, the limit of $\bar{x}$ using L'Hopital's rule is $0.5L$.

Example 11.11

A continuous beam carries a uniform load of 1 k/ft between points A and B and a concentrated load of 20 k placed as shown in Figure 11.20. Find an optimum beam size to carry the given loads.

Solution

From the given information, $n = \frac{2}{3}$. There are two possibilities for the collapse of the system. They are demonstrated in Figures 11.20b and 11.20c.

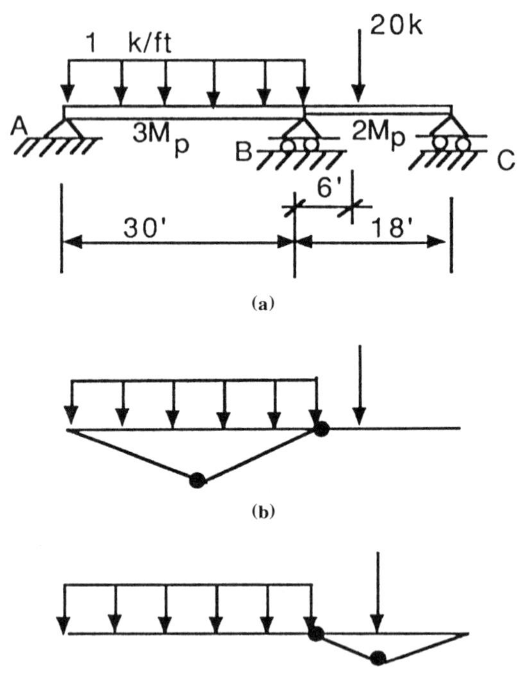

Figure 11.20 Failure mechanism of a continuous beam, two spans.

Considering the case in Figure 11.20b, we obtain the value of $\bar{x}$ from Equation (11.67)

$$\begin{aligned}\bar{x} &= \frac{1}{n}(-1 + \sqrt{(1+n)})L \\ &= 0.436L \\ &= 0.436 \times 30 \\ &= 13.08 \text{ ft}\end{aligned}$$

The collapse load is

$$\begin{aligned}w_c &= 1.70w \\ &= 1.70 \text{ k/ft}\end{aligned}$$

Having obtained the values of n, w_c, and $\bar{x}$, we can determine the moment M_p from Equation (11.66)

$$M_p = 145.8 \text{ ft k}$$

The collapse mechanism in Figure 11.20c requires a different moment. The external and internal work expressions are as follows:

$$W_E = P_c kL\beta \tag{11.68}$$

$$W_I = M_p\beta + M_p(\beta + \alpha) \tag{11.69}$$

$$\alpha = \frac{k}{1-k}\beta \tag{11.70}$$

Combining Equations (11.68), (11.69), and (11.70) and solving for M_p yield

$$M_p = P_c\frac{kL(1-k)}{(2-k)}$$

$$k = \tfrac{1}{3} \tag{11.71}$$

$$L = 18 \text{ ft}$$

$$P_c = 1.7 \times 20 = 34 \text{ k}$$

Substituting the above values in the moment equation results in

$$M_p = 81.6 \text{ ft k}$$

The section capacity on the left is equivalent to $\frac{3}{2}$ of that on the right. The moment capacity for AB is equal to 122.4 ft k. Thus, the first failure mechanism controls.

From the plastic design selection table, we observe that W16 × 31 has a capacity of 162 ft k. This section is the optimum size meeting the load requirement.

The equilibrium of moments will be applied to the same problem given in Example 11.10. The moment diagram for the continuous beam is shown in Figure 11.21.

The moment diagrams above the horizontal line represent the redundant moments. The determinate moments on each span are shown below the line. For the left span,

$$\begin{aligned} M &= \frac{w_c L^2}{8} \\ &= 1.7 \times 1 \times \frac{30^2}{8} \\ &= 191.2 \text{ ft k} \end{aligned}$$

From Figure 11.21, the following moment relationships are established:

$$\begin{aligned} 3M_p &= 191.2 - M_p \\ M_p &= 47.8 \text{ ft k} \end{aligned}$$

and

$$3M_p = 143.4 \text{ ft k}$$

For the right span,

$$\begin{aligned} M &= \frac{P_c ab}{L} \\ &= 20 \times 1.7 \times \frac{6 \times 12}{18} \\ &= 136 \text{ ft k} \end{aligned}$$

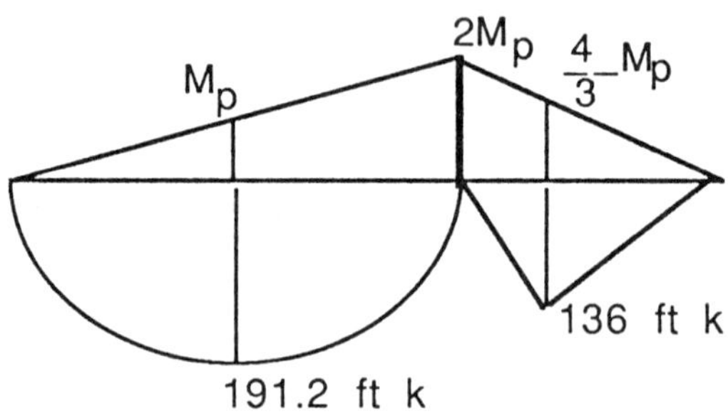

Figure 11.21 Equilibrium of moments approach.

and

$$2M_p - 136 - \tfrac{4}{3}M_p$$

$$M_p = 40.8 \text{ ft k}$$

$$3M_p = 122.4 \text{ ft k}$$

Note that the moment equilibrium on the left controls. It agrees with the results obtained by the failure mechanism method.

Example 11.12

Given the continuous beam shown in Figure 11.22, design the beam for the most economical combination of sizes in the two different spans.

Solution

Apply failure to the left span with $n = 0.50$. From Equation (11.67), the value of $\bar{x}$ is

$$\bar{x} = \frac{1}{n}(\sqrt{1+n} - 1)L$$

$$= 0.450L$$

The expression for the moment is given by Equation (11.66)

$$M_p = \frac{1}{2}wL\frac{\bar{x}(L-\bar{x})}{L+n\bar{x}}$$

$$w = 1.3 \times 1.7 \quad \text{(the factored load for gravity loads)}$$

Substituting the values of n, L, w, and $\bar{x}$ above yields $M_p = 357$ ft k and $nM_p = 178$.

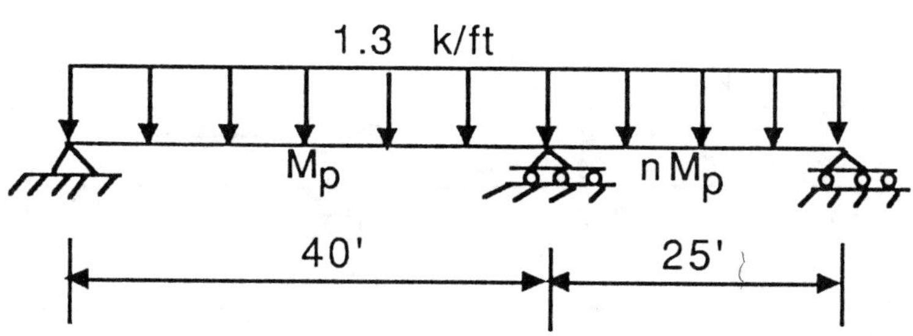

Figure 11.22 Two-span continuous beam with a variable section and uniformly distributed load.

If we consider failure in the right span, $n = 1$ since both plastic hinges appear in that beam. From Equation (11.67), $\bar{x} = 0.414L$. It yields a value of $M_p = 118$ ft k. If the beams are designed for their failure capacity and not any higher, the ratio $n = 118/357 = 0.333$ and not 0.500. Try the solution for the new value of n. The second and third cycle iterations yield values for M_p in the left span of 382 and 384 ft k, respectively. The optimum beam sizes are W24 × 55 for the left span and W14 × 26 for the right span with capacities of 402 and 121 ft k, respectively.

Example 11.13

Design the continuous beam shown in Figure 11.23.

Solution

First, assume that the beam is uniform throughout its length. Consider a value of $n = 1$ for the sections AB and CD.

For AB, we apply Equations (11.66) and (11.67) for M_p and n, respectively. For $n = 1$, the value $\bar{x}$ is $0.414L$

$$M_p = \frac{1}{2} wcL \frac{\bar{x}(L - \bar{x})}{L + n\bar{x}}$$

$$= \frac{1}{2} \times 1.7 \times 3 \times 30^2 \frac{0.414(1 - 0.414)}{1 + 0.414}$$

$$= 393.7 \text{ ft k}$$

Look for the failure in section BC. Equation (11.30) gives the moment as

$$M_p = \frac{P_c L}{8}$$

$$= \frac{1.7 \times 30 \times 20}{8}$$

$$= 127.5 \text{ ft k}$$

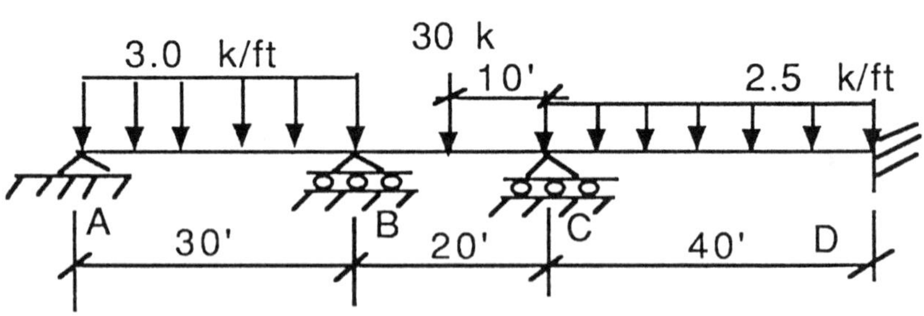

Figure 11.23 Three-span continuous beam.

Check the failure moment in section *CD*. Use Equation (11.60) to show that $\bar{x} = 0.5L$. Then, the value of M_p is determined from Equation (11.37)

$$M_p = \frac{w_c L^2}{16}$$

$$= 1.7 \times 2.5 \times \frac{40^2}{16}$$

$$= 425 \text{ ft k controls}$$

Note that were the beam to have a uniform section, the middle section would be grossly overdesigned. A reasonable approach is to assume that the exterior sections have the same size, whereas the middle section takes on a smaller size. Let the ratio of the moment capacities between the external sections and the middle one be 0.33.

From Equations (11.57) and (11.59) for *CD* and (11.66) and (11.67) for *AB*, the required ultimate moments for the beam are listed as follows:

AB

$$M_p = 495 \text{ ft k}$$

BC

$$M_p = 166 \text{ ft k}$$

CD

$$M_p = 504 \text{ ft k}$$

From the first assumption of a uniform section, $M_p = 425$ ft k. Make the size selection for both cases and compare the overall weight. Use W24 × 68 for the exterior span and W18 × 35 for a total weight of 5464 lb. For a constant section selection of all spans, use W21 × 62 and a total weight of 5580 lb. The first scheme is lighter, but may not be cheaper since the savings in weight are small.

11.8 PORTAL FRAMES

The concept of failure mechanism is extended to frames of which the rectangular frame is but one example. The frame shown in Figure 11.24 represents a portal frame. The moment diagram is drawn on the frame. In the case of continuous beams, the possible failure mechanisms were obvious. However, there are several possibilities for failure mechanisms

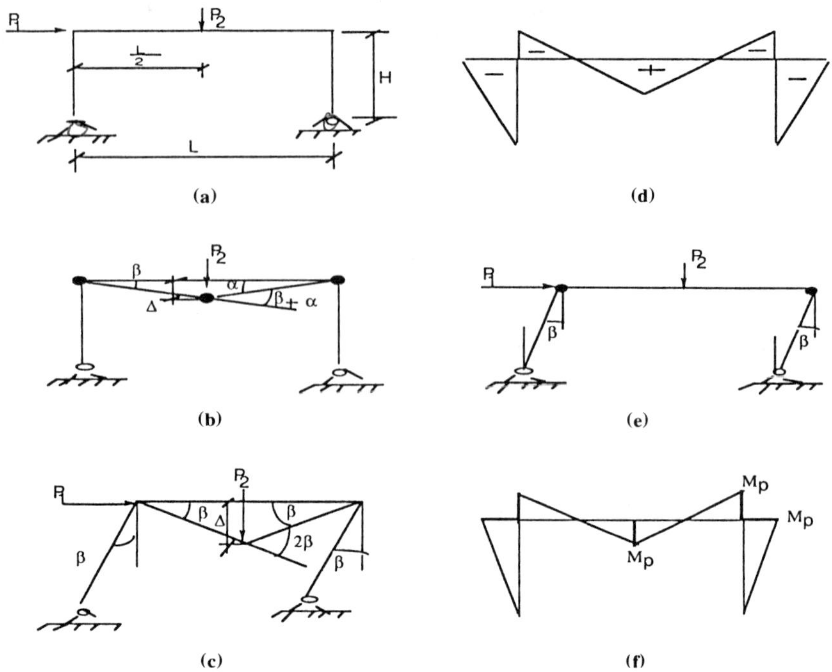

Figure 11.24 Portal frame.

when dealing with frames. There are two primary mechanisms as shown in Figures 11.24c and 11.24d. A third mechanism is a secondary one, and sometimes it is called a combined mechanism. Using the principle of virtual work with each mechanism and solving for M_p is considered the most straightforward method.

Beam Mechanism (P_2 is at $L/2$)

$$P_2\Delta = M_p\beta + M_p 2\beta + M_p\beta \tag{11.72}$$

$$P_2\frac{L}{2}\beta = 4M_p\beta$$

$$M_p = P_2\frac{L}{8} \tag{11.73}$$

Column Mechanism

$$P_1\Delta = M_p\beta + M_p\beta \tag{11.74}$$

$$P_1 H\beta = 2M_p\beta$$

$$M_p = P_1\frac{H}{2} \tag{11.75}$$

Combined Mechanism

$$P_1\Delta_1 + P_2\Delta_2 = M_p 2\beta + M_p 2\beta$$

$$4M_p = P_1 H\beta + P_2 \frac{L}{2}\beta \tag{11.76}$$

$$M_p = \frac{1}{4}\left(P_1 H + P_2 \frac{L}{2}\right)$$

Example 11.14

Consider a portal frame as in Figure 11.24. We are given the following:

$$P_1 = 10 \text{ k}$$
$$P_2 = 30 \text{ k}$$
$$H = 20 \text{ ft}$$
$$L = 40 \text{ ft}$$

The frame has a constant section throughout. Find an optimum size to use in the columns and the beam.

Solution

Beam Mechanism

$$M_p = \frac{1.7 \times 30 \times 40}{8} = 255 \text{ ft k}$$

Column Mechanism

$$M_p = \frac{1.3 \times 10 \times 20}{2} = 130 \text{ ft k}$$

Combined Mechanism

$$M_p = \frac{1.3}{4}\left(10 \times 20 + 30 \times \frac{40}{2}\right) = 260 \text{ ft k controls}$$

Use W24 × 55. The beam should be checked in accordance with AISC Specification N7.

11.9 MINIMUM THICKNESS (WIDTH-THICKNESS RATIO)

When using the plastic method in the design of steel structures, the width-thickness ratio for rolled W, M, or S shapes and similar built-up, single-web shapes subjected to compression involving hinge rotation under ultimate load should be maintained at values not exceeding those listed in Table 11.1.

The width-thickness ratio for a compression flange in a box section should not exceed $190/\sqrt{F_y}$.

For members subjected to ultimate loads, the depth-thickness ratio shall not exceed the values given by Equations (N7-1) or (N7-2) as the case may require (see AISC Manual)

$$\frac{d}{t} = \frac{412}{\sqrt{F_y}}\left(1 - 1.4\frac{P}{P_y}\right) \quad \text{when} \quad \frac{P}{P_y} \leq 0.27 \tag{11.77}$$

$$\frac{d}{t} = \frac{257}{\sqrt{F_y}} \quad \text{when} \quad \frac{P}{P_y} > 0.27 \tag{11.78}$$

Consider the rectangular frame with loads applied as shown in Figure 11.25.

There are four possible failure mechanisms as shown in Figure 11.26. The beam and column mechanisms are simple to predict. Refer to Figures 11.26a, 11.26b, and 11.26c

TABLE 11.1 Limiting flange dimensional proportions

F_y	$b_f/2t_f$
36	8.5
42	8.0
45	7.4
50	7.0
55	6.6
60	6.3
65	6.0

Source: Data on p. 5-96 from Manual of Steel Construction, Ninth Edition, 1989 by American Institute of Steel Construction.

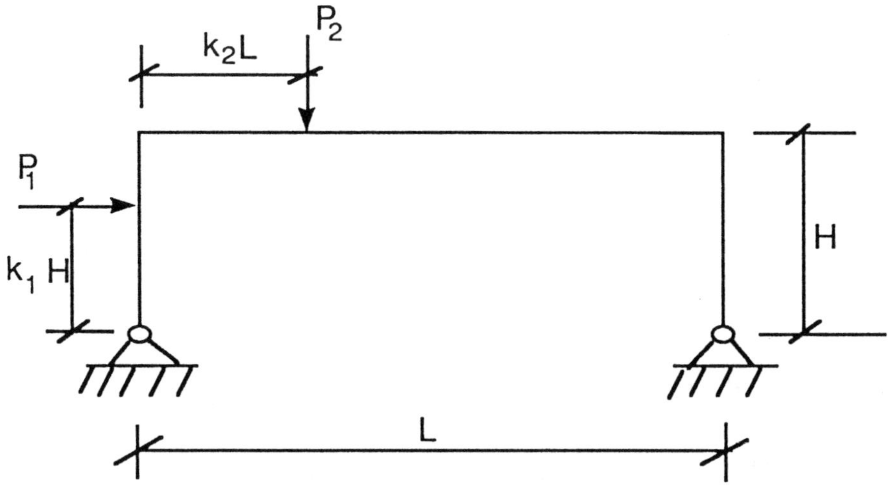

Figure 11.25 Portal frame with concentrated loads at any point on column and girder.

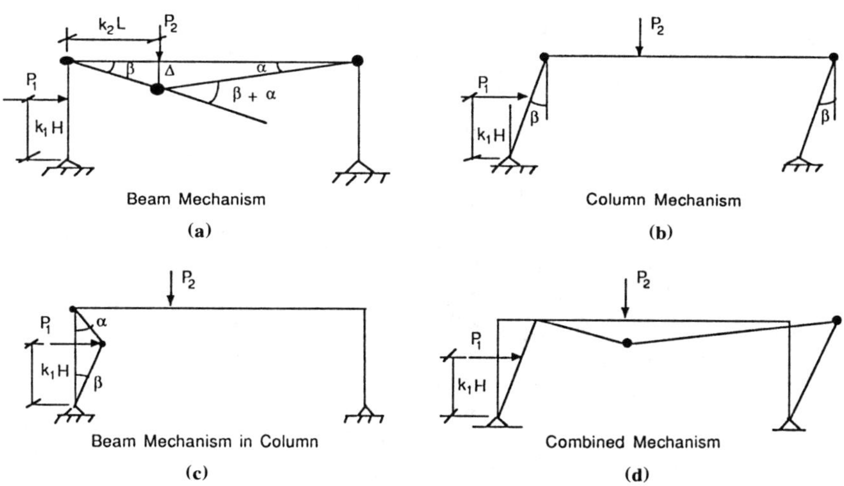

Figure 11.26 Failure mechanism of the portal frame. (a) Beam mechanism. (b) Column mechanism. (c) Beam mechanism in column. (d) Combined mechanism.

$$M_p = P_1 \frac{H}{2} \tag{11.75}$$

Beam Mechanism

$$W_I = 2M_p \beta \frac{1}{1 - k_2} \tag{11.79}$$

and

$$W_E = \frac{P_2 L}{2} \beta \tag{11.80}$$

Equations (11.79) and (11.80) yield

$$M_p = \frac{k_2(1 - k_2) P_2 L}{2} \tag{11.81}$$

Column Mechanism

$$W_E = P_1 k_1 H \beta \tag{11.82}$$

$$W_I = 2M_p \beta \tag{11.83a}$$

and

$$M_p = \frac{P_1}{2} k_1 H \tag{11.83b}$$

Beam Mechanism in the left column

$$W_I = M_p(\alpha + \beta) + M_p \alpha \tag{11.84}$$

$$\alpha = \frac{k_1}{1 - k_1} \beta$$

$$W_I = M_p \frac{1 + k_1}{1 - k_1} \beta \tag{11.85}$$

$$W_E = P_1 k_1 H \tag{11.86}$$

and

$$M_p = \frac{k_1(1 - k_1) P_1 H}{(1 + k_1)} \tag{11.87}$$

To determine the displacements that occur in the combined mechanism, refer to Figure 11.27.

Combined Mechanism

IC = the instantaneous center
$ABECD$ = undeformed structure
$A'B'E'C'D$ = deformed structure

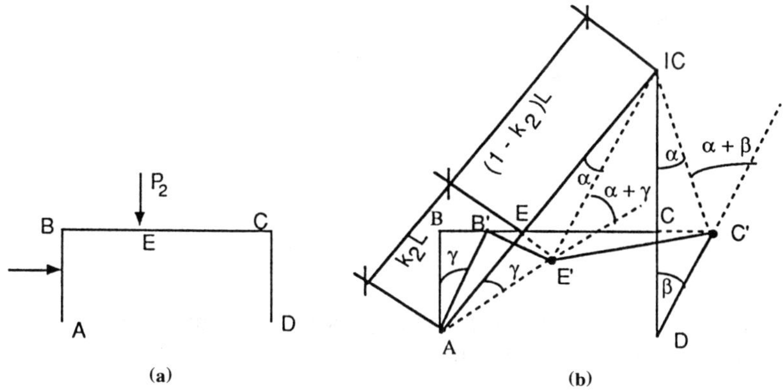

Figure 11.27 Kinematics of the portal frame.

To find the instantaneous center, extend lines AE and DC until they intersect. Their point of intersection represents the instantaneous center for the frame. The diagonal AIC is divided into the ratio k_2L and $(1 - k_2)L$. Writing the expressions for the internal and external work proceeds as follows:

Step 1. Determine line CIC from the similarity of triangles ABE and $ECIC$

$$\frac{H}{k_2L} = \frac{CIC}{(1 - k_2)L}$$

$$CIC = \frac{(1 - k_2)}{k_2}H$$

Step 2. Determine the relationship between the angles α and β from similar triangles DCC' and $CICC'$

$$\alpha = \frac{k_2}{1 - k_2}\beta$$

Step 3. Angle $EE'IC$ is the same as α since triangle $EE'IC$ moves as a plane.

Step 4. Determine the relationship between angles α and γ from similar triangles AEE' and $EE'IC$

$$\gamma = \frac{1 - k_2}{k_2}\alpha$$

$$\gamma = \beta$$

This is true when the height of the columns is the same on both sides.

Step 5. Add the angles α and β and γ and α

$$\alpha + \beta = \frac{1}{1 - k_2}\beta$$

and

$$\gamma + \alpha = \frac{1}{1 - k_2}\beta$$

From the above relationships and by using the concept of virtual work, the expressions for external and internal work are

$$W_I = 2M_p \frac{1}{1 - k_2}\beta \qquad (11.88)$$

$$W_E = P_1 k_1 H_1 + P_2(1 - k_2)L\beta$$

$$M_p = \frac{(1 - k_2)}{2}[k_1 P_1 H + (1 - k_2)P_2 L] \qquad (11.89)$$

When $k_2 = \frac{1}{2}$ and $k_1 = 1$,

$$M_p = \frac{P_1 H}{4} + \frac{P_2 L}{8} \qquad (11.90)$$

Example 11.15

Given a frame that has a span of 75 ft and height of 20 ft with loads $P_2 = 30$ k applied 25 ft from the left column and $P_1 = 10$ k applied 16 ft from the support on the left column, design the frame.

Solution

$$k_1 = 0.80 \qquad \text{and} \qquad k_2 = \tfrac{1}{3}$$

With the values of the k's established, proceed with the application.

Beam mechanism

$$M_p = \frac{k_2(1 - k_2)P_2 L}{2} \qquad \text{[from Equation (11.81)]}$$

$$= \frac{0.333(1 - 0.333)1.7 \times 30 \times 75}{2}$$

$$= 425 \text{ ft k}$$

Column mechanism

$$M_p = \frac{P_1 k_1 H}{2} \quad \text{[from Equation (11.83)]}$$

$$= \frac{1.3 \times 0.80 \times 10 \times 20}{2}$$

$$= 104 \text{ ft k}$$

Beam mechanism in the left column

$$M_p = \frac{k_1(1 - k_1)P_1 H}{2}$$

$$= \frac{0.8(1 - 0.8) \times 1.3 \times 10 \times 20}{2}$$

$$= 22.9 \text{ ft k}$$

Combined mechanism

$$M_p = \frac{(1 - k_2)}{2}[k_1 P_1 H + (1 - k_2)P_2 L] \quad \text{[from Equation (11.89)]}$$

$$= (1 - 0.333)\left(\frac{0.333}{0.667} \times 10 \times 20 + 30 \times 75\right) \times 1.3$$

$$= 679 \text{ ft k controls}$$

Use W27 × 84, M_p = 732 ft k.

11.10 PLASTIC ANALYSIS OF GABLED FRAMES

The geometric relationship between rotations for the various failure mechanisms, displacements, and virtual work is more complicated for gable frames than for continuous beams and portal frames. The method of instantaneous center is used to assist in formulating these relationships. For the structure shown in Figure 11.28, the angle of rotation b will be assigned an arbitrary value and all other rotations and displacements found in terms of it.

The failure mechanisms and principle of virtual work are applied to the gable structure shown in Figure 11.28. The three basic failure mechanisms are presented in Figure 11.29.

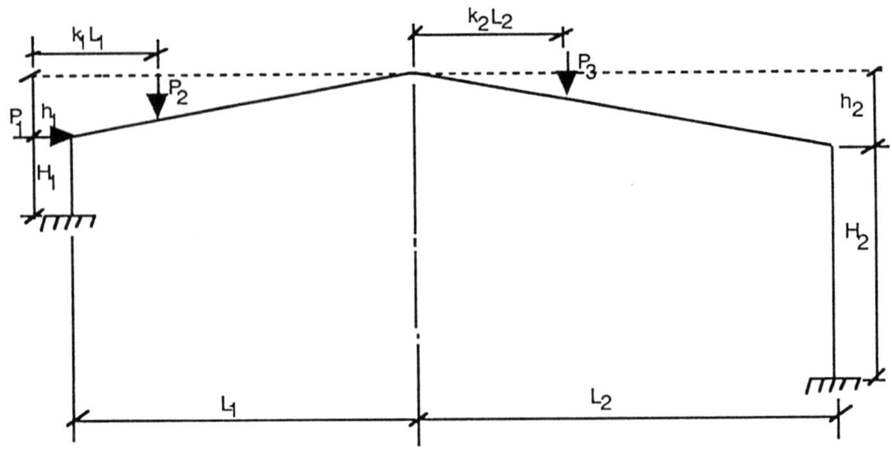

Figure 11.28 Gable frame.

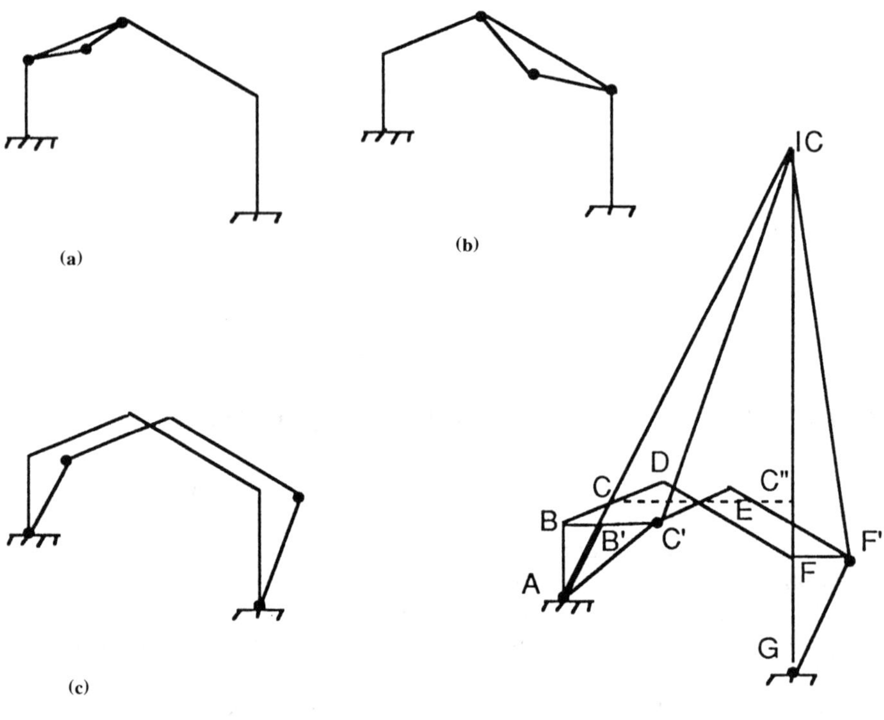

Figure 11.29 Kinematics of the gable frame.

The diagrams in Figures 11.29a and 11.29b represent the beam mechanism in the girders. The expression for the plastic moment is taken from Equation (11.41). For the left girder,

$$M_p = \frac{k_1(1 - k_1)P_2 L_1}{2} \tag{11.91}$$

For the right girder,

$$M_p = \frac{k_2(1 - k_2)P_3 L_2}{2} \tag{11.92}$$

For a symmetrical system in which $L_1 = L_2 = L/2$, $k_1 = k_2 = k$, and $P_2 = P_3 = P$, the plastic moment is

$$M_p = \frac{k(1 - k)PL}{4} \tag{11.93}$$

where

L = the span of the frame

For an unsymmetrical frame, the larger of the moments as given by Equations (11.91) and (11.92) is used in representing the beam failure mechanism.

Since lateral forces are relatively small when compared with gravity loads, the moment from the column mechanism will not control the design.

12

Influence Of Axial Forces on Plastic Moment

12.1 INTRODUCTION

In Chapter 11, the analysis is based on the principle of pure bending. Although this assumption is valid for continuous beams, other secondary effects have to be included in the analysis and design of frames. Secondary effects are attributed to a number of factors such as axial forces, shear forces, buckling (local, lateral, column), repeated loading (fatigue), and brittle fracture. In this text, only the effect of axial loads and shear forces will be included in considering a plastic failure in members of a structural system. The analysis of plastic failure is based on the theory of small deformations.

The procedure for the analysis and design of frames begins with the selection of structural members on the basis of pure bending. The effects of axial and shear forces are introduced, and a check on the adequacy of selected members is made based on AISC specifications. If the initial selection is found inadequate, new sizes are selected and checked. The final selection must satisfy bending, axial, and shear force requirements.

12.2 INFLUENCE OF AXIAL FORCES ON PLASTIC MOMENT CAPACITY

When a member is subjected to axial forces in addition to bending, the magnitude of the plastic moment of that member is reduced. For ordinary portal frames, the axial load is small and its effect may be ignored. However, in cases for which the axial forces are relatively large, their effect must be included in the analysis of the frame.

Figure 12.1 shows the stress pattern under a combination of bending and axial forces. Consider the stress diagram depicted in Figure 12.1f that represents the plastic condition at failure. As illustrated in Figure 12.2b, a part of the stress distribution in this diagram is caused by the axial force. The remainder is due to bending. The stress diagram is divided into two diagrams. The one representing the axial force consists of a core symmetrically distributed about the centroidal axis and is as shown in Figure 12.2d. The second part is presented in Figure 12.2c and is equivalent to the moment M_{pc}, the failure moment under the combined action of bending and axial loads.

For the case when the stress due to P is confined to the web only, the magnitude of P is expressed as

$$P = 2y_0 t_w \sigma_y \tag{12.1}$$

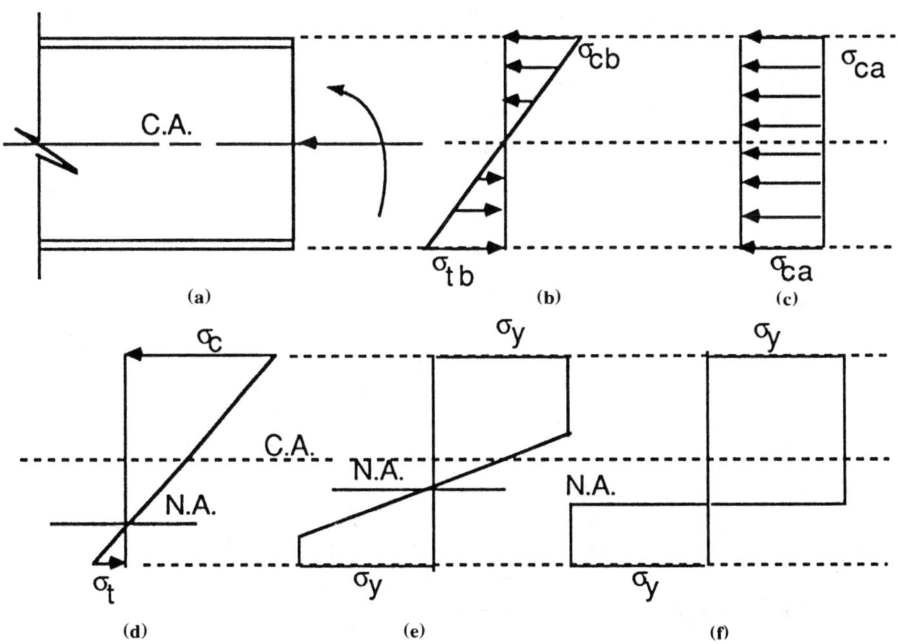

Figure 12.1 Neutral axis falling in the web for combined axial force and bending on beams in the plastic range.

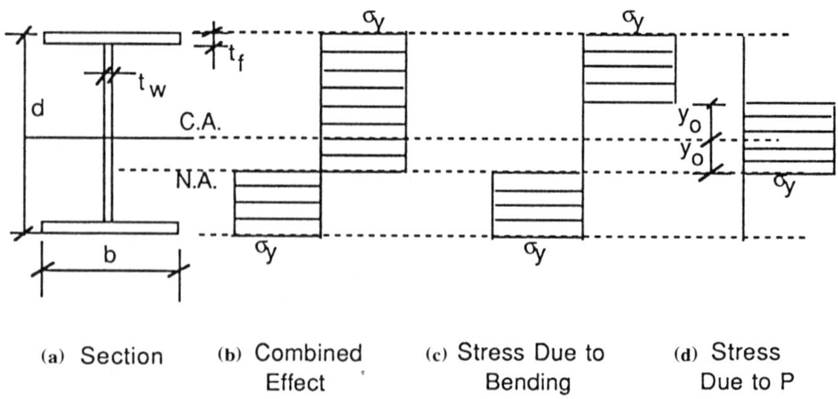

Figure 12.2 Stress distribution due to combined action. (a) Section. (b) Combined effect. (c) Stress due to bending. (d) Stress due to P.

where σ_y is the yield stress, and y_0 the distance from the centroidal axis to the neutral axis. The plastic moment at failure M_{pc} under the combined action of bending and axial loads is obtained from Figure 12.2c. Looking at the section under the bending action alone, we observe that the stress acts on the cross-hatched area as indicated in Figure 12.3a.

The plastic section modulus of the empty section in Figure 12.3a is that of a rectangular section with a height of $2y_0$ and width of t_w. Hence, it will be given by

$$z = \frac{t_w}{4}(2y_0)^2$$

$$= t_w y_0^2 \tag{12.2}$$

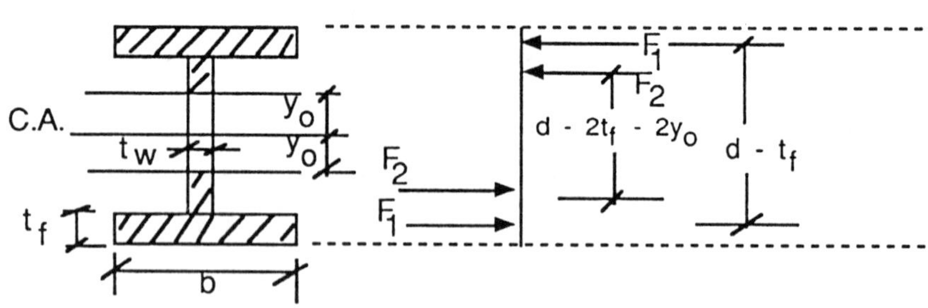

Figure 12.3 Determination of position of the neutral axis in the web.

The plastic section modulus for the equivalent section that is represented by Figure 12.3a is

$$Z_{pc} = Z - z \tag{12.3}$$

where

Z = plastic section modulus of the full section
Z_{pc} = plastic section modulus for the section under the cmbined action of axial and bending forces
z = Plastic section modulus for the empty section shown in Figure 12.3a

The plastic moment can be expressed as

$$M_{pc} = \sigma_y(Z - t_w y_0^2) \tag{12.4}$$

The value of y_0 is obtained from Equation (12.1) and substituted in Equation (12.4)

$$M_{pc} = \sigma_y Z - \sigma_y t_w \left(\frac{P}{2t_w \sigma_y} \right)^2 \tag{12.5}$$

Note that

$$M_p = \sigma_y Z \tag{12.6}$$

Simplifying Equation (12.5) yields

$$M_{pc} = M_p - \frac{P^2}{4t_w \sigma_y} \tag{12.7}$$

Dividing both sides of Equation (12.7) by M_p gives

$$\frac{M_{pc}}{M_p} = 1 - \frac{P^2}{4t_w M_p \sigma_y} \tag{12.8}$$

Substituting Equation (12.6) in Equation (12.8) and noting that $A\sigma_y = P_y$, we can express Equation (12.8) as follows:

$$\frac{M_{pc}}{M_p} = 1 - \frac{A^2}{4t_w Z} \left(\frac{P}{P_y} \right)^2 \tag{12.9}$$

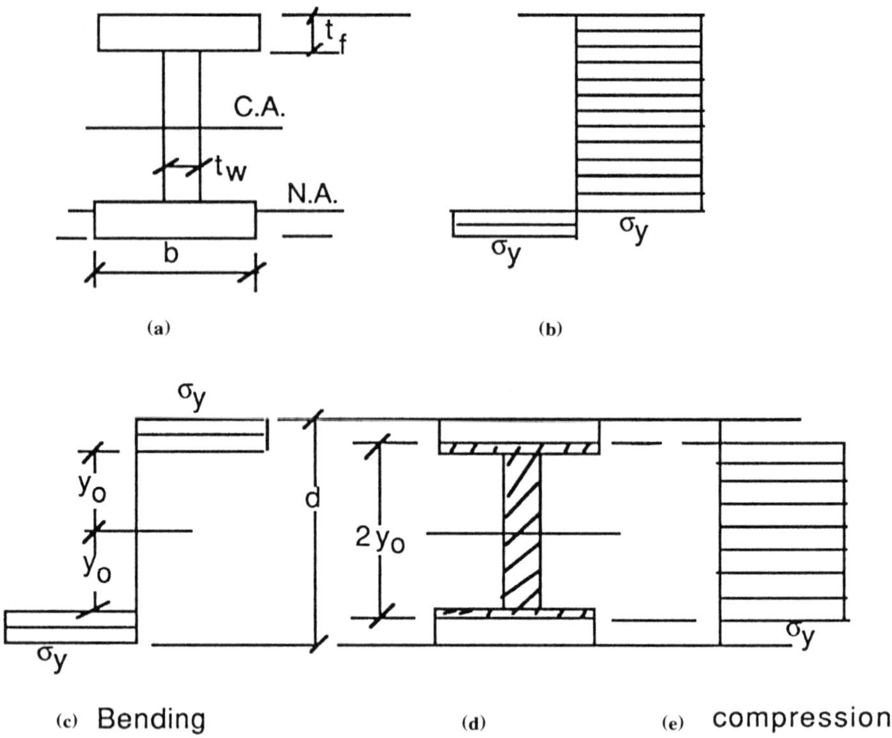

Figure 12.4 Determination of the neutral axis in the flange (a) Beam cross section. (b) Stress diagram. (c) Bending. (d) Area under compression. (e) Compression.

Equation (12.9) applies only when the following equation is satisfied:

$$0 < \frac{P}{P_y} < \frac{t_w(d - 2t_f)}{A} \tag{12.10}$$

When the neutral axis falls in the flange as in Figure 12.4a, a different expression for P is developed.

The value of the axial force can be obtained once the cross-hatched area is defined. Let the equivalent area be denoted by A_c

$$A_c = A - b(d - 2y_0) \tag{12.11}$$

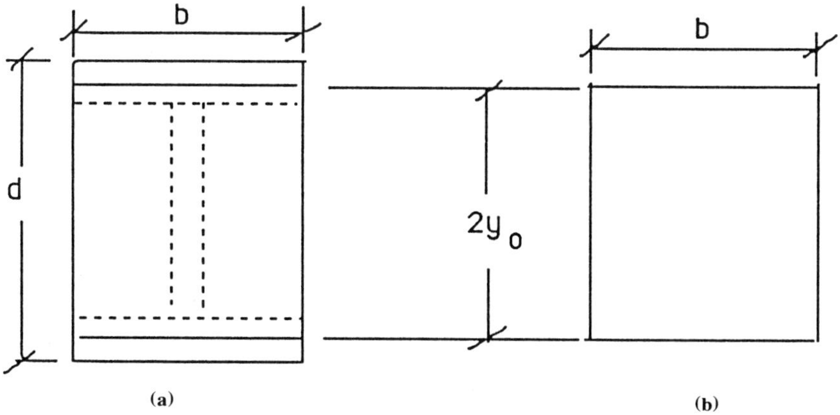

Figure 12.5 Generalized section.

and

$$P = \sigma_y[A - b(d - 2y_0)] \tag{12.12}$$

Consider that Figure 12.4d consists of two rectangles having a width of b and heights of d and $2y_0$, respectively. The inner rectangle is void of any bending stress. Thus, its plastic section modulus can be considered a negative quantity. See Figure 12.5.

The plastic section modulus for the section shown in Figure 12.5 is given by the equation

$$Z_{pc} = Z_1 - Z_2 \tag{12.13}$$

where

$$Z_1 = \frac{bd^2}{4} \tag{12.14}$$

and

$$Z_2 = by_0^2 \tag{12.15}$$

Then,

$$M_{pc} = b\sigma_y\left(\frac{d^2}{4} - y_0^2\right) \tag{12.16}$$

The expression for y_0 is obtained from Equation (12.12)

$$y_0 = \frac{d}{2} - \frac{1}{2b}\left(A - \frac{P}{\sigma_y}\right) \tag{12.17}$$

Combining Equations (12.16) and (12.17) yields

$$M_{pc} = \frac{\sigma_y}{2}\left[d\left(A - \frac{P}{\sigma_y}\right) - \frac{1}{2b}\left(A - \frac{P}{\sigma_y}\right)^2\right] \tag{12.18}$$

Equation (12.18) can be expressed in nondimensional form as

$$\frac{M_{pc}}{M_p} = \frac{A}{2Z}\left[d\left(1 - \frac{P}{P_y}\right) - \frac{A}{2b}\left(1 - \frac{P}{P_y}\right)^2\right] \tag{12.19}$$

where

Z = the plastic section modulus (in.^{3})
$P_y = \sigma_y A$

The above equation should conform to the following condition:

$$\frac{t_w(d - 2t_f)}{A} < \frac{P}{P_y} < 1 \tag{12.20}$$

Example 12.1

Given the frame shown in Figure 12.6, design the columns and girder using the plastic method for A-36 steel. Check for the effect of axial load on the moment capacity.

Solution

For the beam mechanism type of failure, refer to Figure 12.6b. The external work W_E is

$$W_E = 1.7 \times 2 \times 40 \times 20\theta \times \tfrac{1}{2}$$
$$= 13{,}600\theta \text{ kip ft.}$$

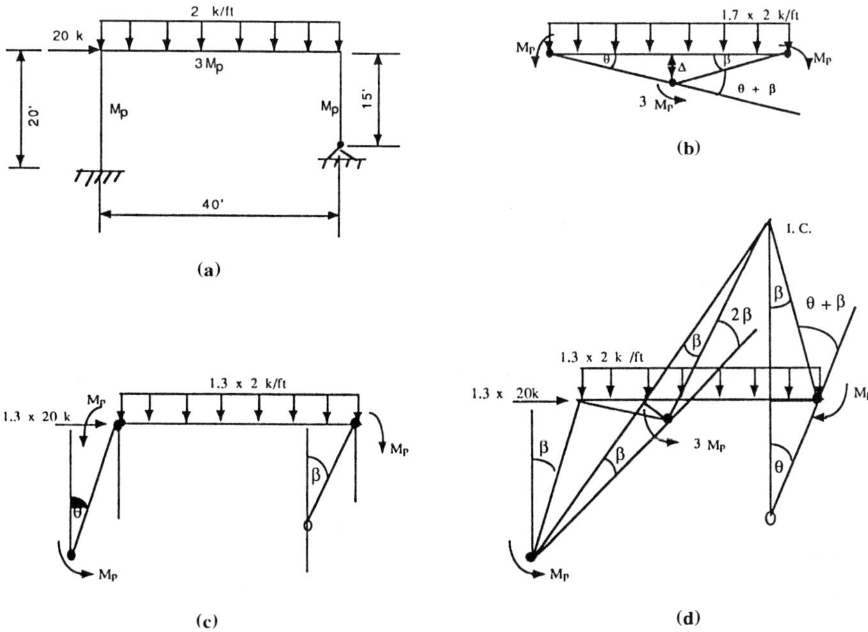

Figure 12.6 Portal frame design example.

The internal work W_I is

$$W_I = M_p\theta + 3M_p(2\theta) + M_p\theta$$
$$= 8M_p\theta$$

Equating external with internal work yields

$$M_p = 170 \text{ kip ft.}$$

Next, check the column mechanism type of failure as shown in Figure 12.6c

$$W_E = 1.3 \times 20 \times 20\theta$$
$$= 520\theta \text{ kip ft.}$$

The expression for internal work is written as follows:

$$W_I = M_p\theta + M_p\theta + M_p\beta$$

From the geometry of the frame in Figure 12.6c, the value of β is determined from the following relationship:

$$\beta = \tfrac{4}{3}\theta$$

Then,

$$W_I = \tfrac{10}{3}M_p\theta$$

and

$$M_p = 156 \text{ kip ft.}$$

The combined mechanism of failure is shown in Figure 12.6d. From the geometry of the frame in its failure position, the relationship between β and θ is established

$$\beta = \tfrac{3}{4}\theta$$

The expressions for external and internal work are

$$\begin{aligned} W_E &= 1.3 \times 2 \times 40 \times 20 \times \beta \times \tfrac{1}{2} + 1.3 \times 20 \times 20\beta \\ &= 1560\beta \text{ k/ft} \\ W_I &= 8M_p\beta + M_p\theta \\ &= \tfrac{28}{3}M_p\beta \end{aligned}$$

Then, the value of the moment is obtained for the combined mechanism

$$M_p = 167 \text{ kip ft.}$$

It is clear from the three mechanisms of failure that the beam mechanism controls design.

Selection of Sizes

The girder size is obtained with

$$\begin{aligned} M_p &= 3 \times 170 \\ &= 510 \text{ kip ft.} \end{aligned}$$

Use W24 × 68; $M_p = 530$ kip ft. for the column size. Try W18 × 35; $M_p = 200$ kip ft. Check for the effect of the axial load on the plastic moment capacity.

The load carried by each column is

$$P = 68 \text{ k}$$

and the ultimate capacity P_y is

$$P_y = 371 \text{ k}$$

The above value for P_y is obtained from the AISC plastic design selection table given in Part 2 of the Manual.

From the section physical properties given in the AISC Manual, Part 1, the following section properties are determined for the column:

$$d = 17.70 \text{ in.}$$
$$t_w = 0.300 \text{ in.}$$
$$t_f = 0.425 \text{ in.}$$
$$Z = 66.5 \text{ in.}^3$$
$$A = 10.3 \text{ in.}^2$$

Use Equation (12.10) to verify whether the neutral axis falls in the web or not

$$\frac{P}{P_y} = \frac{68}{371}$$
$$= 0.187$$
$$\frac{t_w(d - 2t_f)}{A} = \frac{0.300 \times (17.70 - 2 \times 0.425)}{10.3}$$
$$= 0.491 > 0.183$$

From the above result and Equation (12.10), it can be concluded that the neutral axis indeed falls in the web of the column.

Equation (12.9) gives the relationship between the moment capacity without the axial load and that with one

$$\frac{M_{pc}}{M_p} = 1 - \frac{A^2}{4t_w Z}\left(\frac{P}{P_y}\right)^2$$

Substituting the appropriate values of the terms in the above equation yields a value of $M_{pc} = 191$ k/ft.

13

Rigid Connections

13.1 INTRODUCTION

The method of plastic analysis is based on the premise that connections are rigid. Thus, connections play a significant role in the design of indeterminate structures. The basic requirements for the condition of rigid connections are:

1. Sufficient strength of the connections to absorb (resist) the moments in the frame as a result of the given loads.
2. Adequate rotation capacity in the joints to provide sufficient redistribution of the moments.
3. Attainment of an overall stiffness for the structure to keep the displacements relatively small.
4. Cost of fabrication must remain economical to provide an overall reduction in the cost of construction.

There are a variety of connections commonly used in steel construction. These are listed as follows:

1. Corner connection
2. Tee connection
3. Cross connection
4. Column or beam slice connection
5. Column anchorage connection
6. Miscellaneous connections

13.2 STRAIGHT CORNER CONNECTION

In a typical frame, beams and columns are connected at right angles as shown in Figure 13.1. A free-body diagram for the corner, including the points for zero moment, is shown in Figure 13.2. B is chosen at the point of counterflexure in the beam, which is commonly $L/4$ in the fully developed plastic condition. In the absence of any lateral forces on the frame,

$$V_c = N_b \tag{13.1}$$

The shear in the column is assumed to be resisted by the column web

$$V_c = \tau_y t_w d_c \tag{13.2}$$

For a square corner, the column and beam have the same depth. A free-body diagram of the beam alone is presented in Figure 13.3. Summing the moments acting on the free-body diagram in Figure 13.3 to zero yields

$$M_p = V_b L \tag{13.3}$$

where V_b is the shear force in the beam. From the diagram in Figure 13.2b, the shear force in the beam is shown to be equal to the shear in the column

$$V_b = V_c \tag{13.4}$$

and

$$T_0 = \frac{M_p}{d} - \frac{N_b}{2} \tag{13.5}$$

Combining Equations (13.1), (13.3), (13.4), and (13.5) yields

$$T_0 = \frac{M_p}{d}\left(1 - \frac{d}{2L}\right) \tag{13.6}$$

A free-body diagram of the flange at the corner provides the following relationship:

$$T_0 = \tau_y t_w d_c \tag{13.7}$$

Substituting Equation (13.7) in Equation (13.6) and rearranging the terms yield

$$\tau_y t_w d_c = \frac{M_p}{d_c} - \frac{M_p}{2L} \tag{13.8}$$

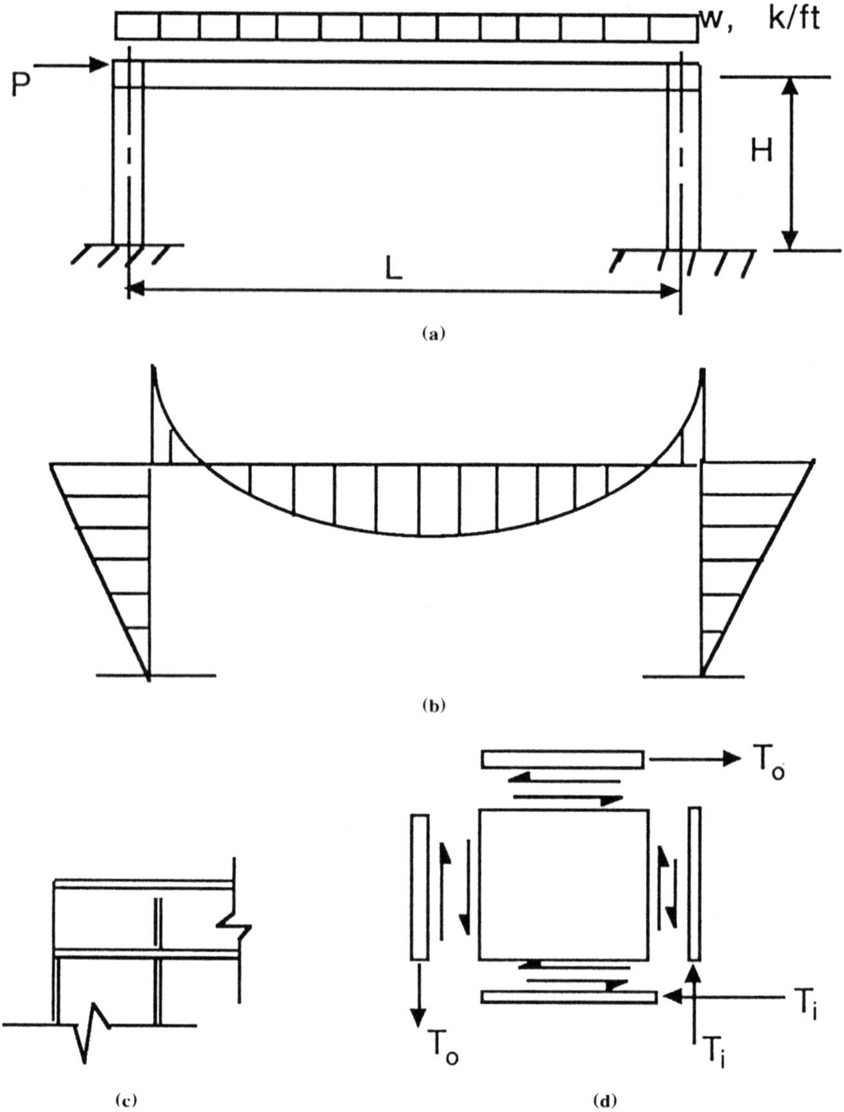

Figure 13.1 Straight corner connection.

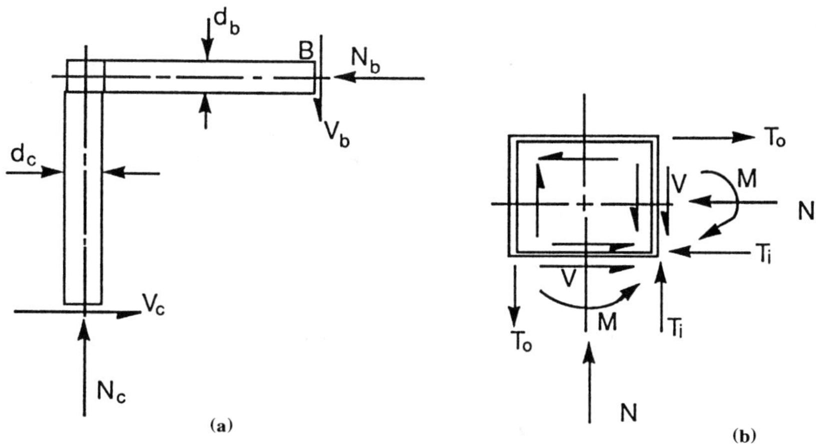

Figure 13.2 Free-body diagram of corner connection.

Simplifying Equation (13.8) results in the following:

$$t_w = \frac{M_p}{\tau_y d_c^2}\left(1 - \frac{d}{2L}\right) \tag{13.9}$$

For a square corner in which $d_c = d_b = d$,

$$t_w = \frac{M_p}{\tau_y d^2}\left(1 - \frac{d}{2L}\right) \tag{13.10}$$

Substituting $\tau_w = \sigma_y / \sqrt{3}$ in Equation (13.10) yields

$$t_w = \frac{\sqrt{3}\, M_p}{\sigma_y d^2}\left(1 - \frac{d}{2L}\right) \tag{13.11}$$

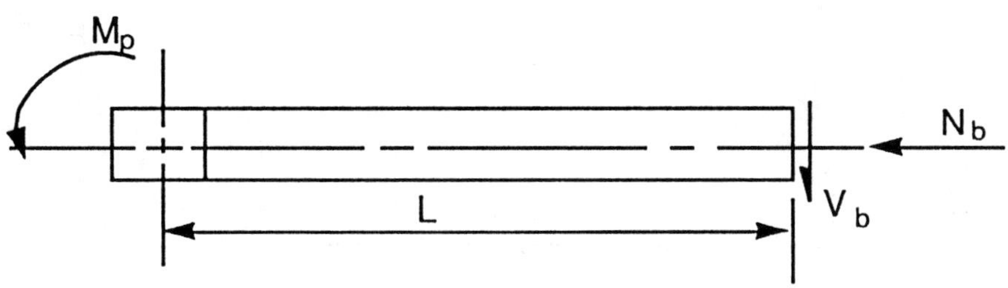

Figure 13.3 Free-body diagram of one leg of a straight corner connection.

Example 13.1

A portal frame with a span of 60 ft and height of 20 ft carries a uniform load of 0.6 k/ft and wind force of 8 k applied at the top of the column. See Figure 13.4. The frame has a constant section throughout its length. Use plastic analysis to design the frame (constructed of A-36 steel).

Solution

The beam mechanism yields $M_p = 230$ ft k and the combined mechanism $M_p = 227$ ft k. It is obvious that the moment for the column mechanism is smaller than both. Hence, the frame must be designed for $M_p = 230$ ft k. Try W18 × 40.

$$M_p = 235 \text{ ft k.}$$
$$d = 17.9 \text{ in.}$$
$$t_w = 0.315 \text{ in.}$$
$$L = \tfrac{1}{4} \times \text{span} = 15 \text{ ft}$$

From Equation (13.11), the required thickness of the web is

$$t_w = \frac{\sqrt{3} \times 230 \times 12}{36 \times 17.9^2}\left(1 - \frac{17.9}{2 \times 15 \times 12}\right)$$
$$= 0.394 > 0.315$$

The size selection is not satisfactory. Either choose a larger section or stiffen the web. Stiffening the web is the subject of the next section.

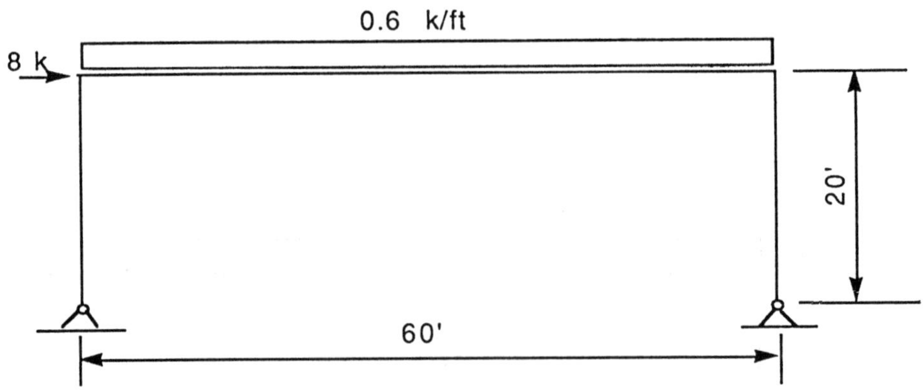

Figure 13.4 Portal frame example.

13.3 STIFFENER FOR A STRAIGHT CORNER CONNECTION

Replacing M_p in Equation (13.11) by $Z\sigma_y$ yields

$$t_w = \frac{\sqrt{3}}{d^2} Z\left(1 - \frac{d}{2L}\right) \tag{13.12}$$

and

$$Z = S_x f \tag{13.13}$$

where

f = shape factor of the section
S_x = section modulus (in.^{3})

Equation (13.12) is rewritten as

$$t_w = \frac{\sqrt{3}}{d^2} fS_x\left(1 - \frac{d}{2L}\right) \tag{13.14}$$

For most commonly used wide-flange sections, the product of $f(1 - d/2L)$ is almost equal to unity. Thus, the required thickness of the web at the corner is expressed by

$$t_w = \frac{\sqrt{3}}{d^2} S_x \tag{13.15}$$

When the beam and column are of different sizes, then the thickness becomes

$$t_w = \frac{\sqrt{3}}{d_c d_b} S_x \tag{13.16}$$

To meet the required strength at the corner without increasing the size of the section, a stiffener is needed at the corner. See Figure 13.5. From a free-body diagram of the top flange as shown in Figure 13.5, the sum of the horizontal forces is expressed as follows:

$$T_0 = dt_w \tau_{\text{web}} + \cos\theta\, T_{sy} \tag{13.17}$$

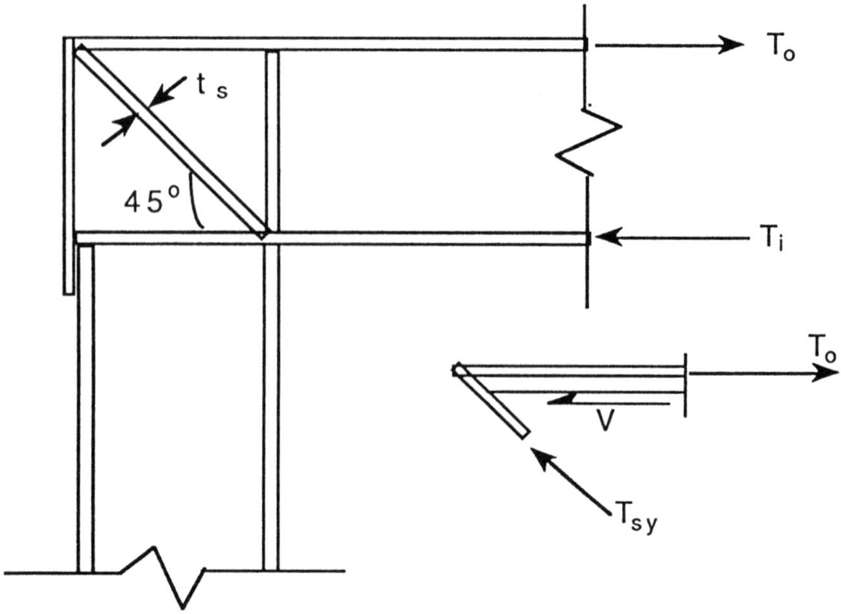

Figure 13.5 Stiffener design for corner connections.

Using the yield strength of the stiffener and a square corner in which $\theta = 45$ deg and $\tau_y = \sigma_y / \sqrt{3}$, we can rewrite the above equation as

$$T_0 = \frac{\sigma_y}{\sqrt{3}} dt_w + \frac{\sigma_y b_s t_s}{\sqrt{2}} \tag{13.18}$$

substituting Equation (13.6) in Equation (13.18) yields

$$\frac{M_p}{d}\left(1 - \frac{d}{2L}\right) = \frac{\sigma_y}{\sqrt{3}} dt_w + \frac{\sigma_y b_s t_s}{\sqrt{2}} \tag{13.19}$$

Representing the left term in Equation (13.19) by

$$\frac{S_x f}{d}\left(1 - \frac{d}{2L}\right)\sigma_y$$

and simplifying as before, the following expression is obtained:

$$\frac{b_s t_s}{\sqrt{2}} = \frac{S_x}{d} - \frac{t_w d}{\sqrt{3}} \tag{13.20}$$

The required area of the stiffener that is needed to strengthen the corner is derived from Equation (13.20)

$$b_s t_s = \sqrt{2}\left(\frac{S_x}{d} - \frac{t_w d}{\sqrt{3}}\right) \tag{13.21}$$

Example 13.2

Consider the frame in Example 13.1. Design the stiffener for the corner. Use W18 × 40.

Solution

$$S_x = 68.4 \text{ in.}^3$$

$$d = 17.9 \text{ in.}$$

$$t_w = 0.315 \text{ in.}$$

$$b_s t_s = \sqrt{2}\left(\frac{68.4}{17.9} - \frac{0.315 \times 17.9}{\sqrt{3}}\right)$$

$$= 0.80 \text{ in.}^2$$

In welding the stiffeners to the web, the thickness of the stiffener must be of the same magnitude as that of the web. For a stiffener thickness of 0.25 in., the width of a double stiffener is 1.6 in. Use a width of 2 in. on each side of the web.

13.4 HAUNCHED CONNECTIONS

Haunches whether tapered or curved are pleasing to the eye, and they modify the section to follow more closely the shape of the moment diagram for the elastic and plastic conditions. The haunch helps to reduce the size of the main member considerably. If the cost of fabricating the haunch is high, it may offset the savings from the reduction of the member size. A feasibility study is recommended before deciding to use a haunched connection. The method of analysis of the frame remains the same with or without haunches.

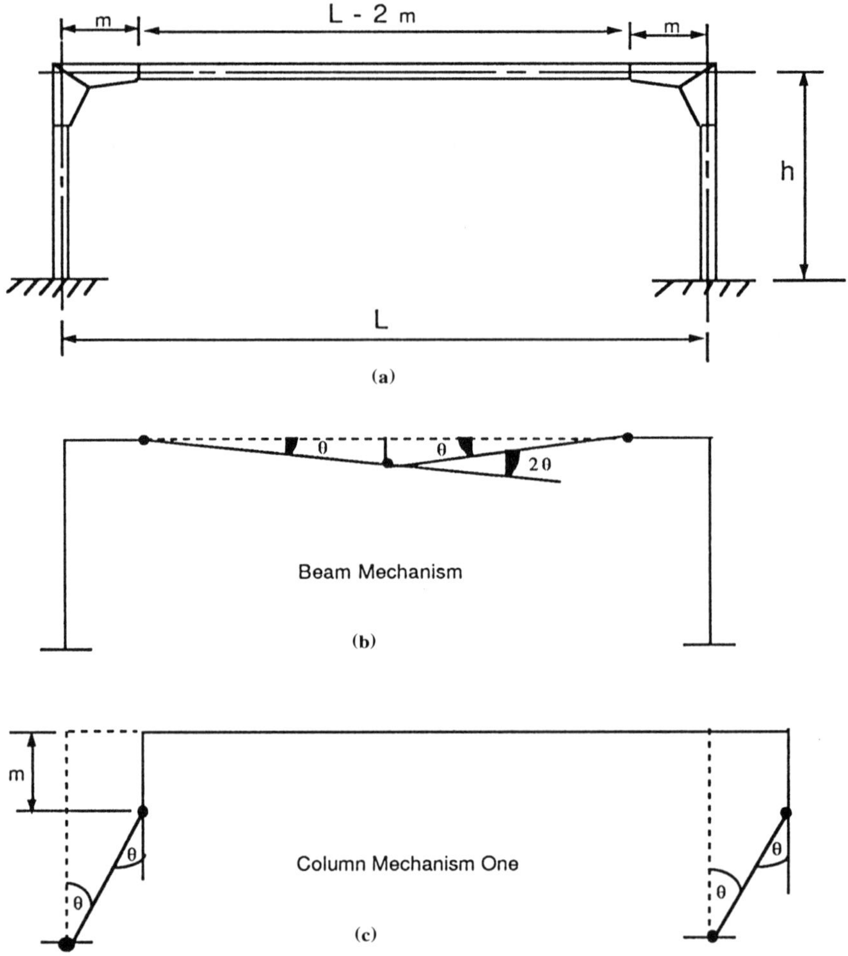

Figure 13.6 Haunched corner connection.

Consider the portal frame with haunched corners as depicted in Figure 13.6.

Beam Mechanism

The beam mechanism is shown in Figure 13.6b. The plastic moment M_p is given by

$$M_p = \frac{w_u(L - 2m)^2}{16} \tag{13.22}$$

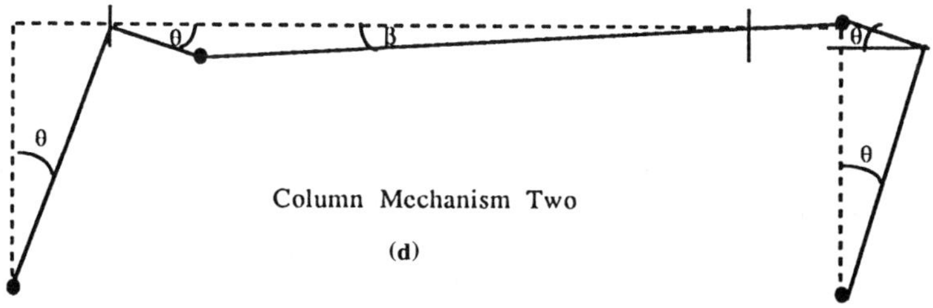

(d)

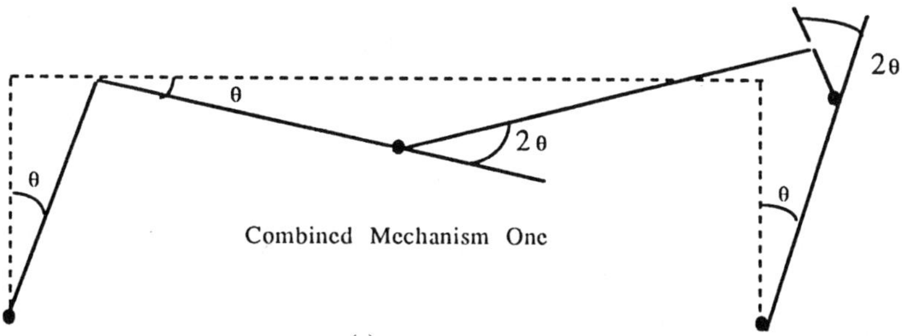

(e)

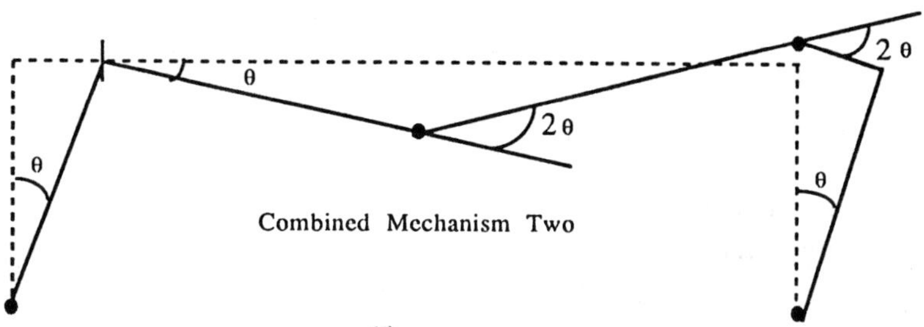

(f)

Figure 13.6 (cont'd.)

Column Mechanism

The external work due to the lateral load consists of two parts. The top m feet displace an equal amount, and the lower $h - m$ feet displace a varying amount as illustrated in Figure 13.6c. The expressions for external and internal work are written as follows:

$$W_E = w_{uL} m h \theta + \tfrac{1}{2} w_{uL} (h - m)^2 \theta \quad (13.23)$$

$$W_I = 4 M_p \theta \quad (13.24)$$

Combining Equations (13.23) and (13.24) yields

$$M_p = \tfrac{1}{8} w_{uL}(h^2 - m^2) \tag{13.25}$$

Another possibility for a column mechanism is shown in Figure 13.6d. The rotation at the bottom of the column is θ, whereas the rotation of the plastic hinge in the beam is $\theta + \beta$. The relationship between θ and β is given by

$$\beta = \frac{2m}{L - 2m}\theta \tag{13.26}$$

and

$$\theta + \beta = \frac{L}{L - 2m}\theta \tag{13.27}$$

Then, the expressions for external and internal work are written as follows:

$$W_I = 2M_p\theta + 2M_p\frac{L}{L - 2m}\theta \tag{13.28}$$

and

$$W_E = \frac{w_{uL}h^2}{2}\theta + 2\frac{w_{uv}m^2}{2}\theta + \frac{w_{uv}(L - 2m)^2}{2}\beta \tag{13.29}$$

where

w_{uL} = factored lateral load
w_{uv} = factored vertical load

Thus,

$$M_p = \frac{1}{4}\left[w_{uL}\frac{h^2}{2} + w_{uv}(L - m)m\right]\left(\frac{L - 2m}{L - m}\right) \tag{13.30}$$

Combined Mechanism

The combined mechanism mode of failure has two possibilities as shown in Figure 13.6e and 13.6f. The plastic moment for each of these two cases is the same and is given by the following:

$$M_p = w_{uL}\frac{h^2}{12} + w_{uv}\frac{L^2}{24} \tag{13.31}$$

Example 13.3

A portal frame carries a gravity load of 0.6 k/ft and may be exposed to a wind force of 0.5 k/ft. The frame has a span of 60 ft and height of 20 ft. Design the frame employing A-36 steel. Use plastic analysis.

Solution

1. Analysis

(a) *Factored loads.*

Gravity Load

$$w_{uv} = 1.7 \times 0.6$$
$$= 1.02 \text{ k/ft}$$

Wind Load

$$w_{uL} = 1.3 \times 0.5$$
$$= 0.65 \text{ k/ft}$$

Combined Load

$$1.3w_{uL}, \qquad 1.3w_{uv}$$

Assume the length of the corner connection to be 5 ft.

(b) *Beam mechanism.*

$$M_p = \frac{w_u(L - 2m)^2}{16}$$
$$= \frac{1.02(60 - 2 \times 5)^2}{16}$$
$$= 159.4 \text{ ft k}$$

(c) *Column mechanism.*

First column mechanism

$$M_p = \tfrac{1}{8}w_{uL}(h^2 - m^2)$$
$$= \tfrac{1}{8} \times 0.65(20^2 - 5^2)$$
$$= 30.5 \text{ ft k}$$

Second column mechanism

$$M_p = \frac{1}{4}\left[w_{uL}\frac{h^2}{2} + w_{uv}(L - m)m\right]\left(\frac{L - 2m}{L - m}\right)$$

$$\times \frac{1}{4}\left[0.65\frac{20^2}{2} + 1.02(60 - 5)5\right]\left(\frac{60 - 10}{60 - 5}\right)$$

$$= 96.2 \text{ ft k}$$

(d) *Combined mechanism.*

$$M_p = w_{uL}\frac{h^2}{12} + w_{uv}\frac{L^2}{24}$$

$$= 0.65\left(\frac{20^2}{12}\right) + 1.02\left(\frac{60^2}{24}\right)$$

$$= 174.7 \text{ ft k control}$$

2. *Selection of Sizes*

Try W18 × 35.

$$M_p = 200 \text{ ft k}$$

$$t_w = 0.300 \text{ in.}$$

$$d = 17.7 \text{ in.}$$

$$S_x = 57.6 \text{ in.}^3$$

$$A = 10.3 \text{ in.}^3$$

(a) *Check the adequacy of the web at the corner.*

$$t_w = \frac{\sqrt{3}}{d^2} S_x$$

$$= \frac{\sqrt{3}}{17.7^2} 57.6$$

$$= 0.318 \text{ in.} > 0.300$$

A stiffener must be used at the corner to strengthen it

$$b_s t_s = \sqrt{2}\left(\frac{S_x}{d} - \frac{t_w d}{\sqrt{3}}\right)$$

$$= \sqrt{2}\left(\frac{57.6}{17.7} - \frac{0.300 \times 17.7}{\sqrt{3}}\right)$$

$$= 0.27 \text{ in.}^2$$

Use a stiffener $1\frac{1}{4}$ in. $\times$ $\frac{1}{4}$ in. on one side only.

(b) *Check the flange.*

$$\frac{b}{2t_f} = 7.1 < 8.5$$

(c) *Check the effect of axial load on the capacity of the column.* The axial load is

$$\begin{aligned} P &= 1.02 \times 30 \\ &= 30.6 \text{ k} \\ P_y &= {}^*371 \text{ k} \end{aligned}$$

An asterisk requires checking shape for compliance with Equations (N7-1) or (N7-2), Sect. N7 AISC ASD Specifications. Assume that girts are placed at 5 ft. on centers; check for the slenderness ratio

$$\begin{aligned} l_y &= 5 \text{ ft} \\ l_x &= 20 \text{ ft} \\ r_y &= 1.22 \text{ in.} \\ r_x &= 7.04 \text{ in.} \\ \frac{l_y}{r_y} &= \frac{5 \times 12}{1.22} \\ &= 49 \\ \frac{l_x}{r_x} &= \frac{20 \times 12}{7.04} \\ &= 34 \end{aligned}$$

The y direction controls

$$\begin{aligned} F_a &= 18.44 \text{ k/in.}^2 \\ P_{cr} &= 1.7F_aA \\ &= 1.7 \times 18.44 \times 10.3 \\ &= 322.9 \text{ k} \\ F_{e'} &= 62.2 \text{ k/in.}^2 \\ P_e &= \tfrac{23}{12}F_{e'}A \\ &= \frac{23 \times 62.2 \times 10.3}{12} \\ &= 1280 \text{ k} \end{aligned}$$

The interaction formulas in the AISC specifications are

$$\frac{P}{P_{cr}} + \frac{C_m M}{\left(1 - \frac{P}{P_e}\right) M_m} \le 1.0$$

$$\frac{P}{P_y} + \frac{M}{1.18 M_p} \le 1.0$$

$$C_m = 0.85$$

$$\frac{30.6}{322.9} + \frac{0.85 \times 174.7}{\left(1 - \frac{30.6}{1280}\right) 200} = 0.86 < 1.0$$

$$\frac{30.6}{371} + \frac{174.7}{1.18 \times 200} = 0.82 < 1.0$$

The selection satisfies the required specifications. The next item to consider is the design of haunched connection.

13.5 HAUNCHED CONNECTIONS WITH CONCENTRATED LOADS

Figure 13.7 shows the various failure mechanisms for haunched portal frames subjected to concentrated loads.

Beam Mechanism

Figure 13.7b illustrates the failure mode for the beam mechanism

$$W_E = P_1 \frac{L - 2m}{2} \theta$$

$$W_I = 4M_p$$

Then,

$$M_p = \frac{P_1}{8}(L - 2m) \tag{13.32}$$

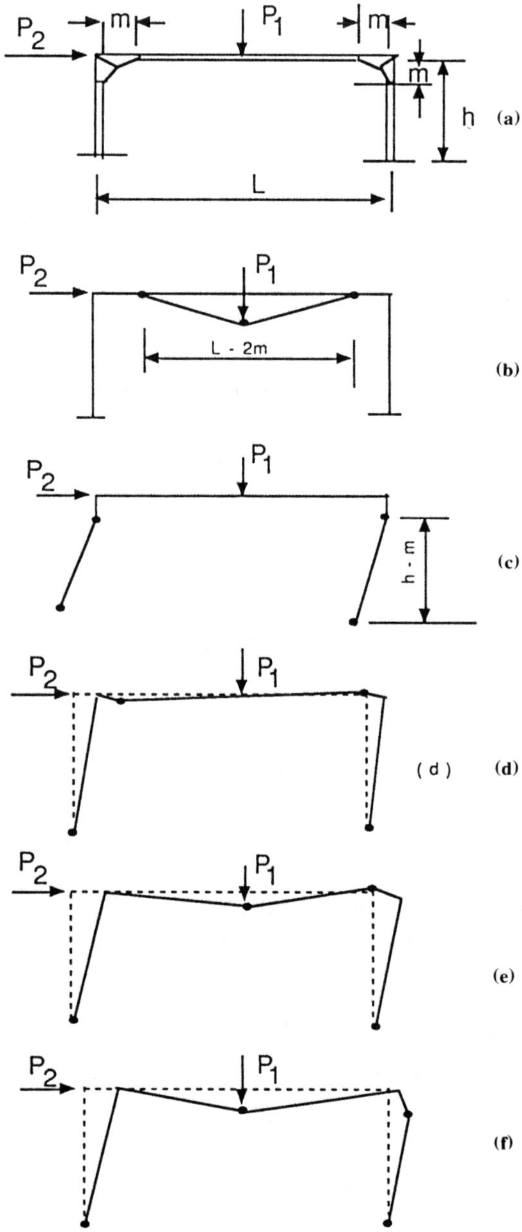

Figure 13.7 Failure mechanism for a haunched corner connection in a portal frame.

Column Mechanism

(a) *Failure in the column.*

$$W_E = P_2(h - m)\theta$$

$$W_I = 4M_p\theta$$

$$M_p = \frac{P_2}{4}(h - m) \tag{13.33}$$

(b) *Failure in the beam.*

$$W_E = P_2 h\theta$$

If the angle swept by the column is considered to be θ, then the angle swept by the beam must be $\theta + \beta$. Then from Figure 13.7d, the relationship between θ and β is established as follows:

$$\beta = \frac{2m}{L - 2m}\theta$$

and

$$\beta + \theta = \frac{L}{L - 2m}\theta$$

The internal work is given by

$$W_I = 2M_p\theta + 2M_p\frac{L}{L - 2m}\theta$$

Then,

$$M_p = \frac{P_1 h}{4}\frac{L - m}{L - 2m} \tag{13.34}$$

Combined Mechanism

If the failure were to occur in the beam as shown in Figure 13.7e, the plastic moment is

$$M_p = \frac{L - 2m}{6L - 8m}(\tfrac{1}{2}LP_1 + P_2 h) \tag{13.35}$$

For the failure mode depicted in Figure 13.7f, the plastic moment is given by the expression

$$M_p = \frac{h - m}{6h - 2m}\left(\frac{P_1 L}{2} + P_2 h\right) \tag{13.36}$$

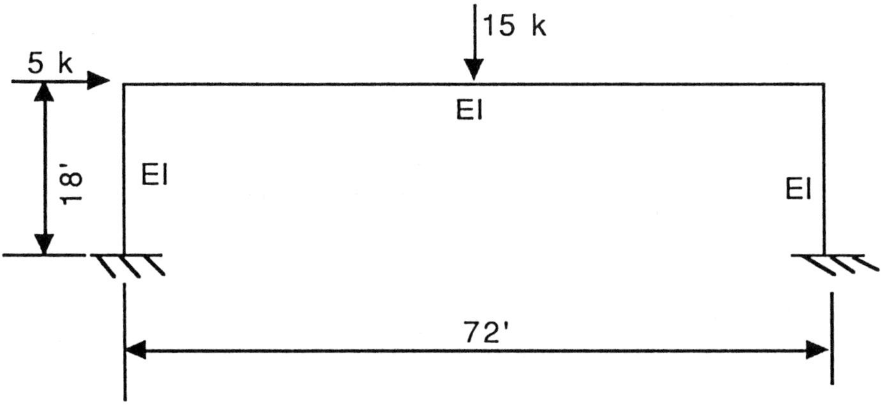

Figure 13.8 Portal frame rigid connection design example.

Example 13.4

A portal frame shown in Figure 13.8 is loaded with a gravity load of 15 k and a wind load of 5 k applied at the top of the column. Use a rigid connection with $m = 5$ ft. Find the required design plastic moment. The given loads are service loads.

Solution

Beam Mechanism

From Equation (13.32), the plastic moment is

$$M_p = \frac{1.7 \times 15 \times (72 - 2 \times 5)}{8}$$

$$= 197.6 \text{ ft k}$$

Column Mechanism

(a) $M_p = \dfrac{1.3 \times 5 \times (18 - 5)}{4}$

$= 21.1$ ft k

(b) $M_p = \dfrac{1.3 \times 5}{4} \times \dfrac{72 - 5}{72 - 10}$

$= 31.6$ ft k

Combined Mechanism

$$\text{(a) } M_p = 1.3 \times \frac{72 - 2 \times 5}{6 \times 72 - 8 \times 5}\left(\tfrac{1}{2} \times 72 \times 15 + 5 \times 18\right)$$
$$= 129.5 \text{ ft k}$$

$$\text{(b) } M_p = 1.3 \times \frac{18 - 5}{6 \times 18 - 2 \times 5} \times \left(\frac{15 \times 72}{2} + 5 \times 18\right)$$
$$= 108.6 \text{ ft k}$$

The beam mechanism controls. Hence,

$$M_p = 197.6 \text{ ft k}$$

13.6 DESIGN GUIDES: CONNECTIONS

To develop full M_p at point H (see Figure 13.9), the required thickness of the web at the corner is

$$t_w \geq \sqrt{3}\,\frac{S}{d^2} \tag{13.37}$$

S being the section modulus of the given section.

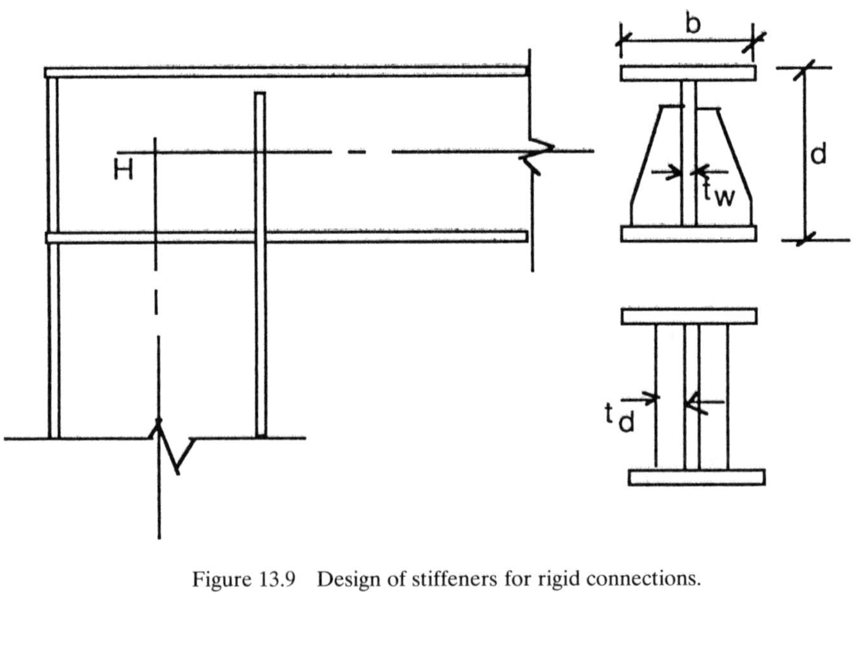

Figure 13.9 Design of stiffeners for rigid connections.

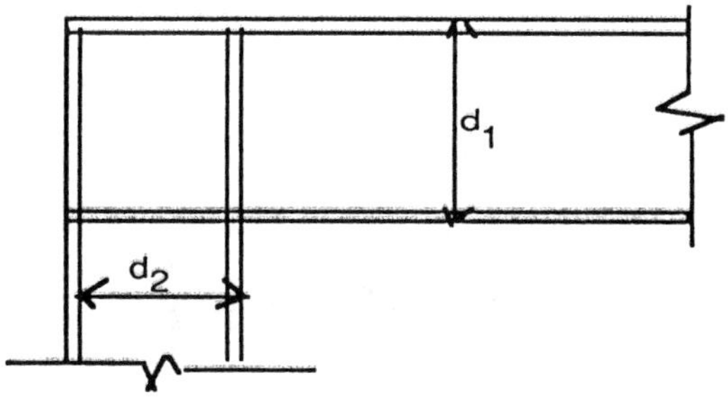

Figure 13.10 Straight corner.

For inadequate connections, use double web plates as shown in Figure 13.9b

$$t_d \geq \sqrt{3}\,\frac{S}{d^2} - t_w \tag{13.38}$$

For a rectangular corner as shown in Figure 13.10, the web requirement for shear capacity is given by

$$t_w \geq \frac{\sqrt{3}\,S_{\min}}{d_q d_2} \tag{13.39}$$

where

$S_{\min}$ = the smaller of the section moduli for the beam and column (in.^{3})
d_1, d_2 = the depth of the beam and column at the corner (in.)

To develop an alternate equation to Equation (13.39), use the following expression for the thickness of the web at the corner:

$$t_w \geq \frac{0.6M}{d_1 d_2} \tag{13.40}$$

When the web of the selected section is not adequate to meet the shear requirement, a stiffener is provided to fulfill it. See figure 13.11

$$t_s = \frac{\sqrt{2}}{b}\left(\frac{s}{d} - \frac{t_w d}{\sqrt{3}}\right) \tag{13.41}$$

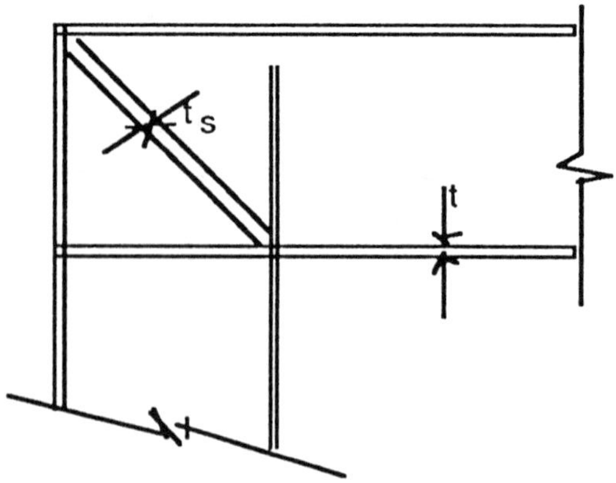

Figure 13.11 Stiffener for a straight corner.

where

$$t_s \le \frac{b}{17} \tag{13.42}$$

A conservative approach is to assume that $t_s = t$, where

t_s = the thickness of the stiffener
t = the thickness of the tension flange for the beam or column

Tapered haunch connections are designed on the same basis. See Figure 13.12. The procedure is a step-by-step method.

1. Select the general layout. Brace at sections R, H, and P. See Figure 13.12.
2. Check the strength at section 1. Develop M_p at point 1

$$Z_1 = bt(d_1 - t) + \frac{t_w}{4}(d_1 - 2t)^2 \tag{13.43}$$

where

$$Z_1 \ge \frac{M_1}{s_y} \tag{13.44}$$

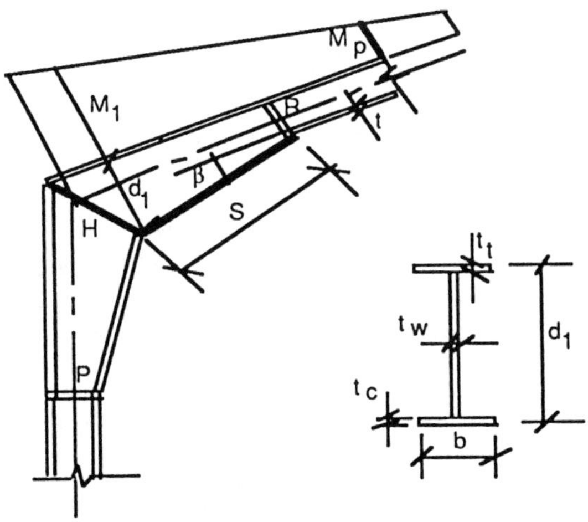

Figure 13.12 Haunched corner rigid connection for a gabled frame.

3. For $Z = M_1/s_y$, check haunch stability according to the following possible alternatives:

(a) For $s/b \leq 4.0$,

$$t_t = t$$

$$t_c = \frac{t}{\cos\beta} \tag{13.45}$$

(b) For $s/b > 4.0$, use Equation (13.45) and provide intermediate bracing or alternatively, the following.

(c) $4.0 < s/b < 17.0$

$$t_t = \left[1 + 0.1\left(\frac{s}{b} - 4\right)\right]t$$

$$t_c = \frac{t_1}{\cos\beta} \tag{13.46}$$

where

t_1 = thickness of tension flange
t_c = thickness of compression flange

(d) At point 1, the relationship $Z_1 \gg M_1/\sigma_y$ must be maintained and $\beta > 24$ deg.

4. For $Z_1 \gg M_1/\sigma_y$ and $s/b < 17$,

$$t_t = t$$

$$t_c = \frac{t_t}{\cos \beta} \tag{13.47}$$

5. The thickness of the web for the connection is equal to that of the member.

6. The selection of bracing is discussed under the following heading.

Stiffeners for Tapered Haunch Connections

Refer to Figure 13.13. The following requirements must be met in the design of haunched and tapered connections:

$$t_s \geq \sqrt{2}\, t_t - 0.82 \frac{t_w d_1}{b} \tag{13.48}$$

$$t_s \geq \sqrt{2}\,(1 - \tan \beta)t \tag{13.49}$$

$$t_s \geq \frac{b}{17} \tag{13.50}$$

$$t_r \geq t_c \tan \beta \tag{13.51}$$

$$t_r \geq \frac{b}{17} \tag{13.52}$$

One can use a conservative approximation by assuming that

$$t_s = t_r = t \tag{13.53}$$

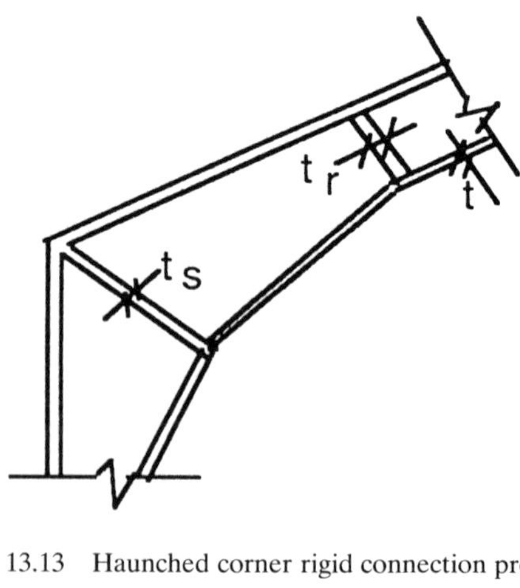

Figure 13.13 Haunched corner rigid connection proportions.

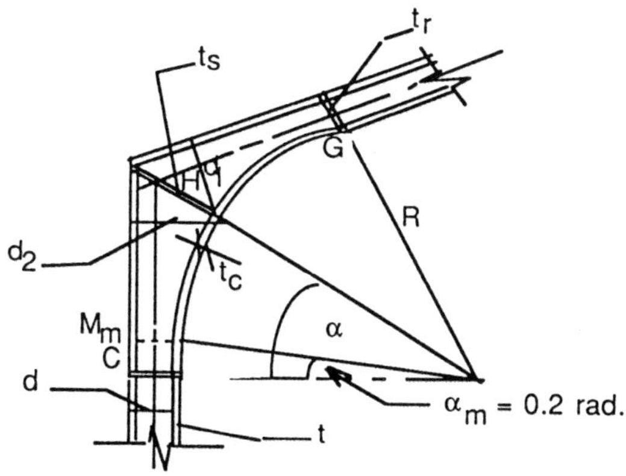

Figure 13.14 Curved haunched rigid connection.

Curved Tapered Connections

Figure 13.14 illustrates a curved connection. The design of a curved haunched rigid connection is outlined as follows:

1. Select the general layout of the connection and brace at points *G*, *H*, and *C*.
2. Check the haunch radius

$$R \leq \frac{4b}{\alpha} \tag{13.54}$$

for $4 < R\alpha/b < 17$, increase the tension and compression flange thickness according to

$$\Delta t = 0.1\left(\frac{R\alpha}{b} - 4\right) \tag{13.55}$$

or provide additional bracing.

3. Select flange thickness to provide plastic moment at *C* or *G*

$$t_t = t_c = \frac{d_m}{2} - \sqrt{\frac{d_m^2 b}{4(b - t_w)} - \frac{\frac{M_m}{d_y}}{b - t_w}} \tag{13.56}$$

where

$$d_m = d + 0.02R$$

$$M_m = \text{moment at } \alpha_m = 0.2 \text{ rad}$$

One approximation that may be used is

$$t_t = t_c = \tfrac{4}{3}t$$

4. Check cross-bending

$$t_c \geq \frac{b^2}{2R} \tag{13.57}$$

5. Use for the haunch web thickness and width of the flange the same dimensions as those of the rolled section.

6. Select stiffeners according to Equations (13.48) through (13.53).

14

Multistory Buildings: Plastic Design

14.1 INTRODUCTION

The use of plastic design concepts in the design of multistory steel building frames provides the structural engineer with yet another powerful tool in analysis and design. The concepts involved in this method have been well documented through several research projects sponsored by the AISC, Canadian Institute of Steel Construction (CISC), and several other national facilities, private as well as military, that are connected to the building industry.

The purpose of this chapter is to acquaint the student and practitioner with the theory behind plastic design methods for braced frames. It is hoped that this introductory chapter may stimulate interest in exploring further the development of a viable procedure in the design of multistory buildings for braced frames, as well as moment-connected ones.

High-rise buildings have been constructed in this country for many years. However, the rapid urbanization of American society over the past few decades has forced developers to construct high-rise buildings in many more locations other than large cities. The continuous increase in the population and the even greater rise in land costs make the use of tall buildings more economical in housing and urban planning. Structural steel frames with the use of plastic design methods may offer savings over buildings utilizing other materials or steel frames designed by the allowable stress methods.

To provide an organic design, the building process at present demands the integrated effort of several engineering teams: architects, and electrical, mechanical, and structural engineers. Each of these groups must understand the other disciplines in order to fully understand the total function of a building. Conduits both vertical and horizontal for the purpose of electrical and mechanical needs in the building must be provided for without adverse effects on the structural integrity of the frame or certain economic and spacial constraints. In the end, the frame must safely support all gravity and lateral forces due to wind, and to a certain extent earthquake forces, and remain aesthetically appropriate and complementary to the existing environment in which the building is located.

Fortunately, the structural engineer has at his or her hands the knowledge of more elaborate methods of analysis, new materials that are stronger than those used in the past, and the availability of computers. New high-strength carbon steel, high-strength bolts, welding equipment, and composite design using steel and concrete in floor construction produce more economical buildings.

14.2 ALLOWABLE STRESS VS. PLASTIC DESIGN METHODS

The AISC has introduced LRFD as a viable method of design, to replace ASD at some point in the future. LRFD uses the elastic method of analysis, thus requiring sophisticated computational procedures for analyzing structural systems. The allowable stress method uses the same approach to analysis as LRFD, but it is based on the assumption that the usefulness of a structural element is reached once a particular characteristic in that element attains its allowable limit. Other points in the structural system remain understressed, which renders this approach uneconomical. When a certain point in an indeterminate structural system reaches yield, redistribution of the internal forces takes place, allowing understressed members in the system to acquire more resistance to the applied loads. When this process is utilized in the design of indeterminate frames, a more economical system is obtained.

The plastic method for the analysis and design of indeterminate frames recognizes the above phenomenon in the behavior of structural systems subjected to external loads. The focus for the usefulness of a structure is shifted to the ultimate load limit that can be reached before a failure mechanism is induced, thus initiating the unrestrained deformation of the structure. The ultimate load represents the strength capacity of the entire structure. The ratio of the ultimate load to the working stress load is known by the load factor F. Let the ultimate moment capacity of the structural system be denoted by M_p, where

$$\begin{aligned} M_p &= \sigma_y Z \\ \sigma_y &= \text{the yield stress} \quad (\text{k/in.}^2) \\ Z &= \text{the plastic section modulus} \quad (\text{in.}^3) \end{aligned} \tag{14.1}$$

and the working stress moment capacity by M_a, where

$$\begin{aligned} M_a &= \sigma_a S_x \\ S &= \text{the yield section modulus} \quad (\text{in.}^3) \\ \sigma_a &= \text{the allowable stress} \quad (\text{k/in.}^2) \end{aligned} \tag{14.2}$$

Since the load factor has been defined by the ratio of the ultimate capacity to the allowable one, it is determined as follows:

$$F = \frac{M_p}{M_a} \tag{14.3}$$

Substituting Equations (14.1) and (14.2) in Equation (14.3) yields

$$F = \frac{\sigma_y Z}{s_a S_x} \tag{14.4}$$

The ratio of the plastic section modulus to the elastic one is known by f, the shape factor. Then, the load factor is expressed by

$$F = \frac{\sigma_y f}{\sigma_a} \tag{14.5}$$

The average shape factor for American wide-flange sections is about 1.12.

According to the AISC specification, the allowable stress for gravity loads is

$$\sigma_a = 0.66\sigma_y \tag{14.6}$$

Then,

$$\begin{aligned} F &= \frac{\sigma_y \times 1.12}{0.66\sigma_y} \\ &= 1.70 \end{aligned}$$

The allowable stress for wind or earthquake forces and a combination of these forces with gravity loads is raised by 33% above that for the gravity loads. Thus, the load factor for lateral loads and a combination of lateral and vertical loads is obtained from Equation (14.5) as follows:

$$F = \frac{\sigma_y \times 1.12}{0.66 \times 1.33\sigma_y}$$

$$= 1.276$$

or

$$F = 1.3$$

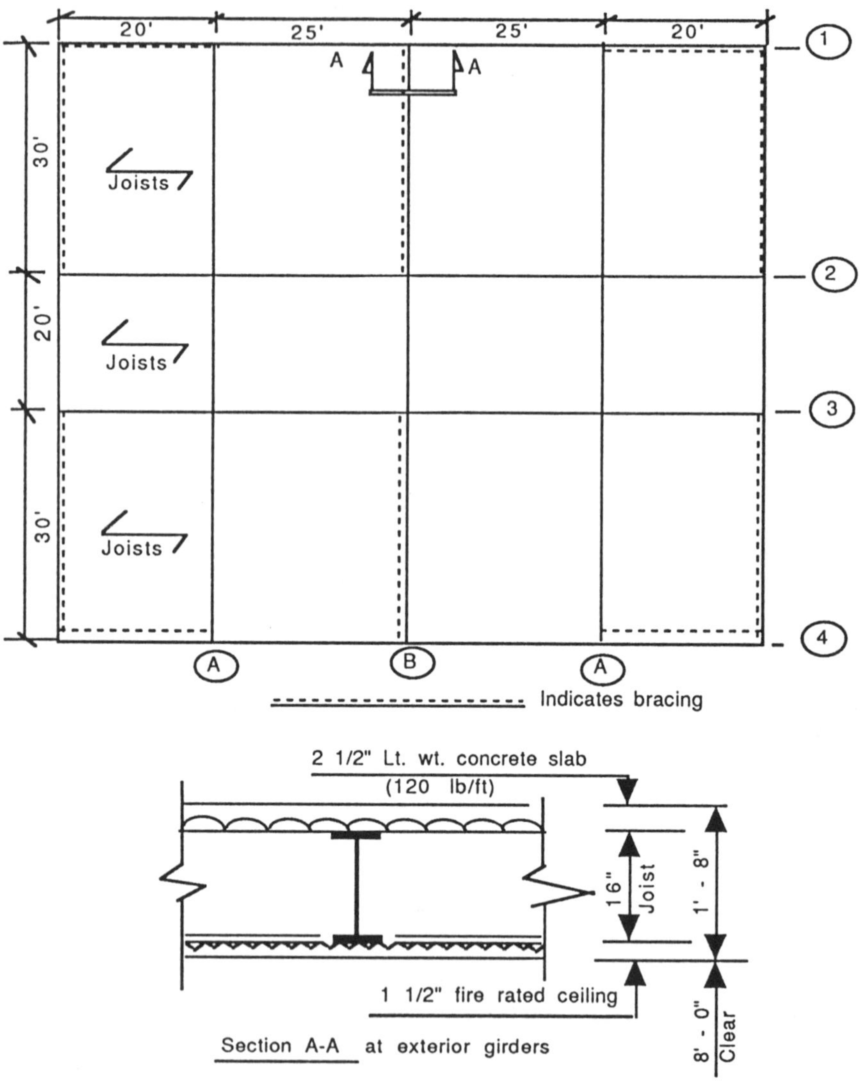

Figure 14.1 Building floorplan and section of an exterior girder.

14.3 APPLICATION TO MULTISTORY BUILDINGS

Consider the building presented in Figures 14.1 and 14.2. The design process involves four stages. Stage 1 considers the design of the unbraced bent for gravity loads only. Stage 2 involves the design of braced bents for combined loads. Stage 3 consists of checking columns, girders, bracing, shear, and uplift. The final stage will include the design of connections.

The structure under consideration is an office building located in downtown, Greensboro, North Carolina. The building loads are tabulated below:

FLOOR LOADS

$3\frac{1}{2}$-in. reinforced concrete slab	43 lb/ft^2
Floor finish	1
Ceiling	5
Partitions	20
Joists	4
MECHANICAL	5 lb/ft^2
Total dead load	78
Live load	50
Total load	128
Exterior walls: 60 lb/ft^2 $\times$ 9.67 ft	580
Parapet: 60 lb/ft^2 $\times$ 4 ft	250

ROOF LOADS

$3\frac{1}{2}$-in. reinforced concrete slab	43
Roofing	3
Insulation	5
Ceiling	5
Joists	2
MECHANICAL	5 lb/ft^2
Total dead load	63
Live load	20
Total load	83
Exterior walls: 60 lb/ft^2 $\times$ 9.67 ft	580
Parapet: 60 lb/ft^2 $\times$ 4 ft	250

For the wind, use the ASCE 7-88 Standard of 10 lb/ft^2 throughout the height of the building.

The design of floor girders is similar to those of the roof. In designing floor girders, the working live loads must be modified by applying the following reduction factor:

$$L = L_0\left(0.25 + \frac{15}{\sqrt{A_I}}\right)$$

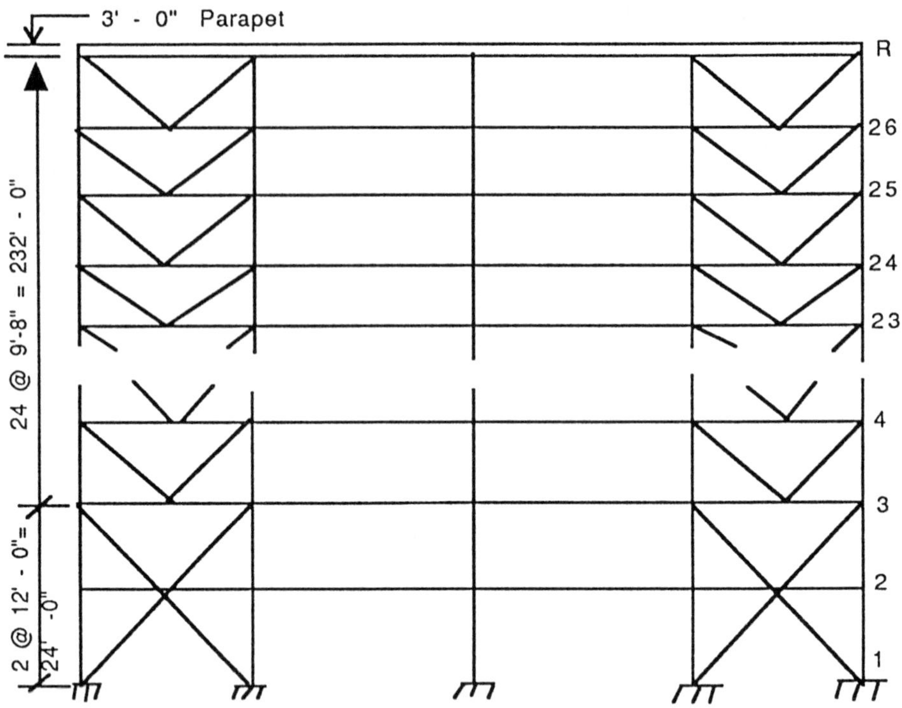

Figure 14.2 Evaluation of 27-story building.

The design of columns in multistory buildings is controlled by the factored load placed on every girder. It is sufficient for a preliminary design to use the axial loads and moments obtained from this loading condition. A "checkerboard" live load condition will produce different moment and end-restraint conditions. Unless the ratio of dead load to total load is less than 0.75 and the adjacent girders have almost equal spans and loads, the checkerboard loading condition can be neglected.

The design of roof and floor girders for the unbraced bent A is given in Tables 14.1 and 14.2, respectively. Table 14.3 presents a summary of the loads from the roof and a typical floor for interior and exterior columns.

The column load calculations are listed in Tables 14.4 and 14.5. The increments for the dead and live loads on an exterior column are 46.3 and 18.75 k, respectively. For an interior column, these increments are 56.4 and 40.63 k, respectively.

TABLE 14.1 Design of Roof Girders: Unbraced Bent A

Line	Item	Unit	Operation	*Bay* Exterior	Interior
1	Bay span	ft		30	20
2	Bent spacing	ft		25	25
3	Unit $DL + LL$ on roof	lb/ft^2		83	83
4	Estimated column depth	ft		1.0	1.0
5	Clear span	ft	(1)–(4)	29	19
6	Roof load on girder	k/ft	(2) × (3)	2.08	2.08
7	Estimated DL of girder	k/ft		0.04	0.03
8	Working load	k/ft	(6) + (7)	2.12	2.11
9	Factored load $F = 1.7$	k/ft	(8) × 1.7	3.60	3.59
10	Required M_p	k/ft	$(9) \times (5)^2/16$	189.2	81.0
11	Required Z	in.^3	(10) × 12/36	63.0	27
12	Section			W18 × 35	W14 × 22
13	Z provided	in.^3		66.5	33.2

Column Moments

The equilibrium of moments at the girder-column connection is expressed as follows:

$$M_{jt} + M_{jb} = -(M_{jR} + M_{jL} + M_{je}) \tag{14.7}$$

where

M_{jt} = moment at the top of the column for joint j
M_{jb} = moment at the bottom of the column for joint j
M_{jL} = moment in the girder at the left end
M_{jR} = moment in the girder at the right end
M_{je} = moment in the column due to eccentric loading

The left-hand side of Equation (14.7) represents the total moment transferred from the girders to the column. The columns will also resist bending moments imposed on them by the girder end moments and shears. From Figure 14.3, the total moment at the centerline of the column from each girder is given by the equation

$$M_{jB} = M_p\left(1 + \frac{4d_c}{L_g}\right) \tag{14.8}$$

$$M_{jt} = -M_p\left(1 + \frac{4d_c}{L_g}\right) \tag{14.9}$$

TABLE 14.2 Design of Floor Girders: Unbraced Bent *A*

Line	Item	Unit	Operation	Bay Exterior	Bay Interior
1	Bay span	ft		30	20
2	Bent spacing	ft		25	25
3	Unit *DL* on floor	lb/ft^2		78	78
4	Unit *LL* on floor	lb/ft^2		50	70
	Live Load Reduction				
5	Floor area	ft^2	(1) × (2)	750	500
6	Minimum reduced live load	lb/ft^2	0.5 × (4)	25	35
7	$L = L_0\left(0.25 + \frac{15}{\sqrt{A_I}}\right)$	lb/ft^2	$L \times (4)$	31.9	50.7
8	Selection of reduced *LL*	lb/ft^2	Largest of (6) or (7)	31.9	50.7
9	Estimated column depth	ft		1.0	1.0
10	Clear span	ft	(1)–(9)	29	19
11	Floor, *DL* on girder	k/ft	(2) × (3)	1.95	1.95
12	Estimated *DL* of girder	ft		0.05	0.04
13	Reduced *LL* on girder	k/ft	(8) × (2)	0.80	1.27
14	Working load	k/ft	(11) + (12) + (13)	2.80	3.26
15	Factored load	k/ft	1.7 × (14)	4.76	5.54
16	Required M_p	k/ft	$(15) \times (10)^2/16$	250.2	125.0
17	Required Z	$in.^3$	(16) × 12/36	83.4	41.7
18	Selected section			W21 × 44	W16 × 26
19	Capacity of girder (given Z)	$in.^3$		95.4	44.2

where

d_c = the depth of the column (in.)
L_g = the length of the girder

M_{jR} and M_{jL} represent the moments at the column ends. They have opposite signs when forming a single curvature. For double curvature in the column, the moment signs are both positive. Equations (14.8) and (14.9) are introduced to the calculations of column moments and are summarized in Table 14.6. The moment diagram for the columns is shown in Figure 14.4.

TABLE 14.3 Column Loads: Unbraced Bent *A*

Line	Item	Unit	Operation	Column Exterior	Column Interior
	Tributary area per floor				
1	From exterior bay (15 × 25)	ft^2		375	375
2	From interior bay (12.5 × 25)	ft^2		—	312.5
3	Total	ft^2	(1) + (2)	375	687.5
4	Unit roof load ($DL + LL$)	lb/ft^2		83	83
	Unit floor loads				
5	Exterior bay: Dead-live	lb/ft^2		78	78
6		lb/ft^2		50	50
7	Interior bay: Dead-live	lb/ft^2		—	78
8		lb/ft^2		—	70
	Loads below roof				
9	$DL + LL$ from roof	k	(3) × (4)	31.1	57.1
10	DL of girder	k		0.53	0.81
11	Estimated DL of column	k		2.0	2.0
12	DL of parapet at 0.25 k/ft	k	0.25 × 25	6.25	—
13	Working load below roof	k	Sum (9) to (12)	39.9	59.9
	Loads per floor				
14	DL from floor: Exterior bay	k	(1) × (5)	29.25	29.25
15	DL from floor: Interior bay	k	(2) × (7)	—	24.38
16	DL of girder	k		0.53	0.81
17	DL exterior wall 0.58 k/ft	k	0.58 × 25	14.5	—
18	Estimated DL of column	k		2.0	2.0
19	Total DL per floor	k	Sum (16) to (19)	46.3	56.4
20	LL from floor: Exterior bay	k	(1) × (6)	18.75	18.75
21	LL from floor: Interior bay	k	(2) × (8)	—	21.88
22	Total LL per floor	k	(20) + (21)	18.75	40.63

EFFECT OF AXIAL FORCE ON PLASTIC MOMENT CAPACITY

Previously, it has been demonstrated that axial force reduces the plastic moment capacity in a structural member under combined effect. This relationship is given by the equation

$$\frac{M_{pc}}{M_p} = 1.18\left(1 - \frac{P}{P_y}\right) \tag{14.10}$$

TABLE 14.4 Exterior Column

Level	*DL* (k)	Tributary Area (ft^2)	Reduction Factor	Sum of *LL* (k)	Reduced Column *LL* (k)	Sum of *DL* + Reduced *LL* (k)	1.7 × Sum *DL* + Reduced *LL* (k)	1.3 × Sum *DL* + Reduced *LL* (k)
Roof								
26	39.9	375	0.80	18.75	15.00	54.9	93.3	71.4
25	86.2	750	0.64	37.50	24.00	110.2	187.3	143.3
24	132.5	1125	0.57	56.25	32.1	164.6	279.8	214.0
23	178.8	1500	0.52	75.00	39.0	217.8	370.3	283.1
22	225.1	1875	0.49	93.75	45.3	271.0	460.7	352.3
21	271.4	2250	0.47	112.5	52.9	324.3	551.3	421.6
20	317.7	2625	0.46	131.25	60.4	378.1	642.8	491.5
19	364.0	3000	0.44	150.0	66.0	430.0	731.0	559.0
18	410.3	3375	0.43	168.75	72.6	482.9	820.9	627.8
17	456.6	3750	0.42	187.50	78.8	535.4	910.2	696.0
16	502.9	4125	0.42	206.25	86.6	589.5	1002.2	766.4
15	549.2	4500	0.41	225.0	92.2	641.4	1090.4	833.8
14	595.5	4875	0.40	243.75	97.5	693.0	1178.1	900.9
13	641.8	5250		262.50	105.0	746.8	1269.6	970.8
12	688.1	5625		281.25	112.5	800.6	1361.0	1040.8
11	734.4	6000		300.00	120.0	854.4	1452.5	1110.7
10	780.7	6375		318.75	127.5	908.2	1543.9	1180.7
09	827.0	6750		337.50	135.0	962.0	1635.4	1250.6
08	873.3	7125		356.25	142.5	1015.8	1726.9	1320.5
07	919.6	7500		375.00	150.0	1069.6	1818.3	1390.5
06	965.9	7875		393.75	157.5	1123.4	1909.8	1460.4
05	1012.2	8250		412.5	165.0	1177.2	2001.2	1530.4
04	1058.5	8625		431.25	172.5	1231.0	2092.7	1600.3
03	1104.8	9000		450.00	180.0	1284.8	2184.2	1670.2
02	1151.1	9375		468.75	187.5	1338.6	2275.6	1740.2
01	1197.4	9750		487.50	195.0	1392.4	2367.1	1810.1
00	1243.7	10,125		506.25	202.5	1446.2	2458.5	1880.1

TABLE 14.5 Interior Column

Level	*DL* (k)	Tributary Area (ft^2)	Reduction Factor	Sum of *LL* (k)	Reduced Column *LL* (k)	Sum of *DL* + Reduced *LL* (k)	1.7 × Sum *DL* + Reduced *LL* (k)	1.3 × Sum *DL* + Reduced *LL* (k)
Roof								
26	59.9	750	0.52	40.63	21.1	81	137.7	105.3
25	116.3	1500	0.44	81.3	35.7	152.0	258.4	197.6
24	172.7	2250	0.4	121.9	48.8	221.5	376.6	288.0
23	229.1	3000	0.4	162.5	65.0	295.1	500.0	382.3
22	285.5			203.2	81.3	366.8	623.6	476.9
21	341.5			243.8	97.5	439.4	747.0	571.2
20	398.3			284.4	113.8	512.1	870.6	665.8
19	454.7			325.1	130.0	584.7	994.0	760.0
18	511.1			365.7	146.3	657.4	1117.6	854.6
17	567.5			406.3	162.5	730.0	1241.0	949.0
16	623.9			446.9	178.8	802.7	1364.5	1043.4
15	680.3			487.6	195.0	875.3	1488.0	1137.9
14	736.7			528.2	211.3	948.0	1611.6	1232.4
13	793.1			568.8	227.5	1020.6	1735.0	1326.8
12	849.5			609.5	243.8	1093.3	1858.6	1421.8
11	905.9			650.1	260.0	1165.9	1982.0	1515.6
10	962.3			690.7	276.3	1238.3	2105.1	1609.8
09	1018.7			731.4	292.6	1311.3	2229.0	1704.7
08	1075.1			772.0	308.8	1383.9	2352.6	1799.0
07	1131.5			812.6	325.0	1456.5	2476.1	1893.5
06	1187.9			853.2	341.3	1529.0	2599.6	1987.9
05	1244.3			893.9	357.6	1601.9	2723.2	2082.4
04	1300.7			934.5	373.8	1674.5	2846.6	2176.8
03	1357.1			975.0	390.0	1747.1	2970.0	2271.2
02	1413.5			1015.8	406.3	1879.8	3195.6	2443.7
01	1469.9			1056.4	422.6	1892.5	3217.2	2460.2
00	1526.3			1097.0	438.8	1965.1	3340.7	2554.6

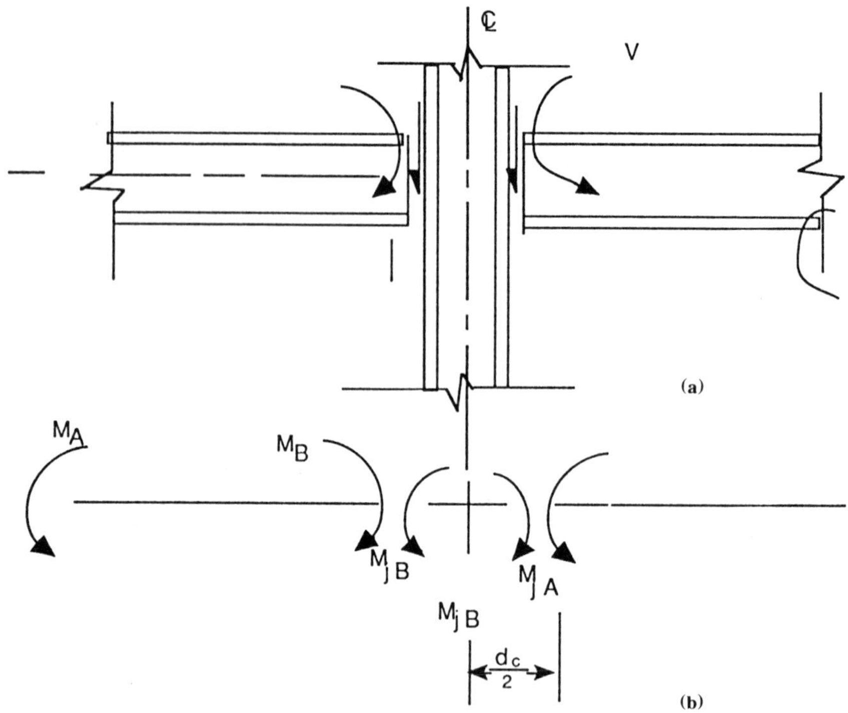

Figure 14.3 Girder-column moment and shear designation.

where

$$M_p = ZF_y \tag{14.11}$$

$$P_y = AF_y \tag{14.12}$$

Equation (14.10) may be expressed as follows:

$$P_y = P + 0.85M_{pc}\left(\frac{P_y}{M_p}\right) \tag{14.13}$$

Combining Equations (14.9) through (14.11) and using the value of $Z/S = 1.12$ for wide-flange sections (average value) and $r_x = 0.43d$ yield the following relationship:

$$P_y = P + 2.1M_{pc} \tag{14.14}$$

TABLE 14.6 Columns Moments, Factored Gravity Loads

				Column	
Line	Item	Units	Operation	Exterior	Interior
1	Left girder required M_p	k ft	Table 14.1 (10)		189.2
2	Estimated d_c/L_g		Table 14.1 (4)/(5)		0.0345
3	At central column $M_jB = M_p(1 + 4d_c/L_g)$	k ft			215.3
4	Right girder required M_p	k ft	Table 14.1 (10)	−189.2	−81.0
5	Estimated d_c/L_g			0.0345	0.0526
6	Column c $M_jA = M_p(1 + 4d_c/L_g)$	k ft		−215.3	−98.0
7	Spandrel (6.2 × 0.5 × 1.7)	k ft	14.3 (12) d_c/L_g	+5.3	
8	Column moment at roof	k ft	−[(3) + (6) + (7)	+210	−117.3
	Moment at Level 26-1				
9	Left girder required M_p	k ft	Table 14.2 (16)		+250.2
10	Estimated d_c/L_g		Table 14.1 (4)/(5)		0.0345
11	Column c $M_jB = M_p(1 + 4d_c/L_g)$	k ft			+284.7
12	Girder right	k ft	Table 14.2	+250.2	+125.0
13	Estimated d_c/L_g			0.0345	0.0526
14	Column c $M_jA = M_p(1 + 4d_c/L_g)$	k ft		−284.7	−151.3
15	Spandrel 14.5 (d_c/c) 1.7		14.5 × 0.5 × 1.7	+12.3	
16	Net girder moment on joint	k ft	−[(11) + (14) + (15)	272.4	−133.4
17	Column moment		0.5 × (16)	+136.2	−66.7

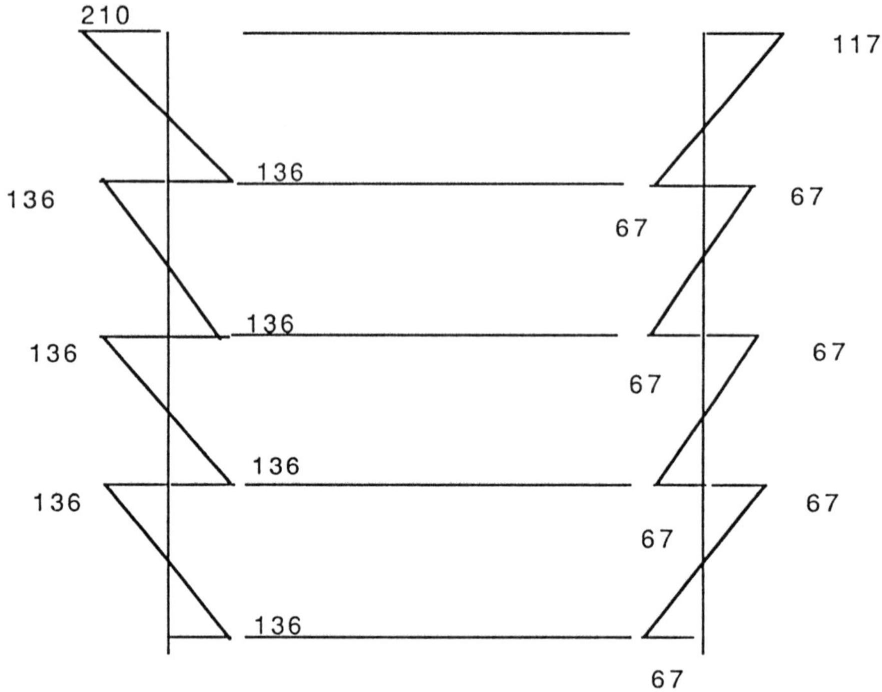

Figure 14.4 Column moment diagrams.

The term P_y in Equation (14.14) is an equivalent axial load that can be computed directly from the axial loads as determined in Tables 14.4 and 14.5. Then, the design of the column can follow the routine procedure outlined in the AISC Manual for concentric loads. Next, a check for wind on the stability of the structure will be conducted. It is not within the scope of this book to accomplish that.

Index

R

S

T

V

W